Dynamic Behavior of Materials

Dynamic Behavior of Materials

Editors

Chuanting Wang
Yong He
Yuanfeng Zheng
Shuhai Zhang
Wenhui Tang

MDPI • Basel • Beijing • Wuhan • Barcelona • Belgrade • Manchester • Tokyo • Cluj • Tianjin

Editors

Chuanting Wang
Nanjing University of
Science & Technology,
China

Yong He
Nanjing University of
Science & Technology,
China

Yuanfeng Zheng
Beijing Institute
of Technology,
China

Shuhai Zhang
North University of China,
China

Wenhui Tang
National University of
Defense Technology,
China

Editorial Office
MDPI
St. Alban-Anlage 66
4052 Basel, Switzerland

This is a reprint of articles from the Special Issue published online in the open access journal *Crystals* (ISSN 2073-4352) (available at: https://www.mdpi.com/journal/crystals/special_issues/dynamic_behaviour_materials).

For citation purposes, cite each article independently as indicated on the article page online and as indicated below:

LastName, A.A.; LastName, B.B.; LastName, C.C. Article Title. *Journal Name* **Year**, *Volume Number*, Page Range.

ISBN 978-3-0365-6942-0 (Hbk)
ISBN 978-3-0365-6943-7 (PDF)

© 2023 by the authors. Articles in this book are Open Access and distributed under the Creative Commons Attribution (CC BY) license, which allows users to download, copy and build upon published articles, as long as the author and publisher are properly credited, which ensures maximum dissemination and a wider impact of our publications.

The book as a whole is distributed by MDPI under the terms and conditions of the Creative Commons license CC BY-NC-ND.

Contents

About the Editors vii

Preface to "Dynamic Behavior of Materials" ix

Chuanting Wang, Yuanfeng Zheng, Shuhai Zhang, Wenhui Tang and Yong He
Editorial for the Special Issue "Dynamic Behavior of Materials"
Reprinted from: *Crystals* **2023**, *13*, 44, doi:10.3390/cryst13010044 1

Chi Yan, Chu Wang, Miaoxia He, Yuecheng Dong, I. V. Alexandrov and Hui Chang Dynamic Behavior of a Novel High-Strength and Ductile Near-α Titanium Ti-Al-Mo-Zr-Fe-B Alloy
Reprinted from: *Crystals* **2022**, *12*, 1584, doi:10.3390/cryst12111584 5

Zezhou Yang, Xiaowei Feng, Xiaoli Zhang, Yongfeng Shen, Xicheng Huang and Ruoze Xie
Study on Dynamic Mechanical Behaviors and J–C Constitutive Model of a Fine-Grained D6A Steel
Reprinted from: *Crystals* **2022**, *12*, 806, doi:10.3390/cryst12060806 15

Guanghui Zhao, Ruifeng Zhang, Juan Li, Cuirong Liu, Huaying Li and Yugui Li
Study on Microstructure and Properties of NM500/Q345 Clad Plates at Different Austenitization Temperatures
Reprinted from: *Crystals* **2022**, *12*, 1395, doi:10.3390/cryst12101395 27

Li Ding, Peihui Shen and Liuqi Ji
Dynamic Response and Numerical Interpretation of Three Kinds of Metals for EFP Liner under Explosive Loading
Reprinted from: *Crystals* **2022**, *12*, 154, doi:10.3390/cryst12020154 39

Xianwen Ran, Xuan Zou, Jingyuan Zhou and Wenhui Tang
Shock Properties of One Unsaturated Clay and Its Equation of State Up to 30 GPa
Reprinted from: *Crystals* **2022**, *12*, 119, doi:10.3390/cryst12010119 59

Jianguang Xiao, Yanxin Wang, Dongmo Zhou, Chenglong He and Xiangrong Li
Research on the Impact-Induced Deflagration Behavior by Aluminum/Teflon Projectile
Reprinted from: *Crystals* **2022**, *12*, 471, doi:10.3390/cryst12040471 71

Chengxin Du, Huameng Fu, Zhengwang Zhu, Kehong Wang, Guangfa Gao, Feng Zhou, et al.
Penetration Failure Mechanism of Multi-Diameter Tungsten Fiber Reinforced Zr-Based Bulk Metallic Glasses Matrix Composite Rod
Reprinted from: *Crystals* **2022**, *12*, 124, doi:10.3390/cryst12020124 89

Anbang Jiang, Yongqing Li, Dian Li and Hailiang Hou
Study on Anti-Penetration Performance of Semi-Cylindrical Ceramic Composite Armor against 12.7 mm API Projectile
Reprinted from: *Crystals* **2022**, *12*, 1343, doi:10.3390/cryst12101343 105

Xiangrong Li, Cong Hou, Huan Tong, Lei Yang and Yongkang Chen
Influential Factors of a Reactive Materials Projectile's Damage Evolution Behavior
Reprinted from: *Crystals* **2022**, *12*, 1683, doi:10.3390/cryst12111683 127

Guancheng Lu, Chao Ge, Zhenyang Liu, Le Tang and Haifu Wang
Study on the Formation of Reactive Material Shaped Charge Jet by Trans-Scale Discretization Method
Reprinted from: *Crystals* **2022**, *12*, 107, doi:10.3390/cryst12010107 145

Mengmeng Guo, Yanxin Wang, Yongkang Chen, Jianguang Xiao and Haifu Wang
The Effect of Aluminum Particle Size on the Formation of Reactive Jet
Reprinted from: *Crystals* **2022**, *12*, 1560, doi:10.3390/cryst12111560 . . . . **163**

Shuai Yue, Yushuai Bai, Zhonghua Du, Huaiwu Zou, Wenhui Shi and Guang Zheng
Dynamic Behavior of Kinetic Projectile Impact on Honeycomb Sandwich Panels and Multi-Layer Plates
Reprinted from: *Crystals* **2022**, *12*, 572, doi:10.3390/cryst12050572 . . . . **181**

Qing Lin, Chunbo Wu, Shuai Yue, Zhonghui Jiang, Zhonghua Du and Mengsheng Li
Dynamic Simulation and Parameter Analysis of Contact Mechanics for Mimicking Geckos' Foot Setae Array
Reprinted from: *Crystals* **2022**, *12*, 282, doi:10.3390/cryst12020282 . . . . **199**

Wenhui Shi, Shuai Yue, Zhiqian Li, Hao Xu, Zhonghua Du, Guangfa Gao, et al.
Dynamic Simulation and Parameter Analysis of Weaved Composite Material for Unmanned Aerial Vehicle Parachute Recovery in Deployment Phase
Reprinted from: *Crystals* **2022**, *12*, 758, doi:10.3390/cryst12060758 . . . . **213**

Zhenze Zhu, Rongfeng Zhou, Xiangming Li, Wentao Xiong and Zulai Li
Flow Field and Inclusions Movement in the Cold Hearth for the Ti-0.3Mo-0.8Ni Alloy
Reprinted from: *Crystals* **2022**, *12*, 1471, doi:10.3390/cryst12101471 . . . . **233**

About the Editors

Chuanting Wang

Chuanting Wang, associate professor of School of Mechanical Engineering, Nanjing University of Science & Technology. He received his PhD degree from University of Southampton, UK, in 2013 and then carried out post-doctoral research at the University of Southern California, USA. He joined the Nanjing University of Science and Technology in 2014. His main research interests are flow behavior of materials under high pressure and the impact-initiated reaction mechanism of reactive materials.

Yong He

Yong He, professor of School of Mechanical Engineering, Nanjing University of Science & Technology. He is Chairman of the Ammunition Professional Committee of the Chinese Military Industry Association, the Deputy Chairman of the Combat Department and Damage Efficiency Professional Committee of the Chinese Aerospace Society, editor board member of Acta Armamentarii, Journal of Ordance Equipment Engineering, Tactical Missile Technology etc. He has published over 200 technical papers in the fields of dynamic behavior of materials and new concept damage.

Yuanfeng Zheng

Yuanfeng Zheng, associate professor of School of Mechatronical Engineering, Beijing Institute of Technology. In 2012, he received his PhD degree from Beijing Institute of Technology in China, and then joined Beijing Institute of Technology to research. His main research interests are reactive materials, terminal effects, impact, shaped charge and damage.

Shuhai Zhang

Yuanfeng Zheng, associate professor of School of Mechatronical Engineering, Beijing Institute of Technology. In 2012, he received his PhD degree from Beijing Institute of Technology in China, and then joined Beijing Institute of Technology to research. His main research interests are reactive materials, terminal effects, impact, shaped charge and damage.

Wenhui Tang

Wenhui Tang, Professor of National University of Defense Technology, China. He is director of the Accounting and Computational Physics Committee of Chinese Society of Nuclear Science, member of the Gas Dynamics Professional Committee, member of the High Pressure Physics Professional Committee of the Chinese Physical Society. He proposed the "Tang theory" for calculation of thermal conductivity of non-metallic crystals at high temperature and high pressure. He published more than 150 peer reviewed articles.

Preface to "Dynamic Behavior of Materials"

The dynamic behavior of materials is a field at the confluence of several scientific disciplines. The processes that occur when materials are subjected to rapid loads can differ significantly from those under static or quasi-static situations. The understanding of the dynamic behavior of materials involves the mechanics of high-strain-rate deformation (elastic, plastic, shock, and detonation waves) and the dynamic responses of materials (constitutive models, shear instabilities, microstructural evolutions, dynamic fractures, and chemical reactions).

This Special Issue on the "Dynamic Behavior of Materials" collected recent research findings on the dynamic behavior of various materials. A collection of fifteen peer-reviewed research articles were included in this Special Issue. The main topics covered include processing technology, state-of-the-art characterization, testing, theoretic modeling, and simulation. We hope that this Special Issue will be interesting for the academic community, and also hope that it will shed light on future research activities in this field.

Chuanting Wang, Yong He, Yuanfeng Zheng, Shuhai Zhang, and Wenhui Tang
Editors

Editorial

Editorial for the Special Issue "Dynamic Behavior of Materials"

Chuanting Wang [1], Yuanfeng Zheng [2], Shuhai Zhang [3], Wenhui Tang [4] and Yong He [1,*]

[1] School of Mechanical Engineering, Nanjing University of Science & Technology, Nanjing 210094, China
[2] State Key Laboratory of Explosion Science and Technology, Beijing Institute of Technology, Beijing 100081, China
[3] School of Environment and Safety Engineering, North University of China, Taiyuan 030051, China
[4] College of Science, National University of Defense Technology, Changsha 410073, China
* Correspondence: yonghe1964@163.com

1. Introduction

The dynamic behavior of materials is a field at the confluence of several scientific disciplines. The processes that occur when materials are subjected to rapid loads can differ significantly from those under static or quasi-static situations. The understanding of the dynamic behavior of materials involves the mechanics of high-strain-rate deformation (elastic, plastic, shock, and detonation waves) and the dynamic responses of materials (constitutive models, shear instabilities, microstructural evolutions, dynamic fractures, and chemical reactions).

This Special Issue on the "Dynamic Behavior of Materials" collected recent research findings on the dynamic behavior of various materials. A collection of fifteen peer-reviewed research articles were included in this Special Issue. The main topics covered include processing technology, state-of-the-art characterization, testing, theoretic modeling, and simulation.

2. Microstructural Evolution of Materials under Dynamic Deformation

Yan et al. [1] investigated the microstructural evolution of a near-αtitanium Ti-6Al-1Mo-2Zr-0.55Fe-0.1B alloy after dynamic deformation. The results indicated that the strength of the alloy increased significantly under dynamic loading. An abundance of deformation twins released the dislocation pile-up and coordinated the plastic deformation of the alloy during dynamic loading. The Johnson–Cook (J–C) constitutive equation of the alloy was obtained. Yang et al. [2] performed uniaxial tensile loading tests on coarse-grained and fine-grained D6A steel, finding that grain refinement effectively improved the strength of fine-grained steel, attributing this to the high density of grain boundaries and cementite particles hindering the movement of dislocations. The parameters of the J–C constitutive model were calibrated. Zhao et al. [3] investigated the microstructure and properties of hot-rolled NM500/Q345 clad plates after heat treatment. They found that the microstructure of the NM500/Q345 clad plate before austenitization was mainly pearlite and ferrite, and both were transformed into lath martensite after austenitization. Ding et al. [4] studied the dynamic response of several metallic materials under explosive loading; the microscopic features of the metals were examined and analyzed with the fracture model.

3. Equations of State and Constitutive Equation of Materials

The complicated composition of unsaturated clay, e.g., solid mineral particles, water, and air, makes it difficult to obtain its precise equation of state (EOS) over a wide pressure range. Ran et al. [5] discussed the high-pressure EOS of unsaturated clay at the mesoscale; a high-pressure EOS of the unsaturated clay up to 30 GPa was developed, and it was in good agreement with the experimental data. Xiao et al. [6] considered the phase change during the impact as initiating a chemical reaction process, introducing three different EOSs to describe the material parameters from a solid reactant state to a solid–gas mixing

state and a gas state. The Johnson–Cook constitutive equation was modified by the adiabatic temperature rise term, and a nonlinear fit was performed to construct a dynamic constitutive relationship at room temperature, which could describe the rheological behavior of the Ti-6Al-1Mo-2Zr-0.55Fe-0.1B alloy under high-strain-rate conditions at room temperature [1]. Similar constitutive equation work was performed on D6A steel, and a corresponding equation was obtained [2].

4. Material and Structure under High-Velocity Impact

Du et al. [7] designed tungsten-fiber-reinforced Zr-based bulk metallic glass rods with gradient structures and conducted penetration experiments of the rods against rolled armor steel. The penetration failure mode of the rods, including bending, the backflow of tungsten fibers, and shear failure, was observed. It was suggested that the bending space and ultimate bending diameter of tungsten fibers should be considered in order to achieve a higher penetration ability of the fiber-reinforced rods. Jiang et al. [8] designed projectile-resistant composite armor consisting of curved ceramic, a steel frame, and a metal back plate. The anti-penetration performance of the armor against 12.7 mm armor-piercing incendiary (API) projectiles was tested. The penetration process was divided into four main stages: the asymmetric erosion of the projectile, ceramic cone squeezing movement, back plate failure, and projectile exit.

Besides the various physical changes, dynamic loading may also trigger a series of chemical reactions of the materials. Xiao et al. [6] proposed a simulation method for the impact-induced deflagration behavior of reactive materials by introducing tunable ignition threshold conditions. The ignition, reaction ratio, pressure, and temperature distribution of an Al/Teflon projectile impacting multilayer targets were analyzed. The increasing trend in the reaction ratio was consistent with the change in radiated flash in the experiments. Li et al. [9] designed a reactive projectile with an Al/PTFE composite as the inner core and a steel shell as the outer case. They studied the penetration and chemical reaction behavior of such a projectile impacting a multilayer target. The influential factors, including impact velocity, target thickness, and projectile structure, on the damage effect were discussed.

5. Explosive–Material Interaction

Dynamic loading on materials could be generated by an explosion; the interaction between explosion products and materials is very quick, and the strain rate could be extremely high. Lu et al. [10] investigated the formation process of reactive-material-shaped charges under an explosion. A two-dimensional finite element model of a shaped charge and a reactive material liner was established; the jet formation process, particle dispersion, and jet particle distribution were analyzed. The results showed that the PTFE matrix accelerated faster than the Al particles under shock loading. The relative displacement results in a density gradient along the axis of the jet, and PTFE becomes the main component of the jet. The initial granule distribution in the liner had a great influence on the particle distribution in the formed jet. Similarly, Guo et al. [11] studied the jet formation process of reactive Al/PTFE liners via X-ray photography. It was observed that the reactive Al/PTFE composite reacted during jet formation, which can be divided into a local reaction stage and an overall reaction stage. Secondary collision occurred in the inner layer of the liner during the jet formation, where the chemical reaction initiated. Moreover, the Al particle size led to a large difference in the jet formation process, showing that the jet formed from liners with larger Al particles was more condensed due to a slower chemical reaction. For tungsten-based materials, the flow behavior of the material under a high strain rate had a significant influence on the final jet formation [4]. Achieving a reasonable coupling effect between the explosive and liner material is a challenge, aiming to achieve a jet/EFP without obvious cracking fracture.

6. Theoretic Modeling and Simulation

Due to the difficulty of experimentally observing dynamic processes, especially those under high strain rates, theoretic modeling and simulation have become very important in explaining and visually displaying dynamic processes that are difficult to observe. In order to study the dynamic response associated with the impact of a kinetic projectile on the internal structure of an artificial satellite, a simulation model of the projectile damaging the multilayer plates was established in ABAQUS by Yue et al. [12]. The influences of initial velocities and incident angles on the damage characteristics were discussed. According to the dynamic characteristics of the adhesion–desorption process between gecko-like polyurethane setae and the contact surface, the microcontact principle of an elastic sphere and plane was established. On this basis, combined with the cantilever beam model, microscale adhesive contact models in the cases of a single and an array of setae were obtained. The contact process was numerically simulated and verified by the adhesion–desorption test. The results showed that the simulation model could reflect the real contact procedure of setae [13]. Shi et al. analyzed [14] the deployment process of an unmanned aerial vehicle's parachute recovery. A dynamics model of the parachute deployment phase was established. The effect of the parachute weight and launch temperature on the dynamic characteristics of the deployment was discussed. Zhu et al. [15] established a three-dimensional numerical model to describe the melt flow field and inclusions movement in the cold hearth for the Ti-0.3Mo-0.8Ni alloy during electron beam cold hearth melting. The solidification and discrete phase models in ANSYS were used to quantitatively analyze the movement of inclusions in the cold hearth. Autodyn was employed to simulate the penetration and reaction processes of reactive projectiles impacting targets, where a modified EOS was embedded via secondary development [6,9]. An adaptive coupling algorithm based on the finite element method and smooth particle hydrodynamics in LS-DYNA was adopted to study the anti-penetration performance of the composite armor structure against the API projectile [8]. LS-DYNA and Autodyn were employed to simulate the jet/EFP formation process of various materials under explosions [4,14,15].

This Special Issue includes a small number of discrete case studies on the "Dynamic Behavior of Materials". We hope that this Special Issue will be interesting for the academic community, and also hope that it will shed light on future research activities in this field.

Funding: This research received no external funding.

Acknowledgments: The contributions of all authors are gratefully acknowledged.

Conflicts of Interest: The authors declare no conflict of interest.

References

1. Yan, C.; Wang, C.; He, M.X.; Dong, Y.C.; Alexandrov, I.V.; Chang, H. Dynamic Behavior of a Novel High-Strength and Ductile Near-α Titanium Ti-Al-Mo-Zr-Fe-B Alloy. *Crystals* **2022**, *12*, 1584. [CrossRef]
2. Yang, Z.Z.; Feng, X.W.; Zhang, X.L.; Shen, Y.F.; Huang, X.C.; Xie, R.Z. Study on Dynamic Mechanical Behaviors and J–C Constitutive Model of a Fine-Grained D6A Steel. *Crystals* **2022**, *12*, 806. [CrossRef]
3. Zhao, G.H.; Zhang, R.F.; Li, J.; Liu, C.R.; Li, H.Y.; Li, Y.G. Study on Microstructure and Properties of NM500/Q345 Clad Plates at Different Austenitization Temperatures. *Crystals* **2022**, *12*, 1395. [CrossRef]
4. Ding, L.; Shen, P.H.; Ji, L.Q. Dynamic Response and Numerical Interpretation of Three Kinds of Metals for EFP Liner under Explosive Loading. *Crystals* **2022**, *12*, 154. [CrossRef]
5. Ran, X.W.; Zou, X.; Zhou, J.Y.; Tang, W.H. Shock Properties of One Unsaturated Clay and Its Equation of State Up to 30 GPa. *Crystals* **2022**, *12*, 119. [CrossRef]
6. Xiao, J.G.; Wang, Y.X.; Zhou, D.M.; He, C.L.; Li, X.R. Research on the Impact-Induced Deflagration Behavior by Aluminum/Teflon Projectile. *Crystals* **2022**, *12*, 471. [CrossRef]
7. Du, C.X.; Fu, H.M.; Zhu, Z.W.; Wang, K.H.; Gao, G.F.; Zhou, F.; Xu, L.Z.; Du, Z.H. Penetration Failure Mechanism of Multi-Diameter Tungsten Fiber Reinforced Zr-Based Bulk Metallic Glasses Matrix Composite Rod. *Crystals* **2022**, *12*, 124. [CrossRef]
8. Jiang, A.B.; Li, Y.Q.; Li, D.; Hou, H.L. Study on Anti-Penetration Performance of Semi-Cylindrical Ceramic Composite Armor against 12.7 mm API Projectile. *Crystals* **2022**, *12*, 1343. [CrossRef]
9. Li, X.R.; Hou, C.; Tong, H.; Yang, L.; Chen, Y.K. Influential Factors of a Reactive Materials Projectile's Damage Evolution Behavior. *Crystals* **2022**, *12*, 1683. [CrossRef]

10. Lu, G.C.; Ge, C.; Liu, Z.Y.; Tang, L.; Wang, H.F. Study on the Formation of Reactive Material Shaped Charge Jet by Trans-Scale Discretization Method. *Crystals* **2022**, *12*, 107. [CrossRef]
11. Guo, M.M.; Wang, Y.X.; Chen, Y.K.; Xiao, J.G.; Wang, H.F. The Effect of Aluminum Particle Size on the Formation of Reactive Jet. *Crystals* **2022**, *12*, 1560. [CrossRef]
12. Yue, S.; Bai, Y.S.; Du, Z.H.; Zou, H.W.; Shi, W.H.; Zheng, G. Dynamic Behavior of Kinetic Projectile Impact on Honeycomb Sandwich Panels and Multi-Layer Plates. *Crystals* **2022**, *12*, 572. [CrossRef]
13. Lin, Q.; Wu, C.B.; Yue, S.; Jiang, Z.H.; Du, Z.H.; Li, M.S. Dynamic Simulation and Parameter Analysis of Contact Mechanics for Mimicking Geckos' Foot Setae Array. *Crystals* **2022**, *12*, 282. [CrossRef]
14. Shi, W.H.; Yue, S.; Li, Z.Q.; Xu, H.; Du, Z.H.; Gao, G.F.; Zheng, G.; Zhao, B.B. Dynamic Simulation and Parameter Analysis of Weaved Composite Material for Unmanned Aerial Vehicle Parachute Recovery in Deployment Phase. *Crystals* **2022**, *12*, 758. [CrossRef]
15. Zhu, Z.Z.; Zhou, R.F.; Li, X.M.; Xiong, W.T.; Li, Z.L. Flow Field and Inclusions Movement in the Cold Hearth for the Ti-0.3Mo-0.8Ni Alloy. *Crystals* **2022**, *12*, 1471. [CrossRef]

Disclaimer/Publisher's Note: The statements, opinions and data contained in all publications are solely those of the individual author(s) and contributor(s) and not of MDPI and/or the editor(s). MDPI and/or the editor(s) disclaim responsibility for any injury to people or property resulting from any ideas, methods, instructions or products referred to in the content.

Article

Dynamic Behavior of a Novel High-Strength and Ductile Near-α Titanium Ti-Al-Mo-Zr-Fe-B Alloy

Chi Yan [1], Chu Wang [1], Miaoxia He [1], Yuecheng Dong [1,*], I. V. Alexandrov [2] and Hui Chang [1,*]

[1] College of Materials Science and Engineering, Tech Institute for Advanced Materials, Nanjing Tech University, Nanjing 211816, China
[2] Department of Materials Science and Physics of Metals, Ufa State Aviation Technical University, 450008 Ufa, Russia
* Correspondence: dongyuecheng@njtech.edu.cn (Y.D.); ch2006@njtech.edu.cn (H.C.); Tel.: +86-25-8358-7270 (Y.D.)

Abstract: In this study, the dynamic compression properties of a new high-strength (>1000 MPa) and ductile (>15%) near-α titanium Ti-6Al-1Mo-2Zr-0.55Fe-0.1B alloy were investigated at high strain rates of 1620 s^{-1}~2820 s^{-1} by a split Hopkinson pressure bar (SHPB). The microstructural evolution of the samples before and after the dynamic deformation was analyzed by electron backscatter diffraction (EBSD). The results indicated that the strength of the alloy enhanced significantly under the dynamic loading compared with the quasi-static compression and increased with the increase in the strain rate. An abundance of deformation twins released the dislocation pile-up and coordinated the plastic deformation of alloy during the dynamic loading. The dynamic plasticity constitutive equation of the alloy was obtained by fitting high strain rate experimental data at room temperature by the Johnson–Cook constitutive equation with the modified temperature term.

Keywords: near-α titanium alloy; dynamic deformation; deformation twin; Johnson–Cook equation

1. Introduction

Near-α titanium alloys have high specific strength, excellent corrosion resistance, and good weldability, making them extensively accepted as admirable structural materials in marine engineering [1–5]. In particular, the Ti80 (Ti-6Al-3Nb-2Zr-1Mo) alloy is widely studied and applied as a typical marine titanium alloy because of its corrosion resistance [6,7], impact toughness [8], low fatigue cracking [9,10], and creep behavior [11] as well as its performance after various welding methods [12,13]. However, the presence of the Nb element makes Ti80 expensive; it also gives it a high melting point of up to 2468 °C. The high melting point makes it possible to generate a high density of inclusions in the casting process, which is harmful to the subsequent forging/rolling process. Moreover, the strength of the Ti80 is approximately equal to 800 MPa, which is not high enough as the pressure shell of the deep-diving submersible dives into the depth of up to 10,000 m. That is why the new type, the 10,000 deep-diving submersible, used α + β alloy in spite of poor weldability.

On the other hand, titanium alloys as structural materials are often subjected to high velocity impact loads, which are required to ensure the structural integrity and continuity with the required dynamic load carrying capacity under the specified impact loads. Since both plasticity and failure behavior of titanium are significantly affected by the strain rate, it is necessary to study the dynamic behavior of titanium at different strain rates [14–16]. Hence, in this work, a novel near-α titanium Ti-6Al-1Mo-2Zr-0.55Fe-0.1B alloy, based on the Ti80 alloy, was designed, in which Nb was replaced by the microalloying elements Fe and B. The aim of the present investigations was to design a novel, low-cost, high-strength, and high-toughness near-α titanium alloy for marine engineering applications and evaluate the dynamic compression properties of this alloy at different strain rates with the split Hopkinson pressure bar (SHPB).

2. Materials and Methods

A near-α titanium Ti-6Al-1Mo-2Zr-0.55Fe-0.1B alloy was designed and melted using vacuum arc fusion (VAR) equipment. The chemical composition of the alloy was investigated by inductively coupled plasma mass spectrometry (ICP-MS) analysis; the actual chemical composition is shown in Table 1. The phase transition temperature of the alloy measured by the metallographic method was 995 °C. After forging, heat treatment [17] was used to optimize the microstructure.

Table 1. Chemical composition of the near-α Ti-6Al-1Mo-2Zr-0.55Fe-0.1B (wt.%) alloy.

Elements	Al	Mo	Zr	Fe	B	Ti
wt.%	5.8	1.07	1.85	0.57	0.044	Bal.

Quasi-static tensile tests were conducted on the MTS 370.10 hydraulic servo fatigue tester. The tensile specimen length was 100 mm, and the diameter and length of the gauge were ϕ5 mm and 25 mm, respectively. The strain rate was 7×10^{-3} s^{-1}. Quasi-static compression experiments were carried out on a universal testing machine with a strain rate of 10^{-3} s^{-1} with the dimension of $\phi 3 \times 6$ mm, which was the sample size. Dynamic mechanical properties were tested by the SHPB equipment at high strain rates of 1620 s^{-1}~2820 s^{-1}, and the sample size was $\phi 6 \times 3$ mm. The diameter of the SHPB compression bar was 14.5 mm, and the input pulse maximum value was 1200 MPa. The samples were prepared by wire cutting, and the end faces were polished by sandpaper grinding.

The Japanese Regulus 8100 cold field emission scanning electron microscope was used to characterize the microstructure of the samples as well as the element distribution; the sample size was 4 mm × 4 mm. After dynamic deformation, EBSD samples were prepared by slicing parallel to the compression direction, and the samples were polished using a Shanghai metallographic PFD-2 electrolytic polishing machine with an electrolytic solution of 12 mL perchloric acid, 120 mL methanol, and 68 mL n-butanol. The polishing temperature was stabilized at -20 °C by adding liquid nitrogen; the polishing current was 0.6~0.7 A, and the polishing time was about 50 s. The surface of the specimen was cleaned with alcohol after polishing. The EBSD was carried out on the field emission scanning electron microscope (SEM, JSM-6700F), equipped with an Oxford Instruments EBSD detector and working at an accelerating voltage of 20 KV. The step size of the samples was 0.05 μm with a scanning area of 1600 μm^2. The HKL Technology Channel 5 system was used to analyze the EBSD test data.

3. Results and Discussions

3.1. Initial Microstructure

The SEM microstructure and composition distribution map of the Ti-6Al-1Mo-2Zr-0.55Fe-0.1B alloy is illustrated in Figure 1. Obviously, the bimodal microstructure with the primary equiaxed α phase and β transformation matrix was achieved by the heat treatment, which was accepted as a result of the simultaneous combination of good strength and ductility [18,19]. The size of the primary equiaxed α phase was about 5.25 μm, occupying 21% of the content. The thickness of the secondary lamellar α phase was about 0.52 μm. On the other side, the distribution of each alloying element showed sufficient homogeneity, as indicated in Figure 1b.

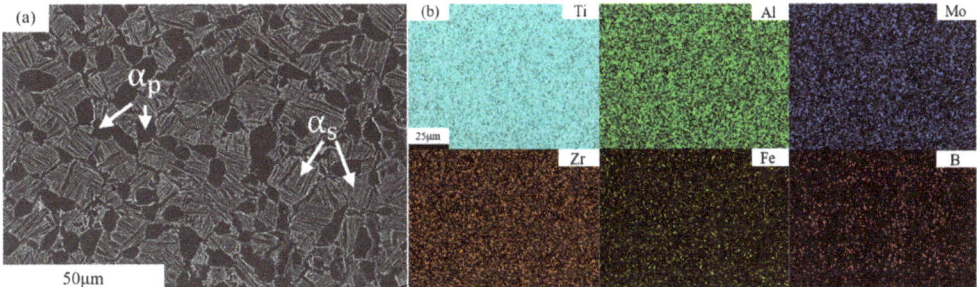

Figure 1. SEM results: (**a**) microstructure and (**b**) composition distribution maps of the Ti-6Al-1Mo-2Zr-0.55Fe-0.1B alloy.

3.2. Quasi-Static Tensile and Compression Experiment

Figure 2a shows the true stress–strain curve for the alloy at the strain rate of 0.007 s^{-1}. Result indicated that the yield stress was 927 MPa; the stress increased slightly with the increase in the strain, and the ultimate tensile stress could reach 1005 MPa. In comparison with a slight increase in the work hardening effect during the tensile test, the alloy showed a significantly higher work hardening during the quasi-static compression experiment (Figure 2b). Simultaneously, the ultimate compression stress increased up to 1370 MPa from 1081 MPa of tensile yield stress. Table 2 shows the quasi-static mechanical property data of the new titanium alloy and the Ti80 alloy with the bimodal microstructure. It can be seen that the strength of the novel alloy is superior to that of the Ti80 alloy. Meanwhile, good ductility is kept at the same level.

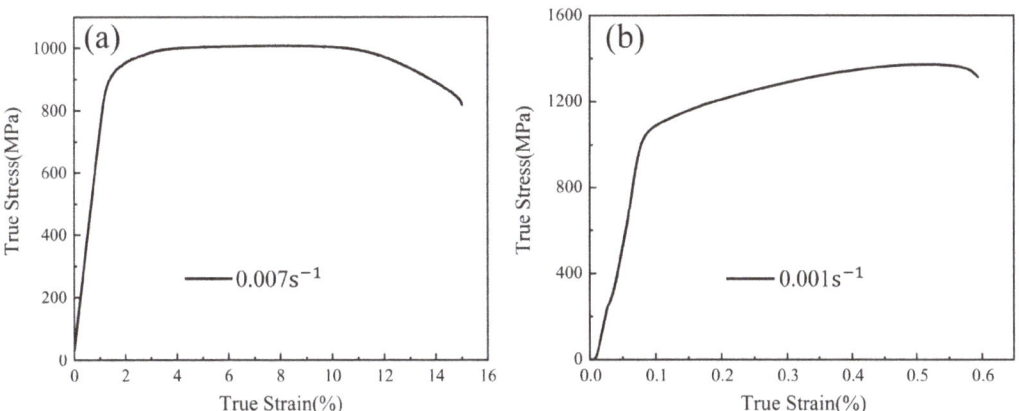

Figure 2. Quasi-static mechanical properties of the Ti-6Al-1Mo-2Zr-0.55Fe-0.1B alloy: (**a**) tension; (**b**) compression.

Table 2. Quasi-static mechanical property data of the Ti-6Al-1Mo-2Zr-0.55Fe-0.1B and Ti80 alloys.

Alloy Composition	Tensile			Compression		
	YS/MPa	TS/MPa	Strain (%)	YS/MPa	UCS/MPa	Strain (%)
Ti-6Al-1Mo-2Zr-0.55Fe-0.1B	927	1005	15.1	1018	1370	60
Ti-6Al-2Zr-1Mo-3Nb [20,21]	842	932	22	881	1046	50

3.3. Dynamic Compression Experiment

Dynamic true stress−strain curves of the Ti-6Al-1Mo-2Zr-0.55Fe-0.1B alloy at different strain rates are shown in Figure 3a. Obviously, compared to the quasi-static stress−strain curve (Figure 2b), they had no obvious yield stage and strain hardening during dynamic loading. The strength was higher during the high strain rate deformation and increased with the increase in the strain rate. Meanwhile, all the dynamic stress−strain curves fluctuated at the plastic deformation stage. The main data extracted from the dynamic stress−strain curves of the novel alloy and the Ti80 alloy are listed in Table 3. Among them, the absorbed impact energy could be considered as the area under the stress−strain curve of the uniform plastic deformation, which could reflect the dynamic mechanical property of materials more accurately by considering the combination of the strength and plasticity. The formula for calculating the absorbed impact energy is as follows [22]:

$$E = \int_{\varepsilon_s}^{\varepsilon_f} \sigma \cdot d\varepsilon \tag{1}$$

where ε_f and ε_s are strains at the beginning and the end of the plastic deformation, respectively. It can be seen that the absorbed impact energy of the novel alloy increases with an increase in the strain rate, exhibiting a higher value than the Ti80 alloy at the similar strain rate.

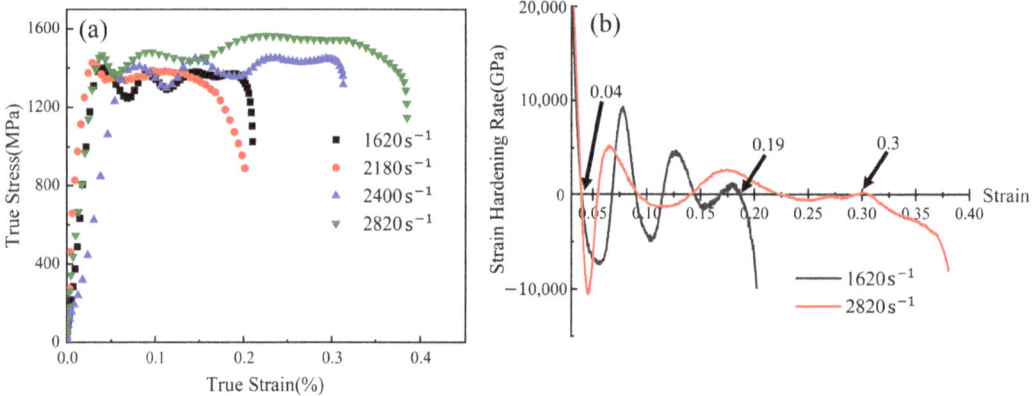

Figure 3. (**a**) Dynamic true stress−strain curves and (**b**) strain hardening rate of the Ti-6Al-1Mo-2Zr-0.55Fe-0.1B alloy.

Table 3. Dynamic mechanical property data of the Ti-6Al-1Mo-2Zr-0.55Fe-0.1B and Ti80 alloys.

Alloy Composition	Ti-6Al-1Mo-2Zr-0.55Fe-0.1B				Ti-6Al-2Zr-1Mo-3Nb [21]		
Strain rate	1620 s^{-1}	2180 s^{-1}	2400 s^{-1}	2820 s^{-1}	1500 s^{-1}	2500 s^{-1}	3500 s^{-1}
UCS /MPa	1398	1438	1454	1567	1348	1418	1478
Strain	0.21	0.21	0.31	0.39	0.13	0.22	0.30
Akv/MJ·m^{-3}	229	240	335	515	158	299	422

The strain hardening rate θ can be used to express the ability of a material to resist plastic deformation, where θ = dσ/dε. Figure 3b shows the strain hardening rate of samples at the high strain rates of 1620 s^{-1} and 2820 s^{-1}. The strain hardening rate curves developed similarly at the beginning of the deformation, and the alloy underwent continuous plastic deformation, which led to the first peak stress at the strain 0.04. After that, the development of the strain hardening rate was non-monotonous. This implies that the thermal softening effect happened during the dynamic loading. It is well known that adiabatic heating is always generated as a result of accumulated plastic work during the dynamic loading when the generated heat cannot dissipate outside in a short time, which increases the local temperature in materials and leads to the dislocation annihilation. The wave-shape curves indicate the competition of strain hardening and thermal softening of the alloy during dynamic loading [23]. When the strain hardening rate θ > 0, the strain hardening effect dominates in the plastic deformation of the alloy, and when θ < 0, it shifts to be dominated by the thermal softening effect. Obviously, for the investigated deformation with two strain rates, the final hardening rate was negative at the strains of 0.19 and 0.3, which shows the advantage of the thermal softening effect.

The inverse pole figures for the initial and dynamically deformed samples are shown in Figure 4a–c. The results indicated that the initial microstructure was homogeneous as a result of heat treatment, and the grain size was about 6.43 μm. After the dynamic compression, the grains suffered fragmentation, and the grain size decreased up to 5.51 μm and 5.3 μm after high strain rates of 1620 s^{-1} and 2820 s^{-1}, respectively. On the other hand, deformation twins activated in the microstructure as a result of dynamic loading and promoted the increase in the strain rate. The morphology of the deformation twins showed diversity under dynamic loading, including twin boundaries terminated at the grain boundaries, those terminated within the grains, and even the generated penetrating twins. Meanwhile, a twin wafer-like microstructure was found in the sample subjected by dynamic loading at 2820 s^{-1}.

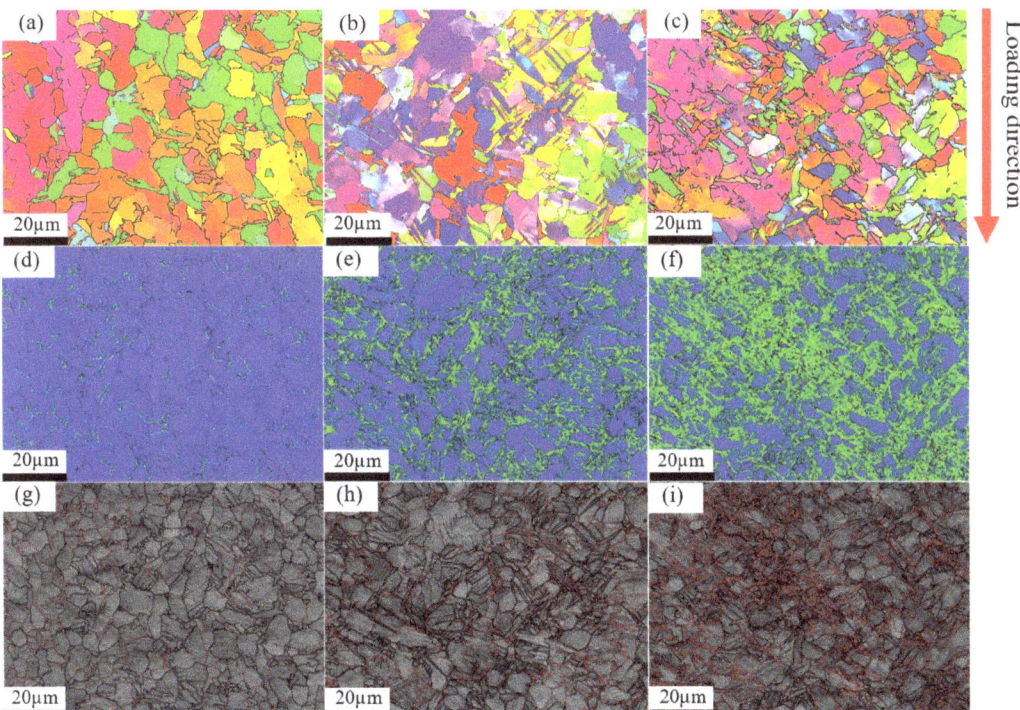

Figure 4. EBSD results: (**a–c**) inverse pole figures; (**d–f**) kernel average misorientation maps; (**g–i**) grain boundary orientations for (**a,d,g**) initial, (**b,e,h**) deformed with strain rate 1620 s^{-1}, and (**c,f,i**) deformed with strain rate 2820 s^{-1} samples.

Figure 4d–f shows the kernel average misorientation (KAM) plot of the new alloy, which can represent the density distribution of the geometrically necessary dislocation (GND) boundaries macroscopically. Those boundaries with KAM > 0.5° are considered as high-density dislocation boundaries (shown in green in the KAM plot), and those with

KAM ≤ 0.5° are considered as low-density dislocation boundaries (shown in blue in the KAM plot). In the initial state, high-density dislocations were rare and only existed as a result of previous heat treatment. However, the dislocation density of the sample increased significantly after dynamic loading, and dislocations were distributed not only at grain boundaries but also in grain interiors, which showed strong strain rate dependence. On the other hand, the fraction of low-angle grain boundaries (LAGBs) increased with the increase in the strain rate, and the fraction of LAGBs increased to 68.84% and 82.37% after the high strain rates of 1620 s^{-1} and 2820 s^{-1}, respectively. Some other relevant parameters of different state samples obtained through EBSD data statistics are listed in Table 4.

Table 4. Microstructure parameters of the Ti-6Al-1Mo-2Zr-0.55Fe-0.1B alloy in the initial and deformed states.

State	Avg·GS (μm)	HAGB (%)	LAGB (%)	The Fraction of Grain in Different States (%)		
				Recrystallized	Substructured	Deformed
Initial	6.43	45.58	54.42	44.4	51.9	3.7
1620 s^{-1}	5.51	68.84	31.16	6.6	67.6	25.8
2820 s^{-1}	5.3	82.37	17.63	5.2	21.7	73.1

Figure 5 shows the pole figures of the initial sample and the samples after the dynamic impact testing at 1620 s^{-1} and 2820 s^{-1}, respectively. The maximum texture intensity of the initial sample was 18.12. The sample after the impact testing with 1620 s^{-1} and 2820 s^{-1} showed a maximum texture intensity of 24.84 and 23.66 along the <0001> pole. Thus, more grains in the deformed samples must be aligned along the <0001> pole direction, demonstrating the large influence of the strain rate on the crystallographic texture evolution.

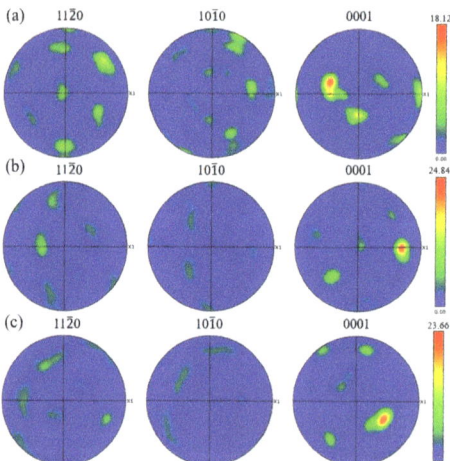

Figure 5. Pole figures of (**a**) initial, (**b**) deformed with strain rates 1620 s^{-1}, and (**c**) 2820 s^{-1} samples.

It is worth mentioning that the high-strength titanium alloy kept integrity and did not show any instability after dynamic loading, showing excellent comprehensive mechanical properties. The new Ti-6Al-1Mo-2Zr-0.55Fe-0.1B alloy is classified as the near-α titanium alloy according to the K$_\beta$ stability coefficient calculation [24]. As is commonly known, the α phase of a titanium alloy has a hexagonal close-packed (HCP) structure, with only three slip systems, but the dislocation slip on prismatic surfaces is facilitated to launch due to the c/a value (1.587) below the ideal HCP value (1.633). Furthermore, the addition of Al elements is considered to reduce the c/a value and benefit the dislocation movement [25–27]. Additionally, an abundance of deformation twins released the dislocation pile-up and coordinated the plastic deformation of the alloy during the dynamic loading [28,29]. On the other hand, the damage of materials under the dynamic loading was always related to the occurrence of adiabatic shear bands, which were considered to accompany the dynamic recrystallization process [30–32]. In our investigation, the adiabatic shear bands have not been observed by optical microscopy and EBSD, and the content of recrystallized grains decreased seriously; even the imposed strain rate was equal to 2820 s^{-1}, as indicated in Table 4. Obviously, the novel titanium alloy has a good potential for the applications due to the simultaneous high strength and ductility, although the deformation mechanism still needs to be revealed further.

4. The Johnson–Cook Constitutive Equation

As mentioned above, when materials suffered high strain rate deformation, the plastic work generated from the deformation could be converted into heat energy, which increased the local temperature of materials. Theoretically, the temperature rise calculation was carried out by the following equation:

$$\Delta T = \beta \frac{W_p}{\rho C_p} \tag{2}$$

where β is the ratio of the mechanical energy transferred into the thermal energy, generally taken as 0.9; W_p is the plastic work; $W_p = \int_0^{\varepsilon_i} \sigma_i d\varepsilon_i$, ρ is the material density; $\rho = 4.45$ g/cm^2; and C_p is the constant pressure specific heat capacity of the material 0.526 J/g·K. The temperature rise of the new alloy at different strain rates can be obtained from the Equation (2); the result is presented in Figure 6. The fitting can be obtained as follows:

$$\Delta T = 26.478 e^{0.0007 \dot{\varepsilon}} \tag{3}$$

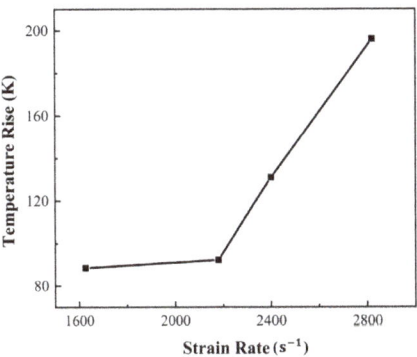

Figure 6. Temperature rises of the Ti-6Al-1Mo-2Zr-0.55Fe-0.1B alloy at different strain rates.

Obviously, the temperature rises of the new alloy increase exponentially under the dynamic loading, especially when the strain rate exceeds 2000 s^{-1}.

The thermal-viscoplastic constitutive model is commonly used to describe the dynamic mechanical response of metallic materials, such as the Johnson–Cook model, the Zerrilli–Armstrong model, and the Steinberg–Guinan model. Among them, the Johnson–Cook model is simple and can better describe the strain hardening, strain rate effects, and thermal softening effect of the material. In this paper, the Johnson–Cook model is used to describe the dynamic intrinsic structure of the new titanium alloy, and the model expression is as follows:

$$\sigma = (A + B\varepsilon^n)\left[1 + C \ln\left(\frac{\dot{\varepsilon}}{\dot{\varepsilon}_0}\right)\right](1 - T^{*m}) \tag{4}$$

where A, B, C, n, and m are coefficients to be determined; σ is the flow stress; ε is the equivalent plastic strain; and $\dot{\varepsilon}_0$ and $\dot{\varepsilon}$ are the reference strain rate and the equivalent plastic strain rate. T^* is the dimensionless temperature ($T^* = (T - T_0)/(T_m - T_0)$), where T_0, T, and T_m are the reference temperature, the deformation temperature, and the melting point temperature of the material, taking $T_0 = 293$ K and $T_m = 1800$ K.

When $\dot{\varepsilon}_0 = 10^{-3} s^{-1}$, $T = T_0$, for the compression experiment performed under quasi-static loading at room temperature, the equation can then be decoupled as follows:

$$\sigma = A + B\varepsilon^n \tag{5}$$

It is rewritten as $\ln(\sigma - A) = \ln B + n \cdot \ln \varepsilon$, substituted into the stress–strain data under the quasi-static compressive loading at room temperature for linear fitting, and the final strain-hardening parameters are obtained as A = 1018 MPa, B = 774 MPa, $n = 0.97$.

The strain rate strengthening parameter C responds to the strain rate effect of the material and can be fitted using stress–strain data for different strain rates at room temperature, where the equation is as follows:

$$\sigma = \left(1018 + 774 \cdot \varepsilon^{0.97}\right)\left[1 + C \cdot \ln\left(\frac{\dot{\varepsilon}}{\dot{\varepsilon}_0}\right)\right] \tag{6}$$

Obviously, the value of C is the slope of $\sigma/(1018 + 777 \cdot \varepsilon^{0.97}) - 1 = C \cdot \ln\left(\frac{\dot{\varepsilon}}{\dot{\varepsilon}_0}\right)$, and after substituting the stress–strain data for different strain rates at room temperature, the value of the strain rate strengthening parameter C is 0.013.

Using the adiabatic temperature rise in Equation (5) plus the reference temperature as the deformation temperature, brought into Equation (6), the value of m is obtained by fitting it to the experimental data and taking the average value of 1.72. Then, the complete J–C constitutive equation is as follows:

$$\sigma = \left(1018 + 774 \cdot \varepsilon^{0.97}\right)\left[1 + 0.013 \cdot \ln\left(\frac{\dot{\varepsilon}}{0.001}\right)\right]\left(1 - T^{*1.72}\right) \tag{7}$$

Figure 7 shows the results of the fitting of the experimental data of the alloy at high strain rates. The J–C constitutive equation constructed in this paper can effectively predict the mechanical response of the alloy under high strain rate deformation, especially in the plastic deformation stage.

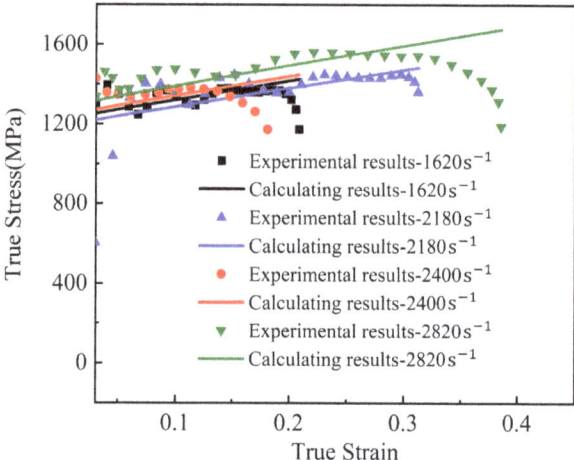

Figure 7. Comparison of the experimental results with the J–C constitutive equation calculations for the Ti-6Al-1Mo-2Zr-0.55Fe-0.1B alloy.

5. Conclusions

(1) A novel near-α titanium Ti-6Al-1Mo-2Zr-0.55Fe-0.1B alloy was developed, which possesses simultaneous high strength (>1000 MPa) and ductility (>15%).
(2) The strength of the alloy under dynamic loading enhanced significantly compared with the quasi-static compression and increased with the increase in the strain rate. An abundance of deformation twins released the dislocation pile-up and coordinated the plastic deformation of the alloy during the dynamic loading.
(3) The Johnson–Cook constitutive equation was modified by the adiabatic temperature rise term, and a nonlinear fit was performed to construct a dynamic constitutive relationship at room temperature, which could describe the rheological behavior of the alloy under high strain rate conditions at room temperature.

The novel near-α titanium Ti-6Al-1Mo-2Zr-0.55Fe-0.1B alloy possesses excellent mechanical properties, a low-cost design, and good weldability, making it a good potential application in marine engineering.

Author Contributions: Author contributions: conceptualization, C.Y. and Y.D.; methodology, Y.D. and C.Y.; validation, C.Y., Y.D. and I.V.A.; formal analysis, C.Y.; investigation, C.Y., H.C. and Y.D.; data curation, C.Y.; writing—original draft preparation, C.Y.; writing—review and editing, C.Y. and Y.D.; supervision, C.W., M.H., Y.D., H.C. and I.V.A.; project administration, I.V.A., H.C. and Y.D.; funding acquisition, Y.D. and H.C. All authors have read and agreed to the published version of the manuscript.

Funding: The authors would like to gratefully acknowledge the support of the National Key Research and Development Program of China (No. 2021YFB3700802), the National Natural Science Foundation of China (No.51931008), the Ministry of Education of the Russian Federation (Project 0838-2020-0006), and the A Project, funded by the Priority Academic Program Development (PAPD) of Jiangsu Higher Education Institutions.

Data Availability Statement: Not applicable.

Conflicts of Interest: The authors declare no conflict of interest.

References

1. Zang, M.C.; Niu, H.Z.; Liu, S.; Yu, J.S.; Zhang, H.R.; Zhang, D.L. Achieving high tensile strength-ductility synergy of a fully-lamellar structured near alpha titanium alloy at extra-low temperatures. *J. Alloy. Compd.* **2022**, *923*, 166363. [CrossRef]
2. Mahender, T.; Anantha Padmanaban, M.R.; Balasundar, I.; Balasundarc, I.; Raghuc, T. On the optimization of temperature and cooling rate to maximize strength and ductility of near α titanium alloy IMI 834. *Mater. Sci. Eng. A* **2021**, *827*, 142052. [CrossRef]
3. Ballat-Durand, D.; Bouvier, S.; Risbet, M.; Pantleonb, W.F. Through analysis of the microstructure changes during linear friction welding of the near-α titanium alloy Ti-6Al-2Sn-4Zr-2Mo (Ti6242) towards microstructure optimization. *Mater. Charact.* **2019**, *151*, 38–52. [CrossRef]
4. Cheng, M.; Yu, B.; Guo, R.; Shi, X.H.; Xu, L.; Qiao, J.W.; Yang, R. Electron beam welding of a novel near α high temperature titanium alloy powder compact: Effect of post-welding heat treatment on tensile properties. *J. Mater. Res. Technol.* **2021**, *10*, 153–163. [CrossRef]
5. Chuvil'Deev, V.N.; Kopylov, V.I.; Nokhrin, A.V.; Tryaev, P.V.; Tabachkovad, Y.N.; Chegurov, M.K.; Kozlova, N.A.; Mikhaylov, A.S.; Ershova, A.V.; Graynov, M.Y.; et al. Effect of severe plastic deformation realized by rotary swaging on the mechanical properties and corrosion resistance of near-α-titanium alloy Ti-2.5Al-2.6Zr. *J. Alloy. Compd.* **2019**, *785*, 1233–1244. [CrossRef]
6. Meng, K.; Guo, K.; Yu, Q.; Miao, D.; Yao, C.; Wang, Q.F.; Wang, T.S. Effect of annealing temperature on the microstructure and corrosion behavior of Ti-6Al-3Nb-2Zr-1Mo alloy in hydrochloric acid solution. *Corros. Sci.* **2021**, *183*, 109320. [CrossRef]
7. Su, B.; Wang, B.; Luo, L.S.; Wang, L.; Su, Y.Q.; Wang, F.X.; Xu, Y.J.; Han, B.S.; Huang, H.G.; Guo, J.J.; et al. The corrosion behavior of Ti-6Al-3Nb-2Zr-1Mo alloy: Effects of HCl concentration and temperature. *J. Mater. Sci. Technol.* **2021**, *74*, 143–154. [CrossRef]
8. Ren, J.Q.; Wang, Q.; Zhang, B.B.; Wang, Y.; Liu, W.F.; Lu, X.F.; Li, S.Z.; Chen, R.J. Charpy impact anisotropy and the associated mechanisms in a hot-rolled Ti-6Al-3Nb-2Zr-1Mo alloy plate. *Mater. Sci. Eng. A* **2022**, *831*, 142187. [CrossRef]
9. Wang, Q.; Ren, J.Q.; Wu, Y.K.; Jiang, P.; Li, J.Q.; Sun, Z.J.; Liu, X.T. Comparative study of crack growth behaviors of fully-lamellar and bi-lamellar Ti-6Al-3Nb-2Zr-1Mo alloy. *J. Alloy. Compd.* **2019**, *789*, 249–255. [CrossRef]
10. Ren, J.Q.; Wang, Q.; Zhang, B.B.; DanYang, D.; Lu, X.F.; Zhang, X.B.; Zhang, X.D.; Hu, J.Y. Influence of microstructure on fatigue crack growth behavior of Ti-6Al-3Nb-2Zr-1Mo alloy: Bimodal vs. *Lamellar Struct. Intermet.* **2021**, *130*, 107058. [CrossRef]
11. Guo, Y.H.; Liu, G.; Huang, Y. A complemented multiaxial creep constitutive model for materials with different properties in tension and compression. *Eur. J. Mech. A Solids* **2022**, *93*, 104510. [CrossRef]

12. Xiong, J.H.; Li, S.K.; Gao, F.Y.; Zhang, J.X. Microstructure and mechanical properties of Ti6321 alloy welded joint by GTAW. *Mater. Sci. Eng. A* **2015**, *640*, 419–423. [CrossRef]
13. Gao, F.Y.; Gao, Q.; Jiang, P.; Liu, Z.Y.; Liao, Z.Q. Microstructure and mechanical properties of Ti6321 alloy welded joint by EBW. *Int. J. Lightweight Mater. Manuf.* **2018**, *1*, 265–269. [CrossRef]
14. Arab, A.; Chen, P.W.; Guo, Y.S. Effects of microstructure on the dynamic properties of TA15 titanium alloy. *Mech. Mater.* **2019**, *137*, 103121. [CrossRef]
15. Chiou, S.T.; Cheng, W.C.; Lee, W.S. The analysis of the microstructure changes of a Fe–Mn–Al alloy under dynamic impact tests. *Mater. Sci. Eng. A* **2004**, *386*, 460–467. [CrossRef]
16. Zhou, T.F.; Wu, J.J.; Che, J.G.; Wang, Y.; Wang, X.B. Dynamic shear characteristics of titanium alloy Ti-6Al-4V at large strain rates by the split Hopkinson pressure bar test. *Int. J. Impact Eng.* **2017**, *109*, 167–177. [CrossRef]
17. Jiang, X.J.; Chen, G.Y.; Men, X.L.; Dong, X.L.; Han, R.H.; Zhang, X.Y.; Liu, R.P. Ultrafine duplex microstructure and excellent mechanical properties of TC4 alloy via a novel thermo-mechanical treatment. *J. Alloy. Compd.* **2018**, *767*, 617–621. [CrossRef]
18. Aguilar, C.; Aguirre, T.; Martínez, C.; Martíneza, C.; De Barbieri, F.; SanMartína, F.; Salinas, V.; Alfonso, I. Improving the mechanical strength of ternary beta titanium alloy (Ti-Ta-Sn) foams, using a bimodal microstructure. *Mater. Des.* **2020**, *195*, 108945. [CrossRef]
19. Chong, Y.; Bhattacharjee, T.; Park, M.H.; Shibata, A.; Tsuji, N. Factors determining room temperature mechanical properties of bimodal microstructures in Ti-6Al-4V alloy. *Mater. Sci. Eng. A* **2018**, *730*, 217–222. [CrossRef]
20. Lv, Y.F.; Fu, H.; Zhang, Y.H.; An, F.B.; Jiang, B. Effects of forging and heat treatment process on microstructure and mechanical properties of Ti80 titanium alloy bars. *Hot Work. Technol.* **2021**, *9*, 81–83.
21. Xu, X.F.; Tayyeb, A.; Wang, L.; Cheng, H.W.; Zhou, Z.; Ning, Z.X.; Liu, X.P.; Liu, A.J.; Zhang, B.B.; Cheng, X.W. Research on dynamic compression properties and deformation mechanism of Ti6321 titanium alloy. *J. Mater. Res. Technol.* **2020**, *9*, 11509–11516. [CrossRef]
22. Dai, J.C.; Min, X.H.; Wang, L. Dynamic response and adiabatic shear behavior of β-type Ti–Mo alloys with different deformation modes. *Mater. Sci. Eng. A* **2022**, *857*, 144108. [CrossRef]
23. Pu, B.; Li, W.B.; Zhang, Q.; Zheng, Y.; Wang, X.M. Research on the dynamic compressive deformation behavior of 3D-Printed Ti6Al4V. *Metals* **2021**, *11*, 1327. [CrossRef]
24. Zhang, T.; Wei, D.X.; Lu, E.; Wang, W.; Wang, K.S.; Li, X.Q.; Zhang, L.Q.; Kato, H.; Lu, W.J.; Wang, L.Q. Microstructure evolution and deformation mechanism of α+β dual-phase Ti-xNb-yTa-2Zr alloys with high performance. *J. Mater. Sci. Technol.* **2022**, *131*, 68–81. [CrossRef]
25. Chai, Y.F.; Shan, L.; Jiang, B.; Yang, H.B.; He, C.; Hao, W.X.; He, J.J.; Yang, Q.S.; Yuan, M.; Pan, F.S. Ameliorating mechanical properties and reducing anisotropy of as-extruded Mg-1.0Sn-0.5Ca alloy via Al addition. *Prog. Nat. Science. Mater. Int.* **2021**, *31*, 722–730. [CrossRef]
26. Qin, D.Y.; Lu, Y.F.; Guo, D.Z.; Zheng, L.; Liu, Q.; Zhou, L. Tensile deformation and fracture of Ti–5Al–5V–5Mo–3Cr–1.5Zr–0.5Fe alloy at room temperature. *Mater. Sci. Eng. A* **2013**, *587*, 100–109. [CrossRef]
27. Guo, Z.; Miodownik, A.P.; Saunders, N.; Schillé, J.P. Influence of stacking-fault energy on high temperature creep of alpha titanium alloys. *Scr. Mater.* **2006**, *54*, 2175–2178. [CrossRef]
28. Xu, F.; Zhang, X.Y.; Ni, H.T.; Cheng, Y.M.; Zhu, Y.T.; Liu, Q. Effect of twinning on microstructure and texture evolutions of pure Ti during dynamic plastic deformation. *Mater. Sci. Eng. A* **2013**, *564*, 22–33. [CrossRef]
29. Liu, Q.; Fang, L.M.; Xiong, Z.W.; Yang, J.; Tan, Y.; Liu, Y.; Zhang, Y.J.; Tan, Q.; Hao, C.C.; Cao, L.H.; et al. The response of dislocations, low angle grain boundaries and high angle grain boundaries at high strain rates. *Mater. Sci. Eng. A* **2021**, *822*, 141704. [CrossRef]
30. Sung-Il, B.; Ratnesh, K.G.; Sharvan, K.; David, N.S. Temperature increases and thermoplastic microstructural evolution in adiabatic shear bands in a high-strength and high-toughness 10 wt.% Ni steel. *Acta Mater.* **2021**, *205*, 116568. [CrossRef]
31. Xu, Y.B.; Yang, H.J.; Meyers, M.A. Dynamic recrystallization in the shear bands of Fe–Cr–Ni monocrystal: Electron backscatter diffraction characterization. *Scr. Mater.* **2008**, *58*, 691–694. [CrossRef]
32. Yang, D.K.; Cizek, P.; Hodgson, P.D.; Wen, C.E. Microstructure evolution and nanograin formation during shear localization in cold-rolled titanium. *Acta Mater.* **2010**, *58*, 4536–4548. [CrossRef]

Article

Study on Dynamic Mechanical Behaviors and J–C Constitutive Model of a Fine-Grained D6A Steel

Zezhou Yang [1], Xiaowei Feng [1,*], Xiaoli Zhang [2], Yongfeng Shen [2], Xicheng Huang [1] and Ruoze Xie [1]

[1] Institute of Systems Engineering, China Academy of Engineering Physics, Mianyang 621999, China; yangzzxx@foxmail.com (Z.Y.); huangxc@caep.cn (X.H.); xierz@caep.cn (R.X.)

[2] Key Laboratory for Anisotropy and Texture of Materials, Northeastern University, Shenyang 110819, China; zhxiaoli001@163.com (X.Z.); shenyf@smm.neu.edu.cn (Y.S.)

* Correspondence: 414fengxw@caep.cn; Tel.: +86-0816-2485474

Abstract: The uniaxial tensile loading tests of coarse-grained D6A steel (CG, d = 20 μm) and fine-grained D6A steel (FG, d = 1.5 μm) were performed using a material testing machine and rotating disk Hopkinson tension bar, respectively. The stress–strain curves of two steels at different strain rates (0.001~1500 s^{-1}) were obtained. Results show that grain refinement effectively improves the strength of FG steel, which is achieved by high-density of grain boundaries and cementite particles hindering the movement of dislocations. In the strain rate range of the test, the strain rate sensitivity of FG D6A steel decreases, which is considered to be the result of the athermal stress that independent of the strain rate becomes the dominant part of the total stress. Combined with the experimental data, the parameters of the Johnson–Cook (J–C) constitutive model were calibrated. The stress–strain curves obtained by simulations are in good agreement with those from tests. These results may provide an important experimental reference and theoretical basis for the application of FG D6A steel in various fields.

Keywords: fine-grained D6A steel; fine grain strengthening; strain rate sensitivity; Johnson–Cook constitutive model; finite element simulation

1. Introduction

The production, use and performance improvement of steel materials are important to develop industrialization. For thousands of years, steel has been widely used in various engineering fields such as civil engineering, chemical engineering, aerospace and weaponry, and has profoundly affected the quality of human life. In recent years, aerospace materials continue to evolve, new materials represented by graphene, carbon fiber and rare earth elements are favored by researchers, but they still cannot surpass high-property steel in terms of material strength, impact toughness and production cost. In the future, ultra-high-strength steel will still be an indispensable material for load-bearing components such as aircraft landing gear, wing spars, and engine crankshafts. In order to meet the rapid development needs of the aviation industry, it is necessary to deeply explore the strengthening mechanism and application prospects of advanced steel materials.

Due to the lightweight and safety requirements of aircraft design, materials with high strength and toughness will be the first choice for airframe components. Ultra-fine-grained (UFG) steel has both ultra-high strength, good toughness [1] and low production cost, which has great application potential in the aviation industry. At present, scholars in China and abroad have carried out extensive research on the preparation process and mechanical properties of UFG steel. Pang [2] investigated the strengthening mechanism of low-temperature annealed UFG steel, and believed that grain refinement was the most effective method to simultaneously improve the strength, plasticity and impact toughness of the material. The strengthening of the experimental steel was not only the result of the interaction between dislocations, but also the interaction of grain boundaries and dislocations. Park [3] found that when the grains are refined to less than a few microns, the strength can be effectively improved. However, the material almost loses the work hardening ability, and its yield ratio is close to 1. Liu et al. [4] studied the effects of strain rate and temperature on the dynamic mechanical behavior of UFG IF steel. They found that UFG IF steel has low strain rate sensitivity and strong temperature sensitivity within a certain temperature range (−70 °C–100 °C). To sum up, the mechanical behavior of UFG steel is closely related to the microstructure (e.g., grain size) and loading conditions (e.g., strain rate, temperature).

Aircraft often face high-speed impact during landing, so it is of great significance to study the dynamic mechanical properties of structural component materials. At present, establishing a macroscopic phenomenological constitutive model or modifying the existing phenomenological model is the main method to describe the mechanical behavior of UFG materials. Khan [5] modified the Johnson–Cook (J–C) to describe the flow stress of nanometallic materials by introducing the Hall-Petch relationship between yield stress and grain size. Zheng et al. [6] introduced the grain size effect on the basis of the J–C model, and then superimposed it with the Armstrong-Frederick nonlinear hardening law. Finally, they proposed a mixed hardening model for UFG materials, which can simultaneously reflect the nonlinear hardening, grain size effect, strain hardening effect and strain rate effect of the material. Based on the modified hyperbolic sine Arrhenius equation, Wang [7] established a constitutive model capable of simulating high temperature micro-compression deformation of UFG pure aluminum. Therefore, building a constitutive model capable of describing large plastic deformation is the basic to promote the practical application of advanced steel materials.

D6A steel is a low-alloy high-strength steel that has been widely used in solid rocket casings, aircraft landing gear, warhead casings and other fields. Shen et al. [8–12] prepared micron-scale fine-grained (FG) D6A steel by using hot rolling and annealing, and analyzed the influence of grain size and nano-precipitation relative to the strength and toughness of FG steel. Grain refinement strengthening, precipitation phase strengthening and texture strengthening are the main mechanisms for the high strength of the investigated steels. Currently, several studies have extensively investigated FG D6A steel, and most of them focus on the preparation process and microstructure evolution. However, its mechanical behavior at dynamic loading was rarely reported. The present study aims to provide experimental and theoretical reference for the use of FG D6A steel in the aviation field.

2. Experimental Materials and Test Methods

The experiment material is D6A coarse-grained (CG) steel, which was produced by Baosteel Group. The D6A steel was forged into a slab with 110 mm in length, 60 mm in width, and 60 mm in thickness. First, in the process of hot rolling, the steel plate was heated to 1100 °C for 3 h, and then hot-rolled to a thickness of 10 mm after six passes. Then, the plate underwent two-phase warm rolling at 760 °C via two passes with ~25% reduction in each pass, and annealing at 660 °C for 20 min after each pass. After hot rolling and two-phase annealing, FG D6A steel samples were obtained. The nominal chemical composition of the steel, measured by using chemical analysis, is listed in as Table 1. The microstructure of FG D6A steel before deformation was characterized by scanning electron microscope (SEM) after polishing and acid leaching. As shown in Figure 1, the investigated steel is mainly composed of gray-black ferrite matrix (F) and white granular cementite (Fe_3C). During the hot rolling process, a large amount of spherical cementite precipitates from the grain boundary and is evenly distributed on the ferrite matrix.

Table 1. Chemical composition of the D6A steel (mass fraction%).

C	Si	Mn	Cr	Mo	Al	Ni	V	Fe
0.43	0.17	0.73	1.05	1.01	0.02	0.61	0.09	Bal

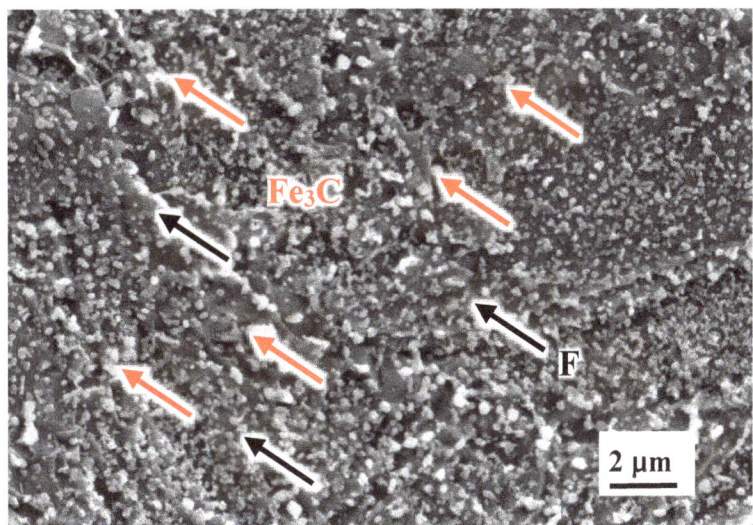

Figure 1. SEM images showing the microstructural characteristics of FG D6A steel.

Figure 2 shows the microstructure of FG steel obtained from electron backscatter diffraction (EBSD) microscopy. At the same time, the EBSD observation data were statistically sorted by Origin software, and the columnar distribution of grain size was obtained, as shown in Figure 3a,b. The grain size of CG and FG D6A steel was 20 μm and 1.5 μm, respectively.

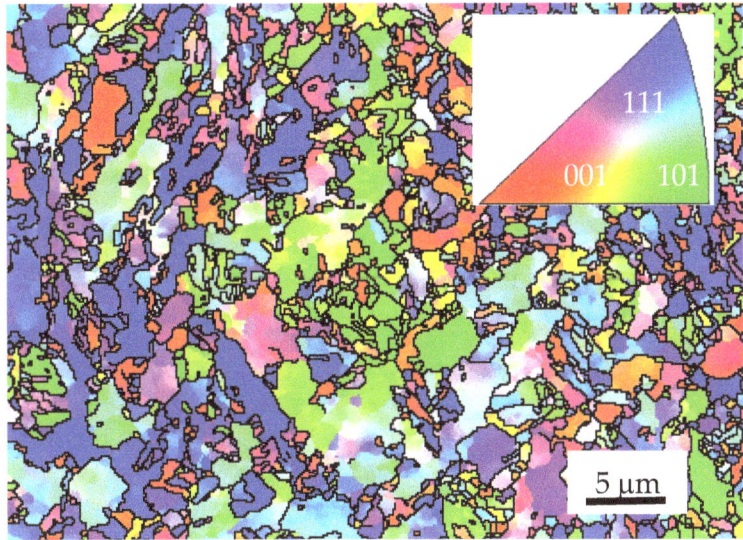

Figure 2. EBSD graph of FG D6A steel.

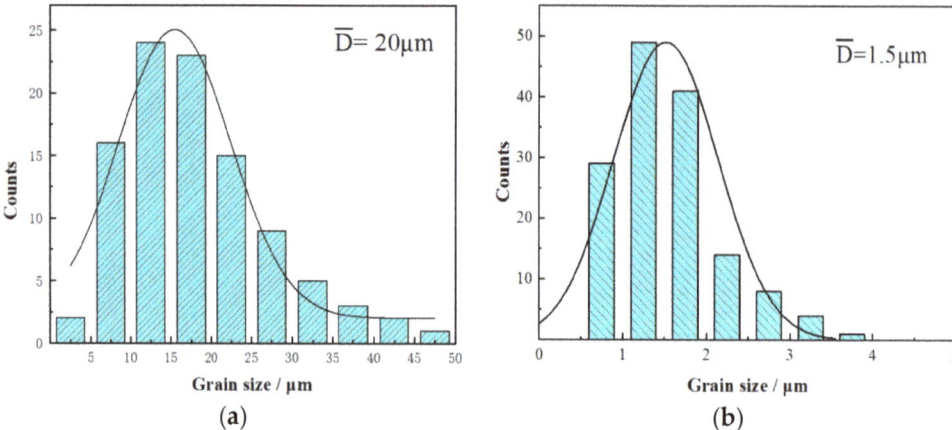

Figure 3. Grain size distributions of D6A steel: (**a**) CG D6A steel; (**b**) FG D6A steel.

Uniaxial tensile tests were carried out on the specimens at room temperature. Quasi-static tensile experiment was performed on a material testing machine and dynamic tensile test was carried out on a rotating disk Hopkinson bar. The steel was prepared into a sheet specimen and connected with the loading device by gluing. Figure 4a,b show the experimental device and specimen. Dynamic uniaxial tensile tests were conducted at various strain rates ($\dot{\varepsilon}$) (400~1500 s^{-1}) by using short metal rods with different sizes (8, 10 and 12 mm). For comparison, similar tensile experiments were performed on CG steel.

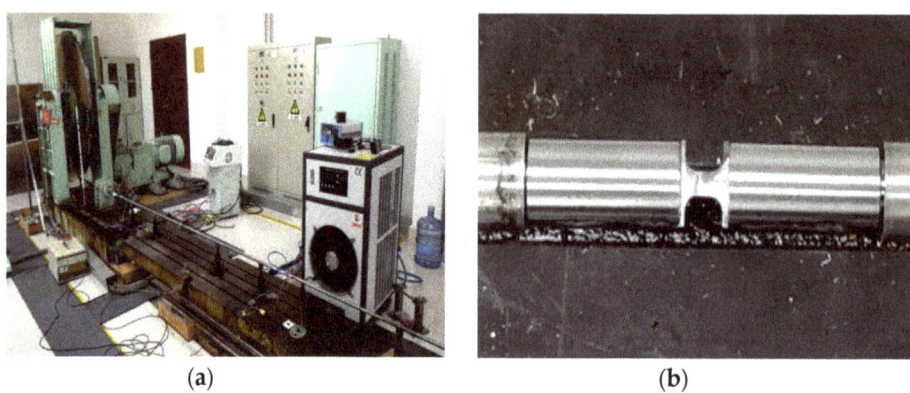

Figure 4. Photographs of Hopkinson tension bar and specimen in this study: (**a**) rotating disk Hopkinson bar; (**b**) steel specimens.

3. Experimental Results and Discussion

The experimental data were processed by the signal acquisition system embedded in the equipment and transformed by Origin software to obtain the stress–strain curves of the steel under different strain rates.

Figure 5a shows the representative stress–strain curves of the as-prepared CG steel, which was tested at room temperatures of 25 °C at a strain rate ($\dot{\varepsilon}$) of 0.001 s^{-1}, and the experiment was repeatable. The quasi-static tensile yield strength of FG D6A steel is approximately 980 MPa (curve turning point in the figure), which is significantly improved compared with that of existing CG steel (390 Mpa [9]).

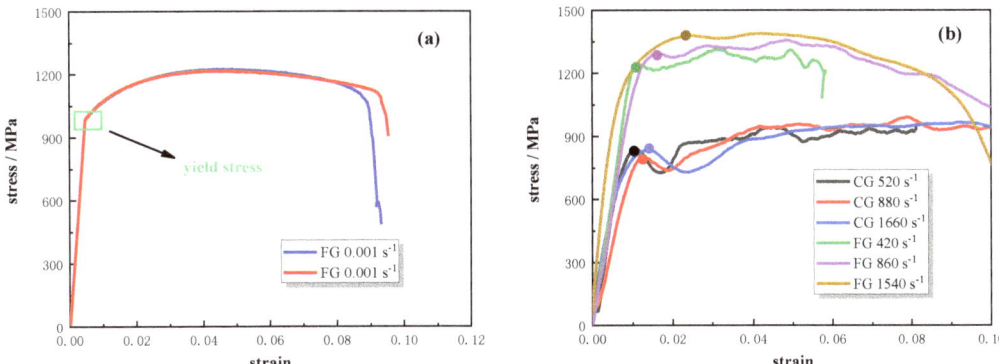

Figure 5. Stress–strain curves of the investigated steel under different strain rates: (**a**) quasi–static loading; (**b**) dynamic loading.

The dynamic tensile tests of the CG ($\dot{\varepsilon}$ = 520, 880, 1660 s^{-1}) and FG ($\dot{\varepsilon}$ = 420, 860, 1540 s^{-1}) D6A steel were carried out, and the obtained stress–strain curves are shown in Figure 5b. It is worth noting that the first peak stress can be regarded as material strength under dynamic loading (as indicated by the dots in the figure). In this experiment, the dynamic strength of CG steel is about 800 MPa, whereas that of FG steel is over 1300 Mpa. The strength of the investigated steel at various strain rates are shown in Table 2. In summary, grain refinement effectively improves the static and dynamic strength of D6A steel.

Table 2. Tensile mechanical properties of D6A steel at room temperature.

Temperature/°C	Grain Size	Strain Rate/s^{-1}	Strength/Mpa
25	FG	0.001	980
		420	1245
		860	1295
		1540	1378
	CG	0.017	390
		520	838
		880	800
		1660	835

Similar to CG steel, the grains of FG steel still exhibit typical crystallographic characteristics, and the microstructure profoundly affects the macroscopic mechanical behavior of materials. High-density grain boundaries have a strong constraint on dislocation slip, which is mainly caused by grain boundary obstacles and the orientation difference between adjacent grains. Due to the disordered arrangement of atoms at grain boundaries, the shear stress required for dislocation slip here increases suddenly. Furthermore, grains in different orientations will restrict mutual deformation, which makes it difficult for slip to proceed along the same crystal axis and eventually increases the critical shear stress of slip. Figure 6 shows that when the external force is not large enough to overcome the obstacle effect of grain boundary, the intragranular dislocation cannot continue to slip through the grain boundary to the adjacent crystal. With the accumulation of deformation, dislocations gather at the grain boundary and tangle with each other to produce dislocation pile-up groups, resulting in local stress concentration (shown in the red symbol), thereby improving the macroscopic strength of the investigated steel.

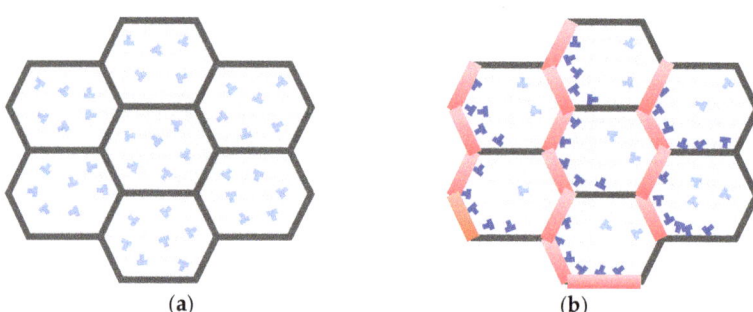

Figure 6. Schematic diagram of grain boundary hindering dislocations: (**a**) initial stage; (**b**) plastic deformation stage.

In addition, the deformation of FG steel can spread to more grains, so the stress distribution is more uniform than that of CG steel. Therefore, it is difficult for micropores and cracks to initiate and propagate, which ultimately increases the strength of the material.

Through microscopic observation, a large number of nano-scale cementite particles are found to be uniformly dispersed on the FG D6A steel matrix, as shown in Figure 2. The strength of the material is effectively improved by introducing nano-scale precipitates, which have been widely used in metal materials. Essentially, precipitation strengthening is achieved by generating small, high-hardness second-phase particles that hinder dislocation motion. At the same time, the interaction between cementite and dislocations is also enhanced, thereby strengthening the metal without greatly affecting the stiffness of the material [13]. Dislocations cannot shear hard cementite particles. In this case, dislocations will bypass particles through Orowan ring or cross-slip [14]. The strength increment is closely related to the volume fractions and size of the precipitated phase.

Figure 7 shows that the extrusion deformation between grains provides more nucleation sites for cementite, and the local temperature rise generated by plastic deformation also accelerates the nucleation process. Fresh dislocations appear around the cementite, and the dislocation density increases uninterruptedly. With the accumulation of plastic deformation, the degree of dislocation pile-up deepens. External force causes the cementite particles to move gradually and gather at the grain boundary. Moreover, cementite has a pinning effect on the grain boundaries, which makes it difficult for the grain boundary slip between grains to proceed and aggravates the stress concentration. Therefore, the interaction of dislocations, cementite and high-density grain boundaries results in FG D6A steel with ultra-high strength.

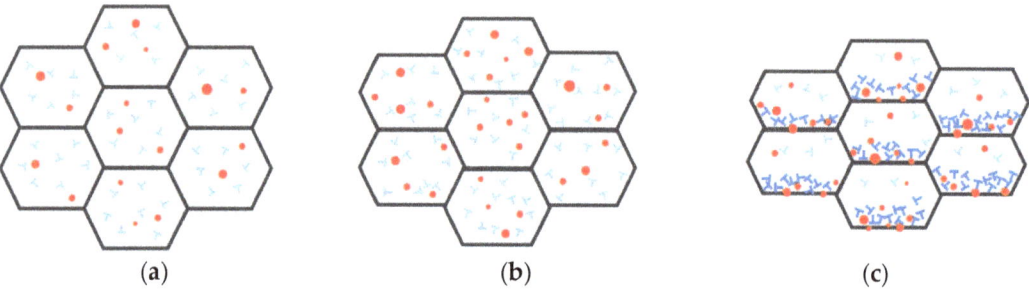

Figure 7. Schematic diagram of grain boundary and cementite hindering dislocations: (**a**) initial stage; (**b**) middle stage; (**c**) final stage.

Figure 8 presents the strength changes of CG/FG steel under static and dynamic tensile loading. D6A steel exhibits remarkable strain rate effect. As the strain rate changes from quasi-static to dynamic, the material strength increases significantly.

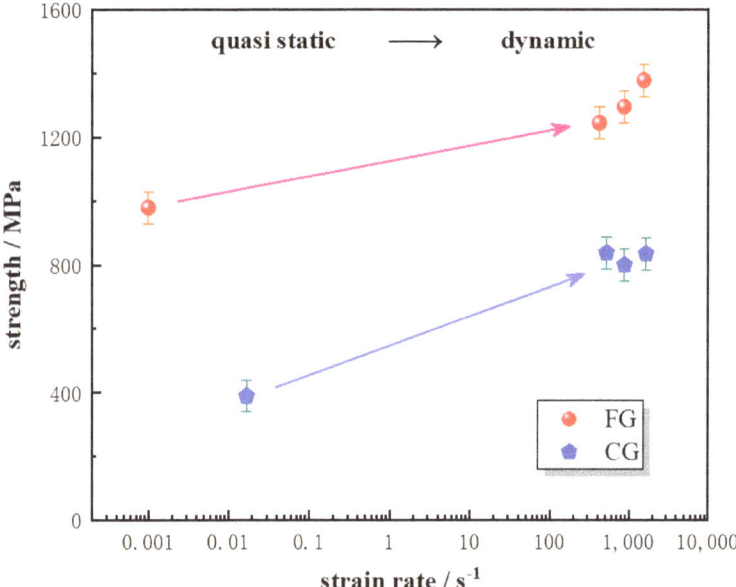

Figure 8. Quasi−static and dynamic tensile strength of D6A steel.

The fracture of the sample was observed by SEM to analyze the microstructure evolution of the investigated steel in the high-speed tensile test. Figure 9 shows the SEM image of the FG D6A steel after dynamic tensile deformation. The comparison shows that the microstructures under different strain rates are basically the same and mainly ferrite and cementite. However, with the increase in strain rate, the deformation of ferrite structure becomes increasingly intense, resulting in the increase in the total number of grains and grain boundary density, which enhances the hindrance of dislocation movement, and thus improves the dynamic strength of FG D6A steel.

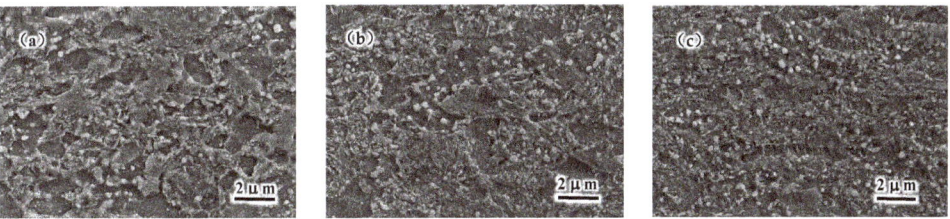

Figure 9. SEM micrograph of D6A steel after deformation: (**a**) 420 s^{-1}; (**b**) 860 s^{-1}; (**c**) 1540 s^{-1}.

Two parameters are generally used to describe the strain rate sensitivity of materials [15], namely the strain rate sensitivity coefficient m and the flow stress increment $\Delta\sigma$. Here, the strain rate sensitivity coefficient can be calculated as follows [16]:

$$m = \left(\frac{\partial \ln \sigma}{\partial \ln \dot{\varepsilon}}\right)_T \approx \frac{\ln(\sigma_2 - \sigma_1)}{\ln(\dot{\varepsilon}_2 - \dot{\varepsilon}_1)} \qquad (1)$$

where $\dot{\varepsilon}$ is the strain rate and σ can be taken as the yield strength corresponding to different strain rates. In addition, the flow stress increment $\Delta\sigma$ can be expressed as follows:

$$\Delta\sigma = \sigma_d - \sigma_s \qquad (2)$$

where, σ_d and σ_s are dynamic and static yield strength, respectively.

As shown in Table 3, the steel exhibits obvious strain rate effect in this experiment. The strain rate sensitivity coefficient m of the original CG steel was about 0.822~0.976 at room temperature, and the flow stress increment $\Delta\sigma$ was about 434 MPa. Compared with the quasi-static yield strength, the dynamic strength more than doubled, and the stress increment ratio reached 113.3%. The strain rate sensitivity coefficient m and flow stress increment $\Delta\sigma$ of FG steel are 0.816~0.923 and 326 MPa, respectively. Although the strength increment of FG steel under dynamic loading is slightly lower than that of the original CG steel, the strength increase ratio is very small, only about 33.2%. Therefore, as the average grain size decreases from 20 µm to 1.5 µm, the strain rate sensitivity of D6A steel shows a weakening trend.

Table 3. Strain rate sensitivity index of D6A steel (0.001~1500 s^{-1}).

Steel	m	$\Delta\sigma$/MPa	Stress Increment Ratio/%
CG	0.822~0.976	434	111.3
FG	0.816~0.923	326	33.2

The flow stress of UFG steel can be mainly divided into two parts [17]: thermal stress and non-thermal stress. Thermal stress corresponds to the thermal activation movement, which mainly overcomes the short-range barrier dependent on temperature and strain rate. Non-thermal stress is related to the long-range obstacles that cannot be overcome in the thermal activation process. The latter is almost independent of strain rate and temperature but largely affected by grain size. Jia established the following mechanical response function [18] in the study of UFG iron with BCC structure:

$$\tau = \tau_0 + \beta d^{-1/2} + g(\gamma) + \overline{\tau_0}\left[1 - \left\{\frac{kT}{\Delta G_{k0}} \ln\left(\frac{\dot{\gamma}_{k0}}{\dot{\gamma}}\right)\right\}^{\frac{2}{3}}\right] \qquad (3)$$

The first three parts of the formula are the non-thermal stress dominated by initial lattice resistance and grain boundary strengthening effect, where $g(\gamma)$ is the strain hardening coefficient. The last part named Peierls-Nabarro stress is related to the short-range barrier, and it can be qualitatively judged that this stress component is related to the strain rate but has no direct relationship with the grain size, that is, the strain rate-dependent stress components have not changed greatly for different grain size steels. By contrast, when the size of D6A steel decreases from 20 µm to 1.5 µm, the grain boundary area increases greatly, and the non-thermal stress will occupy the main part of the flow stress. Therefore, in the same strain rate range, the proportion of strain rate-related thermal stress reduces gradually with the grain size refinement, and macroscopically, the strain rate sensitivity of FG D6A steel is weaker than that of original CG steel.

Furthermore, compared with CG steel, the strain hardening ability of FG D6A decreases under dynamic loading, and plastic instability begins to occur after the stress reaches the peak. This phenomenon deserves further study. Nevertheless, the stress softening phenomenon of some UFG materials [15] did not appear in the D6A steel.

4. Johnson–Cook Constitutive Model and Parameter Fitting

Establishing the constitutive model that can describe the plastic deformation behavior of materials is the premise of accurately simulating severe deformation processes such as impact [19]. In order to promote the application of FG D6A steel in aviation and military fields, it is very important to find a constitutive model that can accurately describe its dynamic deformation behavior. The J–C model has become the most widely used phenomenological viscoplastic model for metallic materials due to its simple form and easy promotion of engineering application. The J–C model well describes the relationship between flow stress and plastic strain, strain rate and temperature. Its classical mathematical expression is as follows [20]:

$$\sigma = \left(A + B\varepsilon_p{}^n\right)\left(1 + C \ln \dot{\varepsilon}^*\right)\left(1 - T^{*m}\right) \tag{4}$$

Therein, σ reflects flow stress; A, B and n are strain-related material parameters at reference strain rates, where A is the yield strength. C is the coefficient describing the strain rate strengthening effect, and m is the factor describing the thermal softening effect. Here, ε_p is the equivalent plastic strain, and $\dot{\varepsilon}^*$ is the dimensionless equivalent plastic strain rate. T^* represents dimensionless temperature. as the temperature rise caused by dynamic loading is small, the thermal softening effect is not considered, so the value of m could be equal to 0. Finally, the simplified J–C model equation is obtained as follows:

$$\sigma = \left(A + B\varepsilon_p{}^n\right)\left(1 + C \ln \dot{\varepsilon}^*\right) \tag{5}$$

The material mechanical properties under different processes are often quite different, so it is necessary to determine the parameters combined with mechanical experiments. When loading at the reference strain rate, the J–C model is further simplified to the Ludwik-Hollomon formula [21]:

$$\sigma = \left(A + B\varepsilon_p{}^n\right) \tag{6}$$

According to the quasi-static loading stress–strain curve (Figure 5a), the yield strength of FG D6A steel is approximately about 980 MPa, that is, $A = 980$. Logarithmic transformation of the above formula can be obtained:

$$\ln(\sigma - A) = \ln B + n \ln \varepsilon_p \tag{7}$$

Fit the curve of $\ln(\sigma - A) - \ln \varepsilon_p$. As shown in Figure 10a, its slope is n, and B is obtained according to the intercept. Finally, the strain-related parameters can be calculated as follows: $B = 493$, $n = 0.625$. After the aforementioned parameters are determined, select the yield strength under different strain rates, and fit the curve. As shown in Figure 10b, the slope is the strain rate hardening coefficient, and $C = 0.026$.

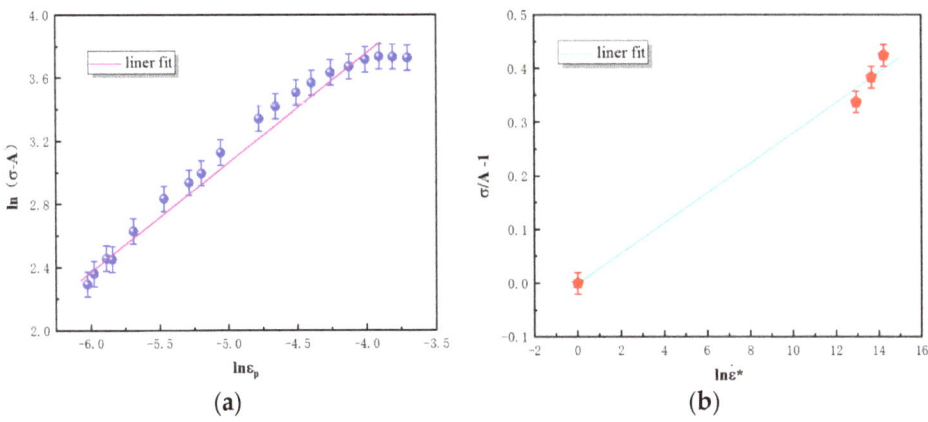

Figure 10. J–C constitutive model constants fitting results: (**a**) $\ln(\sigma - A) - \ln \varepsilon_p$; (**b**) $(\sigma/A - 1) - \ln \dot{\varepsilon}^*$.

In conclusion, the J–C model suitable for FG D6A steel at room temperature can be obtained from the data, as follows:

$$\sigma = \left(980 + 493\varepsilon_p^{0.625}\right)\left(1 + 0.026 \ln \dot{\varepsilon}^*\right) \tag{8}$$

A 3D specimen model is established, the size of which is consistent with the experimental specimen. The left end surface of the specimen is completely fixed, and the right end surface is given velocity boundary conditions to simulate loading at different strain rates. Hexahedral elements were selected for the mesh, and the mesh in the middle was appropriately refined. Select the J–C model that was built in the software and input the above-mentioned relevant parameters. Of note, the elastic modulus of FG D6A steel is 210 GPa. Loading simulations at different strain rates were carried out in ABAQUS software using explicit dynamic analysis type.

Figure 11a shows the nephogram of finite element tensile simulation results. As can be observed, the specimen generates stress concentration in the middle. Calculate the mechanical information of the middle part, plot the stress–strain curve and synchronously compare with the experimental data, as shown in Figure 11b. The results indicate that the stress value of the simulation results is basically consistent with the experimental data. Several stress values under the same strain are selected from the experimental data and the numerical simulation curve, and the comparison shows that the average error in the hardening stage maintains within 15%. Therefore, the parameters accurately describe the dynamic mechanical behavior of FG D6A steel at a certain range of strain rate in our tests.

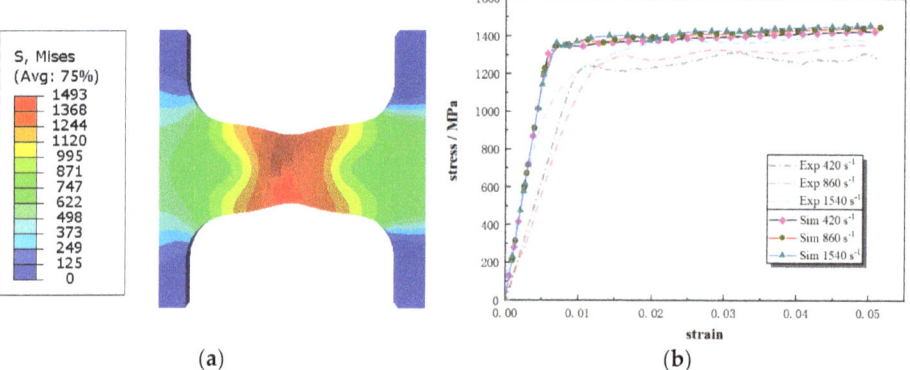

Figure 11. Finite element calculation results of tensile loading of FG D6A steel by using the J–C model: (a) stress cloud; (b) comparison of the result with experimental data.

5. Conclusions

In this study, the quasi-static and dynamic mechanical properties of D6A steel were investigated by using a material testing machine and rotating disk Hopkinson tension bar. From the microscopic point of view, the influence of grain refinement on the macroscopic mechanical properties of materials such as strength and work hardening ability was discussed, and the reasons for the decrease in strain rate sensitivity with the decrease in grain size was analyzed simultaneously. The parameters of the classical J–C model were fitted based on the experimental results. The relevant results can provide certain reference value for promoting the practical application of FG D6A steel, as follows:

1. Grain refinement and cementite precipitation are the main reasons for the strength improvement of FG D6A steel. High-density grain boundaries and hard precipitates improve the strength of the material by hindering dislocation movement. In addition, the investigated steel shows obvious strain rate effect, which is considered to be

the result of grain crushing that further enhances the fine-grain strengthening effect during dynamic loading.
2. Grain refinement reduces the strain rate sensitivity of the investigated steel. According to the existing mechanical response model, the reason is that the non-thermal stress related to grain size in the FG steel increases significantly and occupies the dominant part of the total flow stress.
3. The J–C model suitable for FG D6A steel at room temperature was obtained as follows: $\sigma = \left(980 + 493\varepsilon_p^{0.625}\right)\left(1 + 0.026\ln\dot{\varepsilon}^*\right)$. Combined with the finite element calculation result, this model can effectively predict the dynamic mechanical behavior of FG D6A steel.

Author Contributions: Conceptualization, X.F.; methodology, X.F., X.Z., Z.Y. and Y.S.; software, Z.Y.; investigation, X.F., X.Z., Y.S., R.X. and X.H.; writing—original draft preparation, Z.Y.; writing—review and editing, X.F. and Y.S. All authors have read and agreed to the published version of the manuscript.

Funding: This research was funded by the National Natural Science Foundation of China, grant number 12102413.

Institutional Review Board Statement: Not applicable.

Informed Consent Statement: Not applicable.

Data Availability Statement: Not applicable.

Acknowledgments: The authors gratefully acknowledge the financial support from the National Natural Science Foundation of China with Grant No. 12102413. The authors also gratefully acknowledge the experimental support from the Chinese Academy of Engineering Physics and Northeastern University.

Conflicts of Interest: The authors declare no conflict of interest.

References

1. Calcagnotto, M.; Adachi, Y.; Ponge, D.; Raabe, D. Deformation and fracture mechanisms in fine- and ultrafine- grained ferrite/martensite dual-phase steels and the effect of aging. *Acta Mater.* **2011**, *59*, 658–670. [CrossRef]
2. Pang, Q.H.; Li, W.J.; Cai, M.Y.; Qi, H.; Zhang, C.C.; Wu, J.N. The investigation of strength and plasticity mechanism of low-temperature annealed ultrafine grained stainless steel. *IOP Conf. Ser. Mater. Sci. Eng.* **2018**, *372*, 1–9. [CrossRef]
3. Park, K.T.; Shin, D.H. Microstructural interpretation of negligible strain-hardening behavior of submicrometer-grained low-carbon steel during tensile deformation. *Metall. Mater. Trans. A* **2002**, *33*, 705–707. [CrossRef]
4. Liu, J.J.; Suo, T.; Fan, Y.Q.; Li, J.G.; Zhou, F.H.; Li, Y.L. Dynamic mechanical responses of ultrafine-grained IF steel over a wide range of temperatures. *Int. J. Appl. Mech.* **2018**, *10*, 1–20. [CrossRef]
5. Khan, A.; Chen, X. Nanocrystalline aluminum and iron: Mechanical behavior at quasi-static and hig strain rates, and constitutive modeling. *Int. J. Plast.* **2004**, *22*, 195–209. [CrossRef]
6. Zheng, Z.G.; Xie, C.J.; Sun, T.; Yuan, S. A mixed hardening model of ultrafine-grained materials and numerical simulation. *J. Mech. Eng.* **2014**, *50*, 77–83. [CrossRef]
7. Wang, C.X.; Shi, L.; Xu, J.; Shan, D.; Guo, B. Constitutive relationshi of micro-deformation in an ultrafine-grained pure aluminum. *Mater. Sci. Technol.* **2016**, *24*, 1–7. [CrossRef]
8. Wang, P.J.; Shen, Y.F.; Feng, X.W.; Wang, H.B.; Yang, S.Q. Microstructures and textures of ultrafine grained D6A steel by using rolling and annealing. *J. Iron Steel Res.* **2016**, *28*, 54–59. [CrossRef]
9. Jia, N.; Shen, Y.F.; Liang, J.W.; Feng, X.W.; Wang, H.B.; Misra, R.D.K. Nanoscale spheroidized cementite induced ultrahigh strength-ductility combination in innovatively processed ultrafine-grained low alloy medium-carbon steel. *Sci. Rep.* **2017**, *7*, 2679. [CrossRef] [PubMed]
10. Liang, J.W.; Shen, Y.F.; Zhang, C.S.; Feng, X.W.; Wang, H.B.; Sun, X. In situ neutron diffraction in quantifying deformation behaviors of nano-sized carbide strengthened UFG ferritic steel. *Mater. Sci. Eng. A* **2018**, *726*, 298–308. [CrossRef]
11. Liang, J.W.; Shen, Y.F.; Misra, R.D.K.; Liaw, P.K. High strength-superplasticity combination of ultrafine-grained ferritic steel: The significant role of nanoscale carbides. *J. Mater. Sci. Technol.* **2021**, *83*, 131–144. [CrossRef]
12. Zhang, X.L.; Feng, X.W.; Shen, Y.F. Toughening mechanism of D6A steel during rolling process. *J. Mater. Eng.* **2021**, *50*, 1–15. [CrossRef]
13. Zhang, Z.F.; Lu, L. *Mechanical Behaviour of Materials*; Higher Education Press: Beijing, China, 2017; pp. 596–603.
14. Gladman, T. Precipitation hardening in metals. *Mater. Sci. Technol.* **1999**, *15*, 30–36. [CrossRef]

15. Okitsu, Y.; Takata, N.; Tsuji, N. Dynamic deformation behavior of ultrafine-grained iron produced by ultrahigh strain deformation and annealing. *Scr. Mater.* **2011**, *64*, 896–899. [CrossRef]
16. Liu, X.Y.; Zhang, Q.; Yang, X.R. Deformation, strain rate sensitivity and activation volume of ultrafine-grained commercially pure Ti. *Rare Met. Mater. Eng.* **2020**, *49*, 1867–1872.
17. Meyers, M.A. *Dynamic Behavior of Materials*; John Wiley & Sons: New York, NY, USA, 1994; p. 340.
18. Jia, D.; Ramesh, K.T.; Ma, E. Effects of nanocrystalline and ultrafine grain sizes on constitutive behavior and shear bands in iron. *Acta Mater.* **2003**, *51*, 3495–3509. [CrossRef]
19. Hu, X.; Xie, L.J.; Gao, F.N.; Xiang, J. On the development of material constitutive model for 45CrNiMoVA ultra-high-strength steel. *Metals* **2019**, *9*, 374. [CrossRef]
20. Johnson, G.R.; Cook, W.H. Fracture characteristics of three metals subjected to various strains, strain rates, temperatures and pressures. *Pergamon* **1985**, *21*, 31–48. [CrossRef]
21. Bergstrom, Y.; Aronsson, B. The application of a dislocation model to the strain and temperature dependence of the strain hardening exponent n in the Ludwik-Hollomon relation between stress and strain in mild steels. *Metall. Trans.* **1972**, *3*, 1951–1957. [CrossRef]

Article

Study on Microstructure and Properties of NM500/Q345 Clad Plates at Different Austenitization Temperatures

Guanghui Zhao [1,2], Ruifeng Zhang [1,*], Juan Li [1,2], Cuirong Liu [2], Huaying Li [1] and Yugui Li [1]

1. Engineering Research Center Heavy Machinery Ministry of Education, Taiyuan University of Science and Technology, Taiyuan 030024, China
2. Shanxi Provincial Key Laboratory of Metallurgical Device Design Theory and Technology, Taiyuan University of Science and Technology, Taiyuan 030024, China
* Correspondence: zrf18234222278@163.com

Abstract: In this paper, the change in the mechanical properties of a composite plate was studied using the heat treatment method, and it was found that the performance of the composite plate was greatly improved under the process of quenching at 900 °C and tempering at 200 °C. The hot-rolled NM500/Q345 clad plates were subjected to heat treatment tests of 860 °C, 900 °C, and 940 °C austenitization + 200 tempering. With the help of an optical microscope, scanning electron microscope, EBSD, and transmission electron microscope, the microstructure, interface element distribution, and defect composition at the composite bonding interface of hot rolling and heat treatment were analyzed. An analysis and friction and wear tests were carried out on the wear resistance of the clad NM500. It was found that the microstructure of the NM500/Q345 clad plate before austenitization was mainly pearlite and ferrite, and both were transformed into lath martensite after austenitization. As the austenitization temperature increased, the size of the martensitic lath bundle also became coarse. After austenitization at 900 °C and tempering at 200 °C, the lath-like martensite structure of NM500 contained high-density dislocations between the laths. With the increase in the austenitization temperature, the surface Rockwell hardness showed a trend of first increasing and then decreasing. The wear was the worst when the material was not quenched. When the clad plate was quenched at 900 °C and tempered at 200 °C, the wear of NM500 was the lightest; the maximum depth of the wear scar was 14 μm; the width was the narrowest, 0.73 mm; and the wear volume was the smallest, 0.0305 mm^3.

Keywords: NM500/Q345 clad plate; austenitization; microstructure; properties

Citation: Zhao, G.; Zhang, R.; Li, J.; Liu, C.; Li, H.; Li, Y. Study on Microstructure and Properties of NM500/Q345 Clad Plates at Different Austenitization Temperatures. *Crystals* **2022**, *12*, 1395. https://doi.org/10.3390/cryst12101395

Received: 2 September 2022
Accepted: 28 September 2022
Published: 1 October 2022

Publisher's Note: MDPI stays neutral with regard to jurisdictional claims in published maps and institutional affiliations.

Copyright: © 2022 by the authors. Licensee MDPI, Basel, Switzerland. This article is an open access article distributed under the terms and conditions of the Creative Commons Attribution (CC BY) license (https:// creativecommons.org/licenses/by/ 4.0/).

1. Introduction

With the advancement of science and technology, the operation speed of mechanical equipment has accelerated, and the requirements for the wear resistance of parts have also became increasingly higher. When a piece of equipment and its components are severely worn, not only may it greatly reduce its production efficiency, but it may also cause a huge waste of funds and materials. Serious wear and tear may directly lead to serious casualties and cause disastrous consequences [1–3].

Wear-resistant cast iron steel, high-manganese steel, and low-alloy wear-resistant steel, as traditional wear-resistant materials, all have good wear resistance, and they can withstand a large external force impact and high stress at the same time, making them suitable for bad conditions. Manufacturing components made from these materials can significantly increase the life of machinery and equipment. High-manganese steel can produce work hardening under the working conditions of high-impact load, but under low-impact conditions, the work hardening ability of high-manganese steel cannot be fully exerted [4–6]. NM500 is the most commonly used material in low-alloy, high-strength, wear-resistant steel. It has the characteristics of high strength, good toughness, a simple production process, a low production cost, and strong weldability and machinability, and it is widely used in the fields of construction machinery, mining, and metallurgy [7–10].

Traditional wear-resistant steel adds more alloying elements, which greatly increases the production cost. Q345 steel is a low-alloy steel (C < 0.2%) with good comprehensive properties, and good plasticity and weldability, and it is widely used in buildings, bridges, vehicles, ships, pressure vessels, etc. NM500 is a high-strength, wear-resistant steel plate commonly used in critical wear-resistant contact surfaces for friction and impact. Therefore, the relatively cheap Q345 is used as the base layer, and the low-alloy steel NM500 is used as the cladding layer to prepare high-strength, wear-resistant clad plates by vacuum hot rolling. Wear-resistant steel clad plates make full use of the performance advantages of the base cladding metal, make up for their respective deficiencies, and achieve complementary performance. At the same time, they also save a lot of scarce precious metal resources and reduce energy consumption. Construction is of great significance [11].

At present, there are not many studies on as-rolled, wear-resistant, steel–carbon clad plates. Cheng Muhua et al. [12] studied the effect of rolling reduction on the microstructure and tensile fracture of vacuum hot-rolled NM360/Q345R clad sheets. Ref. [13] found that an increase in rolling reduction can promote the metallurgical bonding of NM500 steel and Q345 steel. Qiu Jun et al. [14] found that, after austenitization at 900 °C and tempering at 250 °C, the composite interface of wear-resistant steel and carbon steel Q345 was in good contact. Lan Kun [15] used a thermal simulation tester to study the deformation resistance of the wear-resistant steel–carbon steel clad plate interface under a single pass. Li Jin et al. [16] studied hot-rolled NM360/Q345R clad plates under compression. When the tensile strength reached 70%, the tensile fracture surface was smooth, and no delamination occurred. Liu Pengtao et al. [2] studied the effect of rolling reduction on the microstructure and wear resistance of BTW wear-resistant steel clad plates. At present, there are few studies on the effects of the austenitization process on the microstructure and properties of NM500 clad plates.

In this paper, a hot-rolled NM500/Q345 clad plate was used as the material, austenitization heat treatment at different temperatures was used, and then OM, EBSD, and TEM were used to further study the NM500/Q345 clad plate microstructure. Its mechanical properties were analyzed using a Shimadzu tensile machine, and a corresponding analysis and research on the hardness and wear resistance of the NM500 steel were conducted. Through the use of the heat treatment method in this paper, the performance of the material was greatly improved, providing certain theoretical support for the practical application of wear-resistant steel cladding.

2. Experimental Materials and Procedures

The material used was a vacuum hot-rolled NM500/Q345 composite sheet, and its main components are shown in Tables 1 and 2. A cuboid sample of 20 mm × 15 mm × 20 mm (length × width × height) was taken by wire cutting, of which the NM500 layer was 8 mm, and the Q345 layer was 12 mm. They were heated to 860 °C, 900 °C, and 940 °C for 20 min; then quenched in a water bath; and finally tempered at 200 °C for 30 min. According to the different austenitization temperatures, the samples were marked as original samples and 860 °C + 200 °C, 900 °C + 200 °C, and 940 °C + 200 °C samples in turn. The experimental process is shown in Figure 1.

Table 1. Chemical composition of Q345 (mass fraction) %.

C	Si	Mn	P	S	Cr	Ni	Mo	Ti	V	Fe
0.13	0.44	1.52	0.013	0.004						Bal

Table 2. Chemical composition of NM500 (mass fraction) %.

C	Si	Mn	P	S	Cr	Ni	Mo	Ti	B	Al
0.38	0.40	1.70	0.020	0.010	1.20	1.00	0.65	0.050	0.00045	0.010

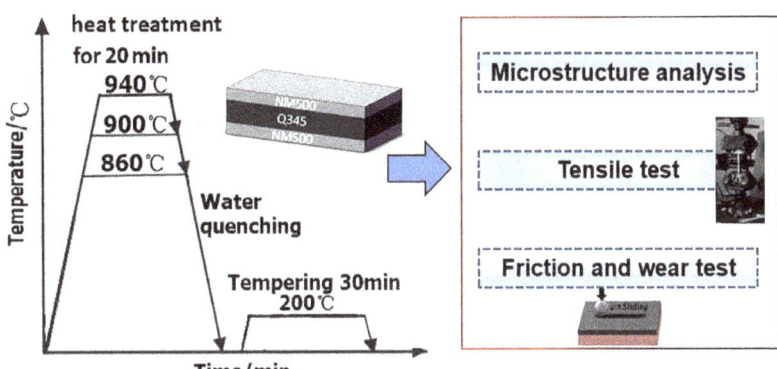

Figure 1. Schematic diagram of the experimental flow.

The metallographic samples in an unused state were prepared by wire electric discharge cutting, and the observed surface was the RD-ND surface. Sanding was carried out with 400–2000 grit sandpaper, and metallographic mechanical polishing was conducted. Metallographic etching was carried out using 4% nitric acid ethanol solution. Optical microscopy (OM) and a ZIESS SIGMA-300 scanning electron microscope (SEM) were used for microstructure observation. The elemental distribution was analyzed by energy spectroscopy (EDS). Using a 10% perchloric acid alcohol solution, the polished metallographic samples were electropolished at 25 V for 60 s at room temperature, and the grain properties of the clad plates were investigated by electron backscatter diffraction (EBSD; Oxford). Moreover, with TEM preparation, 5% perchloric acid alcohol solution double-jet thinning, a JEOL JEM-F200 transmission electron microscope (TEM) was used to analyze their micro-domain fine structure.

The composite NM500 surface of the quenched + tempered sample was polished and removed, sanded with sandpaper, and metallographically polished to ensure the same roughness, and then it was rinsed with alcohol and dried for use. The test was carried out on the reciprocating sliding friction and wear module of an RTEC (MFT-5000) friction and wear testing machine. Silicon nitride ceramic balls with a diameter of 6 mm were used, the friction test time was 30 min, the frequency was 1 Hz, and the experimental load was 150 N. The friction method was linear reciprocating motion, and the stroke was 6 mm (that is, 12 mm reciprocating in 1 s for a total of 1800 cycles). To reduce errors, all experiments were performed three times.

3. Results and Discussion

3.1. Microstructural Analysis

Figure 2a–d show the microstructure changes in the NM500/Q345 clad plate wear-resistant layer (NM500) without austenitization and after austenitization at different temperatures. In Figure 2, the middle layer is the composite interface, the lower layer of the composite interface is carbon steel Q345, the upper layer is NM500, and its microstructure distribution is uniform. At the interface, there are no defects, such as large holes and cracks. Moreover, no Si and Mn segregations are observed near the interface, and the elements are uniformly distributed. The microstructure before austenitization is mainly pearlite and ferrite, and both are transformed into lath martensite after austenitization. The structures obtained at each austenitization temperature are all martensite structures, and the martensite is in the shape of lath. Due to the different austenitization temperatures, the size of the martensite obtained is also different. With an increase in the austenitization temperature, the size of the martensite lath bundle also becomes coarse. Too-high austenitization temperature easily makes the austenitized grains grow, the martensite Ms temperature decrease, and the martensitic structure become coarse. Quenched at 940 °C, the martensitic laths are somewhat coarse.

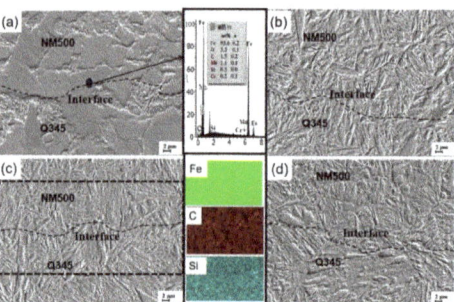

Figure 2. Microstructure and EDS of NM500/Q345 composite board in different states: (**a**) original; (**b**) 860 °C + 200 °C; (**c**) 900 °C + 200 °C; (**d**) 940 °C + 200 °C.

Figure 3 presents microstructure IPF images of the NM500/Q345 composite board in different states. As shown in Figure 4, after austenitization, the clad plate has no defects, such as cracking, and the grains of the clad plate become very fine. The refined grains help to significantly increase the properties of the clad plate. After austenitization and the tempering of the clad plate, the obtained microstructure is a lath-like martensite structure. The austenitization heating temperature affects the austenite grain size. By adjusting the austenitization temperature, the martensitic lath size of the steel after phase transformation can be changed. At 860 °C, the martensite arrangement is uneven, and the microstructure cannot be completely transformed into austenite. When the austenitization temperature is 900 °C, the structure has relatively clear prior austenite grain boundaries, and the martensite arrangement is relatively neat. When the austenitization temperature is 940 °C, the austenitized grains grow, and large martensite laths are formed during austenitization. Excessive martensite affects the toughness and plasticity of the material, and it affects the mechanical properties of the material.

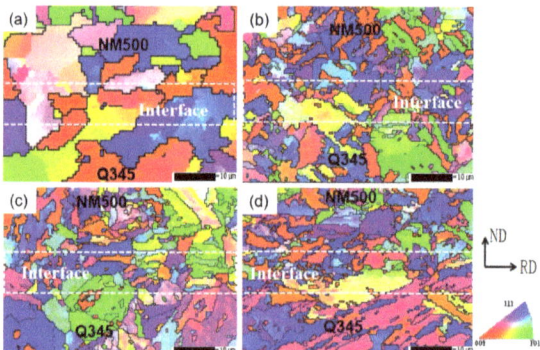

Figure 3. Microstructure IPF of NM500/Q345 composite board in different states: (**a**) original; (**b**) 860 °C + 200 °C; (**c**) 900 °C + 200 °C; (**d**) 940 °C + 200 °C.

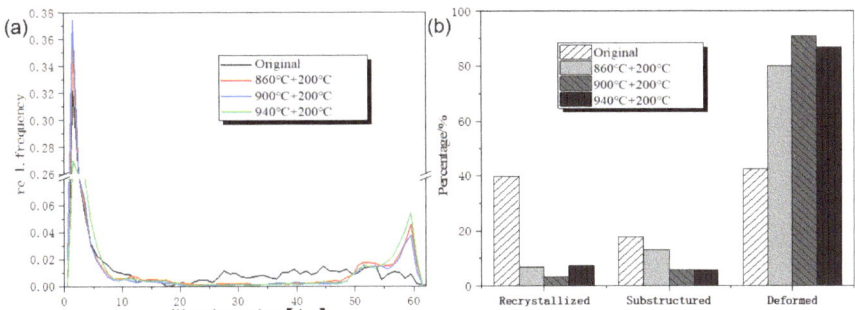

Figure 4. Grain boundary dislocation distribution and recrystallization: (**a**) misorientation scale diagram under different conditions; (**b**) grain scale diagram in different states.

Figure 4 is a statistical graph of the grain boundary misorientation lines and recrystallization statistics of EBSD at different heat treatment temperatures. Compared with the original specimens after rolling, the NM500/Q345 composites at different quenching temperatures have two discontinuous orientation peaks. The average grain boundary orientation difference is 53°. At 940 °C, the small-angle grain boundaries account for less, and there are more twins. However, the proportion of the small-angle grain boundaries at 900 °C is significantly higher than that at other temperatures, and there are relatively few twins at Σ60°. As shown in Figure 5b, a large amount of recrystallization occurs in the structure after rolling, and the deformed grains in the structure after austenitization heat treatment increase, especially at 900 °C.

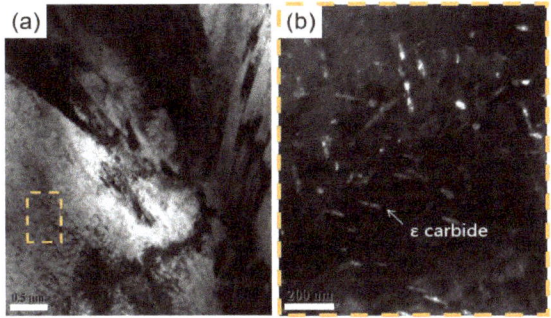

Figure 5. TEM image of NM500 of NM500/Q345 composite board at 900 °C + 200 °C: (**a**) high density dislocations in martensite; (**b**) acicular and rod-shaped precipitation in martensite.

Figure 5 shows a TEM image of the microstructure of the NM500/Q345 clad plate after quenching at 900 °C + tempering at 200 °C. In Figure 5, it can be observed that, after austenitization at 900 °C + tempering at 200 °C, the microstructure of NM500 presents a typical lath-like martensite structure (see Figure 5a), and there are many substructures with a high density of dislocations between the laths. Moreover, it can be seen in the TEM pictures that there are a certain number of needle-like or rod-like precipitates in the lath martensite (see Figure 5b). The length of these precipitates is 50–100 nm; the width is about 10 nm; and the spatial shape is flake-like, in a coherent relationship with the matrix, and usually precipitates on a specific habit surface. Some research data show that such precipitates are ε carbides, namely, $Fe_{2.4}C$ [17]. In Figure 5, spherical precipitates with a size of less than 10 nm can also be observed, and these precipitates are dispersed in the lath. Such precipitates have a certain positive effect on the precipitation strengthening of the material [18,19].

3.2. Mechanical Property Analysis

Figure 6 shows the mechanical properties of the clad plates in different states. In Figure 6, the tensile strength and elongation of the NM500 wear-resistant steel clad plate show a trend of first increasing and then decreasing with the increase in the austenitization temperature. When the austenitization temperature is 900 °C + tempering is 200 °C, its elongation and tensile strength are the largest, at 22.68% and 1432.28 Mpa, respectively. As the austenitization temperature increases to 940 °C, the tensile strength and elongation decrease slightly compared with those of the samples at 900 °C. However, the elongation of the specimens at each austenitization temperature is greatly improved compared with that of the single wear-resistant steel. Moreover, the tensile strength exceeds 1250 MPa, which meets the requirements of the national standard for high-strength, wear-resistant steel plates for construction machinery. The hardness of the NM500 surface of the clad plate after austenitization and that of the hot-rolled clad plate are compared and analyzed, as shown in Figure 6b, and the surface hardness is greatly improved. With the increase in the austenitization temperature, the surface Rockwell hardness shows a trend of first increasing and then decreasing. When the austenitization temperature is 900 °C, the hardness reaches the maximum, which is 47.2 HRC. A large number of microscopic defects (such as dislocations, twins, and stacking faults) are generated in it, which strengthens the martensite, so the hardness of the sample after austenitization is greatly improved [20–22].

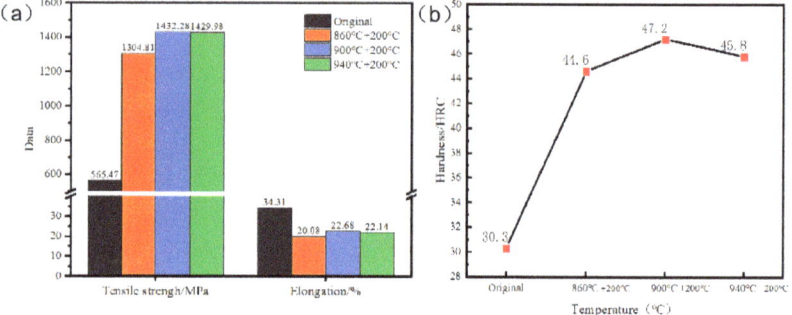

Figure 6. Mechanical properties of clad plates tempered at different austenitization temperatures + 200 °C tempering: (**a**) the data of tensile strength and elongation of clad plate; (**b**) the hardness of clad plates.

Figure 7 shows a tensile fracture diagram of the samples tempered at different austenitization temperatures. There is obvious necking at the tensile fracture site, and different degrees of cracking and delamination appear at the bonding interface of the clad plate. After quenching at 900 °C, delamination does not occur in any small part of the interface. The reason for delamination is because the elongation of NM500 and Q345 is different. With coordinated deformation during tensile deformation, the composite interface generates a certain shear force, and when the shear force exceeds the bonding force at the interface, delamination of the two metals occurs. With the help of the SEM analysis, the tensile fracture morphology of NM500 and Q345 near the interface is determined. The distribution of dimples is uneven. The dimples are deformed along the necking direction, and there are still white spherical second-phase particles in some dimples, which are related to wear resistance. The dimples on the steel side are smaller in size and shallower in depth. When the quenching temperatures are 860 °C and 940 °C, the fracture morphology has more tear edges besides dimples. When the austenitization temperature is 900 °C, the dimple distribution is relatively uniform. To sum up, it can be judged that the tensile fracture mode of the clad plate after heat treatment is ductile fracture. When the austenitization temperature is 900 °C, each clad plate shows good toughness.

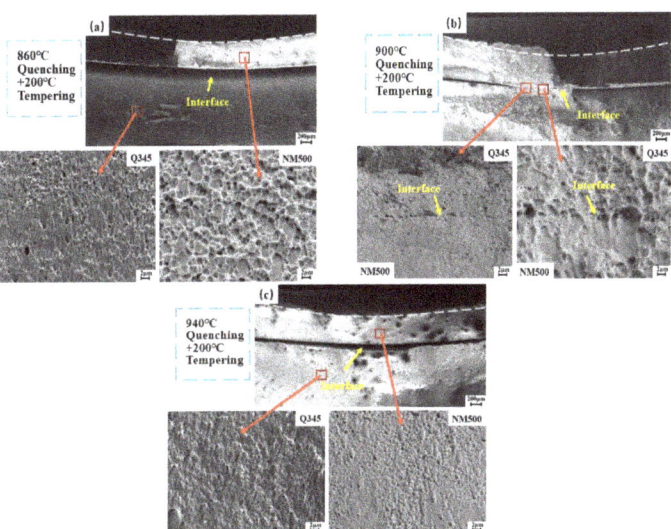

Figure 7. Fracture morphology of clad plate: (**a**) 860 °C + 200 °C; (**b**) 900 °C + 200 °C; (**c**) 940 °C + 200 °C.

3.3. Friction and Wear Analysis of Clad NM500

The friction coefficients of the different samples under the same test conditions are shown in Figure 8a for one test cycle. The friction coefficient curve of the sample shows a trend of rapid increase, then decreases to a certain extent, and then slowly increases and gradually stabilizes. In the initial wear stage, the overall friction coefficient shows an upward trend and fluctuates greatly. This is because, at the beginning of the experiment, the surface of the friction pair is uneven, and the contact surface of the friction pair is point contact at the beginning of the sliding phase, the load per unit area is very large, the friction is dry, and the sliding friction surface is not lubricated. Hence, the friction coefficient rises rapidly [23]. During the wear process, the surface temperature of the material rises sharply as the wear continues. Affected by the surface temperature of the material and the direction of the temperature gradient along the depth direction, adhesive wear occurs on the contact surface, and the wear debris sticks to the scratch due to the influence of the temperature. On the marks, the friction coefficient rises but fluctuates greatly. As the wear continues, the adhesive wear decreases, the friction coefficient of the sample is stabilized in a certain range, and the wear experiment enters the stable wear stage.

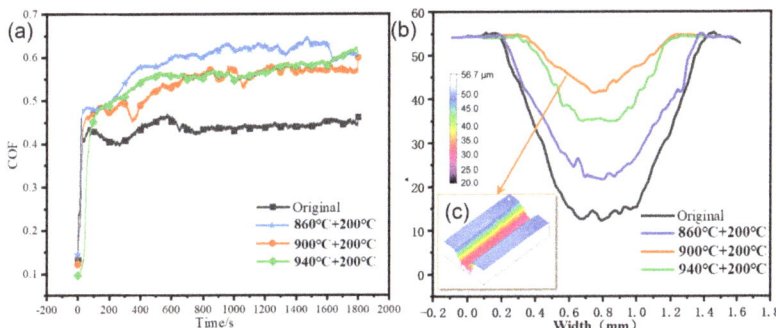

Figure 8. Friction and wear experimental data of NM500/Q345 clad plate wear—resistant layer (NM500) after austenitization at different temperatures: (**a**) friction coefficient; (**b**) wear scar section curve; (**c**) 3D morphology at 900 °C.

The friction coefficient of the NM500/Q345 clad plate before and after NM500 austenitization is quite different. The original sample before austenitization has a soft material, and it is easy to produce wear debris on the surface, which reduces the surface roughness and minimizes the friction coefficient. After austenitization + tempering, the hardness of the wear-resistant steel increases, the wear resistance also increases, and the friction coefficient increases with it. After austenitization at 860 °C, the friction coefficient is the largest. This may be due to incomplete austenitization at this temperature, resulting in an uneven structure of the wear-resistant steel and an increased surface roughness during wear. When the austenitization temperatures are 900 °C and 940 °C, the difference in the friction coefficient is not big, and the friction coefficient is slightly smaller at 900 °C. When the austenitization temperature is high (900 °C and 940 °C), austenitization is complete, the structure is a lath-like martensite structure, and the friction coefficient is relatively consistent.

During the sliding friction and wear process, the sample and the counter-abrasive silicon nitride ceramic are subjected to the combined action of shear stress and compressive stress. Due to the different austenitization conditions of the material, the material removal mechanism during the friction and wear process is also different, resulting in a different macroscopic appearance of the wear scar. Figure 8b shows the scratch cross-section curves of the wear surface scratches of the NM500/Q345 clad plate wear-resistant layer (NM500) unquenched and quenched at different temperatures, and the original surface is selected as the reference surface [24]. Figure 8b shows the macroscopic morphology of the middle part of the scratch, and the width and depth of the different samples are quite different. It can be seen in Figure 8b that the maximum depth of the scratches on the unquenched specimen is 43 μm, and the width is 1.30 mm. When the austenitization temperature is 860 °C, the maximum depth of the scratch is 33 μm, decreasing by 23.26%, and the width is 1.15 mm, decreasing by 11.54%. When the austenitization temperature is 900 °C, the maximum depth of the scratch is 14 μm, decreasing by 57.58%, and the width is 0.73 mm, decreasing by 36.52%. When the austenitization temperature is 940 °C, the maximum depth of the scratches is 20 μm, increasing by 42.86%, and the width is 0.78 mm, increasing by 6.85%. When the material is not quenched, the wear is the worst, the value of the maximum depth is the largest, and the width is the widest. After austenitization at 900 °C, the wear condition is greatly improved, the value of the maximum depth is the smallest, the width is the narrowest, and the wear condition is relatively the best.

The amount of wear is one of the important manifestations of wear resistance [25]. The scratch wear volume can be calculated using Gwyddion software. Figure 9 shows the wear volume values for the wear layer of the NM500 raw material and three quenched temperatures. When the austenitization temperature is 860~940 °C, with the increase in the austenitization temperature, the wear volume first decreases and then increases. When the austenitization temperatures are 860 °C, 900 °C, and 940 °C, the average wear volumes are 0.0701 mm^3, 0.0305 mm^3, and 0.1051 mm^3, respectively. Compared with the average wear volume before austenitization of 0.12821mm^3, when the austenitization temperature is 900 °C, the decrease is the largest, and the wear amount is the least, which also means the best wear resistance.

The wear amount is the result of the combination of the hardness difference between the friction pairs, the structure of the material, the actual contact area, and other factors. In general, the higher the hardness of a material, the less likely it is for a relatively soft material to have scratches on its surface. In general, the higher the hardness of the material, the better the wear resistance of the material. Therefore, the hardness value is usually used as one of the important indicators to measure the wear resistance of materials. When the austenitization temperature is 900 °C, the wear resistance of the wear-resistant layer NM500 is the best. This may be because the austenitization temperature is not high at this temperature, and the original austenitized grains are small. After austenitization, the strip martensite block is small, so the surface of the wear-resistant layer NM500 has a certain strength and toughness, and the wear resistance is the best.

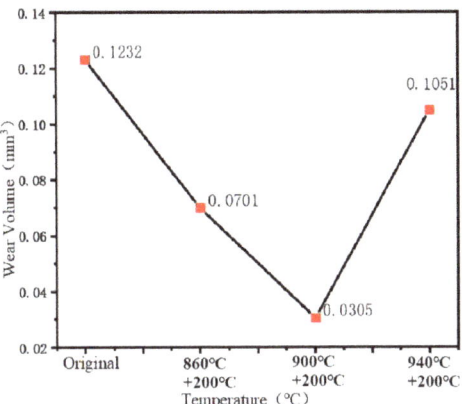

Figure 9. Wear volume of NM500/Q345 clad plate wear-resistant layer under different states of NM500.

4. Conclusions

In this paper, a hot-rolled NM500/Q345 clad plate was used as the material, and austenitization heat treatment at different temperatures was used. The microstructure of the NM500/Q345 clad plate was studied using OM, EBSD, and TEM, and its mechanical properties were analyzed using a Shimadzu tensile machine and by examining its hardness and wear resistance. The following conclusions were drawn:

1. The microstructure of the NM500/Q345 clad plate before austenitization was mainly pearlite and ferrite, and both were transformed into lath martensite after austenitization. The structures obtained at each austenitization temperature were all martensite structures, and the martensite was in the shape of lath. As the austenitization temperature increased, the size of the martensitic lath bundle also became coarse. After quenching at 900 °C + tempering at 200 °C, in the lath-like martensite structure of NM500, there were many substructures between the laths, and a certain number of needles were of precipitated shape or rod-shaped ε precipitates.
2. When quenched at 860 °C + tempered at 200 °C, the martensite arrangement was uneven, and the microstructure could not be completely transformed into austenite. When quenched at 900 °C + tempered at 200 °C, the structure had relatively clear prior austenite grain boundaries, and the martensite arrangement was relatively neat. When quenched at 940 °C + tempered at 200 °C, the austenitized grains grew, and large martensitic laths formed during austenitization. Therefore, the trend of the tensile strength and elongation of the composite plate first increased and then decreased with the increase in the quenching temperature.
3. When quenched at 900 °C + tempered at 200 °C, its elongation and tensile strength were the largest, which were 22.68% and 1432.28 Mpa, respectively. As the austenitization temperature increased to 940 °C, the tensile strength and elongation decreased slightly compared with those of the samples at 900 °C. Due to the martensite structure that formed after quenching, the hardness of the wear-resistant layer of the composite plate first increased and then decreased with the increase in the quenching temperature. The Rockwell hardness of the original sample was 30.3 HRC. When the quenching temperature was 860 °C, the hardness increased to 44.6 HRC. When the quenched temperature was 900 °C, the hardness reached the maximum, which was 47.2 HRC, When the quenching temperature was 940 °C, the hardness decreased to 45.8 HRC.
4. The original sample before austenitization had a soft material, and it was easy to produce wear debris on the surface, which reduced the surface roughness and min-

imized the friction coefficient. After austenitization + tempering, the hardness of the wear-resistant steel increased, the wear resistance also increased, and the friction coefficient increased with it. After quenching at 860 °C and tempering at 200 °C, the friction coefficient was the largest. When the austenitization temperatures were 900 °C and 940 °C, the friction coefficient was not much different. The wear was the worst when the material was not quenched. After austenitization at 900 °C + tempering at 200 °C, the wear of NM500 was the lightest; the maximum depth of the wear scar was 14 μm; the width was the narrowest, 0.73 mm; and the wear volume was the smallest, 0.0305 mm^3.

Author Contributions: Conceptualization, G.Z. and J.L.; methodology, G.Z. and R.Z.; formal analysis, G.Z.; investigation, R.Z., H.L. and C.L.; resources, G.Z.; data curation, R.Z., Y.L. and C.L.; writing—original draft preparation, R.Z.; writing—review and editing, G.Z., R.Z. and J.L.; visualization, J.L.; supervision, G.Z. and H.L.; funding acquisition, G.Z., J.L., H.L. and Y.L. All authors have read and agreed to the published version of the manuscript.

Funding: This research was funded by the National Key Research and Development Program of China (2018YFA0707305), the Scientific and Technological Innovation Programs of Higher Education Institutions in Shanxi (2021L292), the Fundamental Research Program of Shanxi Province (20210302124009 and 20210302123207), the Taiyuan University of Science and Technology Scientific Research Initial funding (20212026), and the Shanxi Outstanding Doctorate Award Funding Fund (20222042).

Institutional Review Board Statement: Not applicable.

Informed Consent Statement: Not applicable.

Data Availability Statement: Not applicable.

Acknowledgments: The authors would like to thank the Provincial Special Fund for Coordinative Innovation Center of Taiyuan Heavy Machinery Equipment for providing the facilities for the experimental works.

Conflicts of Interest: The authors declare no conflict of interest.

References

1. Zhao, D. Study on Tribological Performance of Three Commonly Used Wear-Resistant Steels. Master's Thesis, China University of Mining and Technology, Xuzhou, China, 2017.
2. Liu, P.T.; Ma, L.F.; Zhi, C.C.; Ma, Z.; Zhou, C.; Zhao, G.; Fan, Q.; Jia, W. Effect of Annealing on Heavy-Load Wear Performance of Wear Resisting Steel-Carbon Steel-Cladded Plate. *Tribol. Trans.* **2020**, *64*, 101–110. [CrossRef]
3. Ye, J.; Zhang, H.; Liu, X.; Liu, K. Low Wear Steel Counterface Texture Design: A Case Study Using Micro-pits Texture and Alumina–PTFE Nanocomposite. *Tribol. Lett.* **2017**, *65*, 165. [CrossRef]
4. Liang, L.; Yan, L.; Li, G.; Yi, C.; Xiangtao, D.; Zhaodong, W. Effect of Heat Treatment on Microstructure and Mechanical Properties of Low-Alloy Wear-Resistant Steel NM450. *Mater. Res. Express* **2021**, *8*, 045606. [CrossRef]
5. Hussein, A.; Ahmed, M.R.; Abid, H.; Alzaid, M.; Farhat, L.B.; Badruddin, I.A. Effects of gating design on structural and mechanical properties of high manganese steel by optimizing casting process parameters. *J. Mech. Sci. Technol.* **2022**, *36*, 3931–3937.
6. Pei, W.; Zhang, Y.; Yang, S.; Li, X.; Zhao, A. Study of work-hardening behavior of high manganese steel during compression. *Mater. Res. Express* **2022**, *9*, 066503. [CrossRef]
7. Wu, X. Study on Mechanical Properties and Wear Resistance of Microalloyed Low Alloy Wear Resistant Steel. Master's Thesis, Zhengzhou University, Zhengzhou, China, 2020.
8. Li, H.; Zhao, G.; Ma, L.; Li, J.; Zhi, C. Microstructure analysis of hot-rolled NM500/Q345/NM500 composite interface. *Mater. Res. Express* **2019**, *6*, 016548. [CrossRef]
9. Zhao, G.; Zhang, J.; Song, Y.; Li, J.; Li, H.Y.; Shuai, M.R. Study on friction and wear properties of wear-resistant layer of NM500/Q345 composite plate at different quenching temperatures. *Precis. Form. Eng.* **2021**, *13*, 35–41.
10. Bao, S.; Yang, G.; Xu, Y.; Han, N.; Zhu, X. Austenite grain growth behavior of medium manganese martensitic NM500 steel. *Steel* **2022**, *57*, 152–159.
11. Cheng, M.H. Research on NM500/Q345/NM500 Three Layer Rolling Composite and Heat Treatment Process. Master's Thesis, Taiyuan University of Science and Technology, Taiyuan, China, 2018.
12. Cheng, M.; Huang, Q.; Zhao, G.; Ma, L.F. Effect of reduction ratio on microstructure and tensile fracture of vacuum hot rolled NM360/Q345R clad plate. *Acta Mater. Metall.* **2017**, *16*, 286–292.

13. Li, J.; Liu, C.R.; Song, Y.H.; Zhao, G.; Ma, L.; Huang, Q. Influence of Hot Rolling + Heat Treatment on Microstructure and Mechanical Properties of NM500/Q345/NM500 Composite Plate. *J. Mater. Sci.* **2021**, *56*, 6016–6030. [CrossRef]
14. Qiu, J.; Cheng, X.R.; Lan, K.; Lan, D.; Wang, Y.; Huang, D.J. Effect of Rolling and Heat-Treatment on Structure and Properties of Wear-Resistant Steel NM450D-Carbon Steel Q235B Clad Plate. *Spec. Steel* **2017**, *38*, 56–59.
15. Lan, K.; Cheng, X.R.; Qiu, J.; Lei, D.; Peng, Y. Influence of Reduction Ratio on Interfacial Bonding Properties of Wear-Resistant Steel and Carbon Steel Composite Plate. *Hot Work. Technol.* **2017**, *46*, 121–124.
16. Li, J.; He, Q.Z.; Li, H.; Zhao, G.H.; Cheng, M.H. Research on mechanical properties and interface microstructure of hot-rolled NM360/Q345R composite plate. *Hot Work. Technol.* **2018**, *47*, 118–121.
17. Zhu, C.; Cerezo, A.; Smith, G.D.W. Carbide characterization in low-temperature tempered steels. *Ultramicroscopy* **2009**, *109*, 545–552. [CrossRef] [PubMed]
18. Ceschini, L.; Marconi, A.; Martini, C.; Morri, A.; Di Schino, A. Tensile and impact behaviour of a microalloyed medium carbon steel: Effect of the cooling condition and corresponding microstructure. *Mater. Des.* **2013**, *45*, 171–178. [CrossRef]
19. Jia, T.; Li, M.; Pei, X.; Wang, Z. On the Spheroidizing Annealing Behavior in Cr/Nb Microalloyed Medium Carbon Steels. *Steel Res. Int.* **2018**, *90*, 1800353. [CrossRef]
20. Dewangan, S. Effect of heat treatment into tensile strength, hardness and microstructural attributes of TIG welded Ti-6AL-4V titanium alloy. *Aust. J. Mech. Eng.* **2022**, 2059906. [CrossRef]
21. Babaei, H.; Shafyei, A.; Amini, K. The Effect of Heat Treatment on Mechanical Properties and Microstructure of the AISI 422 Martensitic Stainless Steel. *Mechanika* **2017**, *22*, 13599. [CrossRef]
22. Dwangan, S.; Nemade, V.V.; Nemade, K.H.; Bohra, P.M.; Kartha, S.R.; Chowrasia, M.K. A discussion on mechanical behaviour of Heat-Treated low carbon steel. *Mater. Proc.* **2022**, *63*, 362–367. [CrossRef]
23. Mo, D.Y.; Gong, M.F.; Wu, H.B.; Song, R. Surface Modification of 5Cr5MoWSi Steel and Tribological Properties. *Mach. Des. Manuf.* **2020**, *12*, 95–99.
24. Li, Y.L.; Bao, H.S.; Li, Q. Effect of Hardness on Rolling Wear Properties of 95Cr18 Steel. *Met. Funct. Mater.* **2020**, *27*, 38–45.
25. Jiang, M.M.; Cao, P.; Zhao, Y.H.; Geng, H.R.; Li, J.; Gong, Z.Y. Effect of Different Load and Rotating Speed on Wear Properties of Vacuum-Quenched Cr12MoV Steel. *Hot Work. Technol.* **2020**, *1*, 143–146,150.

Article

Dynamic Response and Numerical Interpretation of Three Kinds of Metals for EFP Liner under Explosive Loading

Li Ding [1,*], Peihui Shen [1] and Liuqi Ji [1,2]

1. School of Mechanical Engineering, Nanjing University of Science & Technology, Nanjing 210094, China; sphjy8@njust.edu.cn (P.S.); jiliuqi@163.com (L.J.)
2. Xi'an Modern Control Technology Institute, Xi'an 710065, China
* Correspondence: dingli@njust.edu.cn

Abstract: In order to study the dynamic response of tungsten heavy alloy materials under explosive loading, two kinds of typical tungsten alloys for explosively formed projectile (EFP) liner and one kind of existing EFP liner were tested in a flash X-ray experiment, with copper liner as a reference. Results showed that copper liner could form a coherent EFP, while 90W–9Ni–Co and W–25Re liners fractured to different extents. The microscopic features of the three kinds of metals were examined and compared with the original liner, and the microstructure evolutions under explosive loading were analyzed with the fracture model and mechanism of the two kinds of tungsten alloys' fracture determined. Associated with the stress and strain conditions by numerical simulation, the fracture mechanism of tungsten heavy alloys can be analyzed. The crack-tip plastic zones of 90W–9Ni–Co and W–25Re are much smaller than copper, and due to the severe stress concentration at the tip of cracks, it is easy for cracks to propagate and trigger the cleavage in tungsten alloys. The value of a crack-tip plastic zone $r(\theta)$ can be used to explain the fracture phenomenon in explosive loading, which can be an alternative guideline for the material selection criteria of the EFP liner. The research results are significant in understanding the dynamic forming, microstructure evolution, and fracture mechanism of tungsten heavy alloys.

Keywords: explosively formed projectile; tungsten heavy alloy; liner material; dynamic response; fracture mechanics

1. Introduction

According to the shaped-charge mechanism, the explosively formed projectile (EFP) makes full use of explosive blasting to form the liner into a preferred penetrator without breaking [1–3]. The liner of EFP should undergo extremely, yet controlled, plastic deform, which makes designing an optimal EFP a very complicated task [4]. The preferable properties of EFP liner material are high density, high ductility, high strength, and a high enough melting point to avoid melting in the liner due to adiabatic heating under explosive loading. The most common liner materials for EFP are copper (Cu), ferrum (Fe), tantalum (Ta), and Ta-W alloys.

Manfred Held [5] showed a comparison of liner materials with their densities, bulk-sound velocities, possible maximum jet tip velocities, and a ranking based on the product of possible jet tip velocities and square root of density. The ranking results clearly shows that tungsten has particularly good potential for a shaped-charge liner. However, whether tungsten can be used as an EFP liner has not been discussed.

The good mechanical properties of tungsten and its alloys have drawn much attention in recent years, especially in military applications. The high density (19.3 g/cm^3), high strength, high sound speed, high melting point (3410 °C), and excellent corrosion resistance make tungsten alloys desirable materials for use as a shaped-charge liner or ballistic penetrator [6,7]. Tungsten in its pure state has limitations, especially the low-temperature brittleness, which restricts its application. Alloys of tungsten with nickel, cobalt, ferrum, or rhenium have witnessed tremendous improvement in mechanical properties at lowering the ductile-to-brittle transition temperature (DBTT). Therefore, it is essential to combine tungsten with other alloying elements to form an alloy for its applications in Tungsten heavy alloys (WHA), defense, and other military applications [8–11].

Michael T. Stawovy [12,13] submitted patents explaining a single-phase tungsten alloy that could be used for forming a shaped-charge liner for a penetrating jet or explosively formed penetrator. The alloy is essentially composed of cobalt, tungsten, and nickel. One preferred composition is, by weight, from 16% to 20% cobalt, from 35% to 40% tungsten, and the balance is nickel and inevitable impurities. The tungsten alloy is worked and recrystallized and then formed into the desired product. However, relevant experimental results were not provided to verify his viewpoint.

S. Rolc et al. [14] conducted a numerical simulation of linear EFP with different materials (copper, iron, tungsten, MONEL Alloy 400, INONEL Alloy 600, INCONEL Alloy 625, INCO Alloy HX, INCOLOY Alloy 800 HT, nickel 200, and Hadfield steel) and found they could all form complete EFPs in the simulation. C. G. Bingol et al. [15] carried out a numerical investigation of the thickness and radius of curvature of the liner on the formation of EFP and found that tungsten, nickel 200, and molybdenum could form EFP as well as copper and tantalum. However, only an experiment of Cu EFP was presented to testify his conclusion. Among their research, whether liners of tungsten and its alloys could form EFPs and the dynamic response of tungsten liner under explosive loading have not yet been studied in experiments. Weibing Li et al. [16] investigated the effect of liner material density and elongation on the shape of dual mode penetrators. However, as a promising liner material, tungsten alloy was studied only in numerical simulation. Yajun Wang et al. [17] carried out a numerical simulation on the relationship between structural characteristics and materials of liner on EFP formation. The liner of tungsten alloy could form complete EFPs with low solidity in the numerical simulation. However, their numerical simulation results have not yet been validated in experiments.

Robert P. Koch et al. [18] carried out experiments to study the performance of a two phase tungsten tungsten–nickel–iron heavy alloy (W–Ni–Fe) and a single phase nickel–tungsten alloy (Ni–W) when used as MEFP liner. They found that W–Ni–Fe heavy alloy liner with a density of 17.1 g/cm^3 fractured into pieces in the charge structure designed for minimal deformation, while both the W–Ni–Fe heavy alloy liner and the Ni–W alloy liner with a density of 11.1 g/cm^3 fractured into pieces, but the potential reasons for the fracture phenomenon have not been analyzed. The radiograph of the X-ray experiment and soft recovered tungsten alloy penetrators from the tests clearly showed the fracture characteristics with charge structure for severe deformation.

In this paper, three kinds of metal materials for EFP liner are selected to be tested under explosive loading, within which 90W–9Ni–Co and W–25Re are tested as two kinds of typical tungsten heavy alloy and copper as a validated EFP liner material reference. The dynamic response and the formation characteristic were observed by a flash X-ray experiment.

In order to explain the different responses of the three kinds of metals, the microscopic features were examined and compared in the original liner and recovered fragment. The fracture models were determined and the microstructure evolutions under explosive loading were analyzed. Associated with numerical simulation results and fracture mechanics, the fracture mechanism of the two kinds of tungsten alloys were analyzed.

2. Experimental Details and Results
2.1. Experimental Details

Parameters of the three kinds of metal material for the EFP liner are listed in Table 1. As two kinds of typical tungsten alloy with good plasticity, 90W–9Ni–Co and W–25Re alloy have much higher density and yield stress compared with copper (OFHC, oxygen-free high-conductivity copper). However, copper has better ductility than the two kinds of heavy tungsten alloys. Powder metallurgy is used to produce the 90W–9Ni–Co and

W–25Re alloy liner. According to the structure of liner, preparation procedures of tungsten alloy liner are set below: a. mixing of powder; b. isostatic pressing of performs; c. sintering; d. rolling to sheet; e. annealing; and f. stamping or machining to liners.

Table 1. Parameters of three kinds of metal materials for EFP liner.

Materials	Density ρ/g·cm^{-3}	Melting Point T_m/°C	Elongation δ/%	Yield Stress σ_s/MPa	Tensile Stress σ_b/MPa
Copper	8.96	1356	30–40	120	245–315
90W–9Ni–Co	17.10	3200	35	600	930
W–25Re	19.50	3100	15–18	/	1600–1950

Figure 1 shows the EFP charge structure. As shown in Figure 1a, the EFP charge structure is composed of a detonator, booster pellet, casing, charge, liner, and retaining ring. The charge is made of explosive 8701, which is a kind of RDX-based explosive, with a density of 1.71 g/cm^3. The detonation velocity of the explosive is 8315 m/s. The length to diameter ration of the charge is 0.8. The length and diameter of the charge are denoted as l and CD. For the hemispherical liners, R_i is the liner's inner curvature, while R_o is the outer curvature (next to the charge), and h is the thickness of the liner. For the constant thickness of the liner used in this paper, h equals $R_o - R_i$. The mass of the liner for the three kinds of metal material stays the same. The material of casing is steel #45, with a thickness denoted as δ which equals to 0.045 CD. The retaining ring is also made of steel #45. Figure 1b shows the 3D geometric sketch of the EFP charge structure. The components structure is shown in Figure 1c and the assembly status of the EFP structure in the experiment is presented in Figure 1d.

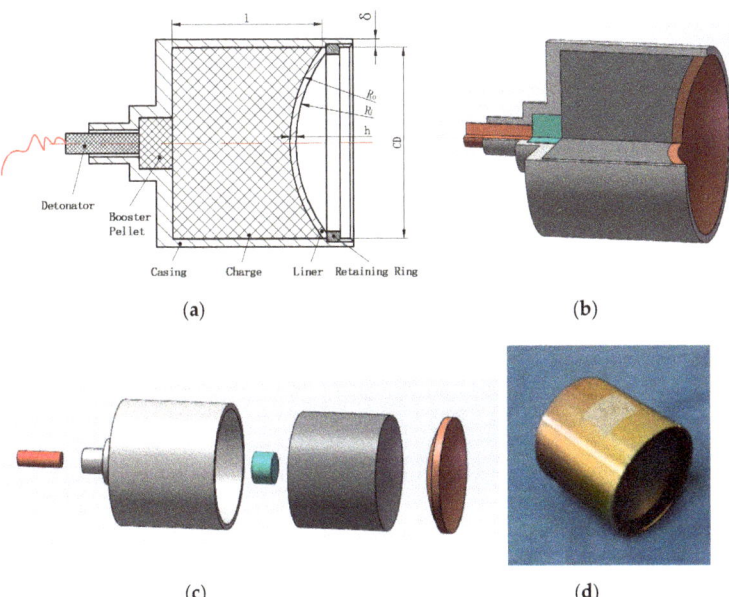

Figure 1. EFP charge structure. (**a**) 2D geometric sketch. (**b**) 3D geometric sketch. (**c**) Components structure. (**d**) Assembly structure.

Then flash X-ray experiment was carried out to observe the dynamic response and formation characteristic of liners. The schematic diagram of the flash X-ray experiment of the EFP is shown in Figure 2. Scandiflash-450 system is used for the flash X-ray experiment, which is designed by Scandiflash AB Company in Sweden and is widely used in ballistics and hypervelocity impact studies. The center point initiation method is applied in the experiment. The material of the target is Q235 steel and the thickness of the target is 40 mm which equals to 0.714 CD. The protective box, made of Q235 steel plates, is used to protect the X-ray film from the fragments generated by the steel casing in explosion. Steel cables are used to connect the protective box to the upper stand, and strings are used to connect the EFP charge structure to the upper stand.

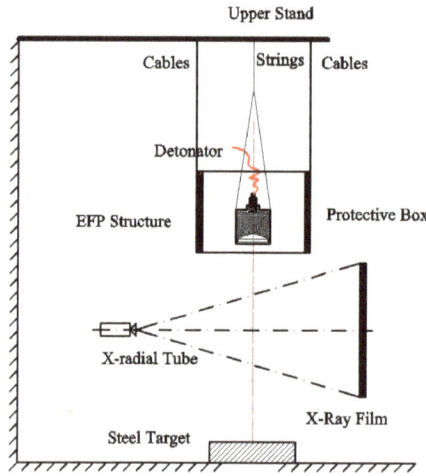

Figure 2. Schematic diagrams of the flash X-ray experiment of a typical EFP.

2.2. Experimental Results

Figure 3 shows the dynamic response and the formation status of three liners in the flash X-ray experiment. For each liner of different material, the images were captured 220 μs and 250 μs after detonation. An EFP with a clear outline was observed in the picture in the experiment of the Cu liner, while tungsten heavy alloy liners did not form an intact EFP. Broken pieces of fragments were observed in the experiment of the 90W–9Ni–Co liner after 220 μs. The W–25Re liner broke into parts of fragments at 250 μs. For the 90W–9Ni–Co and W–25Re liners, both vertical and horizontal fractures could be observed in the X-ray picture [19].

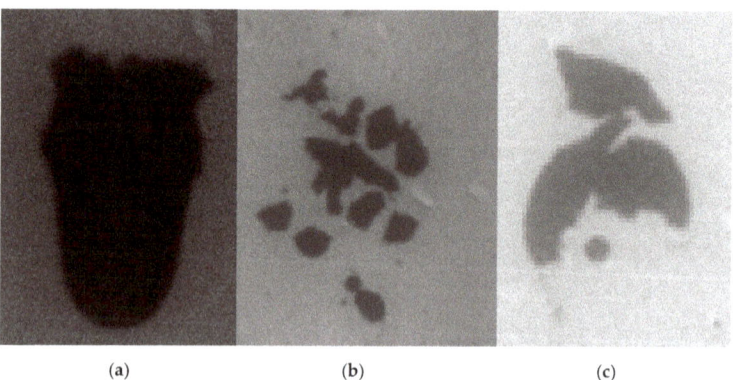

Figure 3. X-ray photographs of liners for different metal materials in the experiment. (**a**) Copper. (**b**) 90W–9Ni-Co. (**c**) W–25Re.

3. Microstructure Analysis

As there are different mechanical properties of the three kinds of metal liners, there could be a remarkable difference among the dynamic responses under explosive loading. In order to explain the different responses of the three kinds of metal materials, the microscopic features were examined and compared in the original liner and recovered fragments, and the microstructure evolutions under explosive loading were analyzed.

3.1. Copper Liner

Figure 4 compares the microstructures in the original liner and recovered residual of the copper liner. The liner's original microstructure is shown in Figure 4a,b. The average diameter of grain varies from 3 μm to 5 μm with equiaxed crystal structure, which are uniformly distributed. In Figure 4c,d, the stretched grain structure, dimples, and slip surfaces can be observed. The length of the grain can be longer than 30 μm and the length of plastic deformation zone can be 30–50 μm. The grain size has grown with the adjustment of grain boundaries, which means it has experienced tremendous plastic deformation and dynamic recrystallization occurs. The ductile fracture surface observed in Figure 4 can be summarized as dimple fracture and be used as assertive evidence to explain the dynamic macroscopic response of the copper liner under explosive loading.

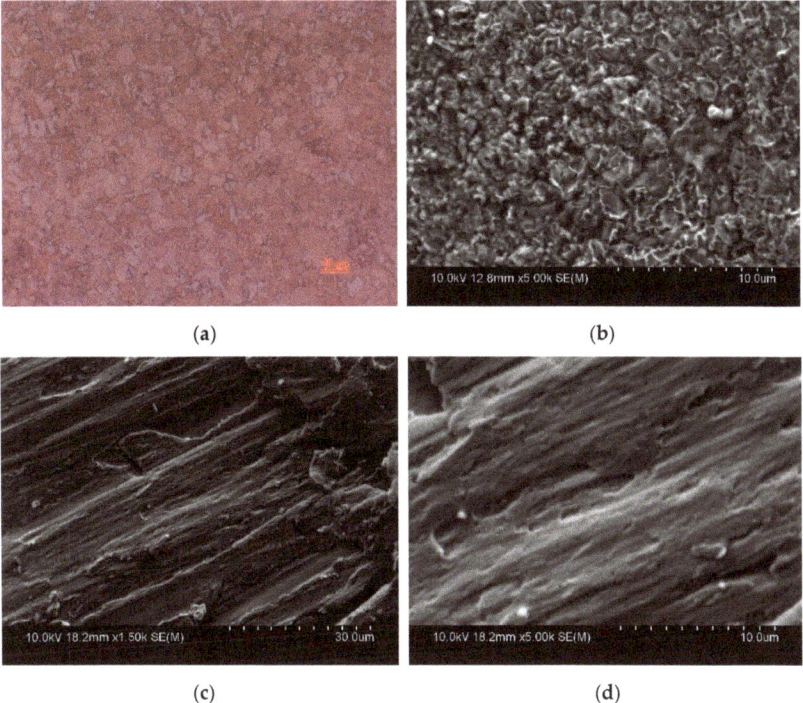

Figure 4. Comparison of microstructures in the original and recovered copper liner. (**a**) Original structure captured by metalloscope. (**b**) Original structure captured by SEM. (**c**) Fracture surface of recovered residual. (**d**) Enlarged view of fracture surface.

With the characteristics of dynamic recrystallization and ductile fracture surface observed in the copper liner in Figure 4, it can be concluded that the copper liner has the ability to sustain a large amount of plastic deformation without rupture in the forming process of an EFP under explosive loading.

3.2. Tungsten Heavy Alloy Liners

Figure 5 presents the microstructures of the original liner and the retrieved residual of the 90W–9Ni–Co alloy after explosive loading. As shown in Figure 5a, the tungsten particles and matrix can be observed in the two-phase compound. The average diameter of the tungsten particles is about 10~50 μm, and the tungsten particles are evenly distributed in the Ni–Co–W alloy. Moreover, the recrystallization and crystal twin can be observed, which indicate the grain growth of tungsten particles in the processing of manufacture. In Figure 5b, tungsten particles and the Ni–Co matrix, which is abnormally line-shaped with white color, can be observed. With no slip surface, cleavage steps are observed and only a small amount of plastic deformation occurs in the Ni–Co matrix. As the average diameter of the tungsten particles is about 2~5 μm, associated with cracks, it can be concluded that cleavage is the main mechanism in the microstructure evolution under explosive loading.

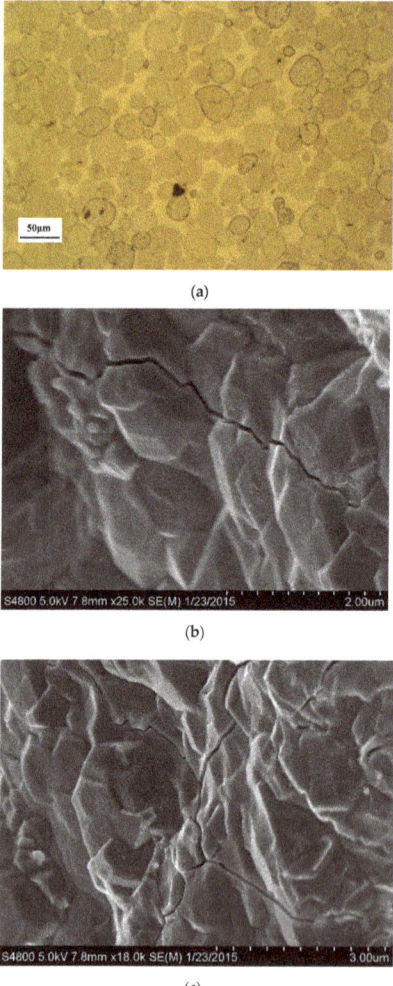

Figure 5. Microstructures in recovered 90W–9Ni–Co alloy liner. (**a**) Original structure captured by metalloscope. (**b**) Crack distribution on the fracture surface. (**c**) Crack pattern around tungsten particles.

Figure 6 makes a comparison of microstructures between the original liner and retrieved residual of the W–25Re alloy. Being a solid solution, as shown in Figure 6a, the fibrous grains dominate in the original microstructure of the W–25Re alloy. After explosive loading, pores and cracks appear and the grains refine. The average diameter of the grain is about 5–10 μm in the recovered residual, as shown in Figure 6b,c. Slip surfaces with a small amount of plastic deformation occur in the fibrous grains. From the fracture surface, it can be inferred that though there is a small amount of plastic deformation, transgranular cleavage is the major cause of the fracture of the W–25Re alloy liner under explosive loading.

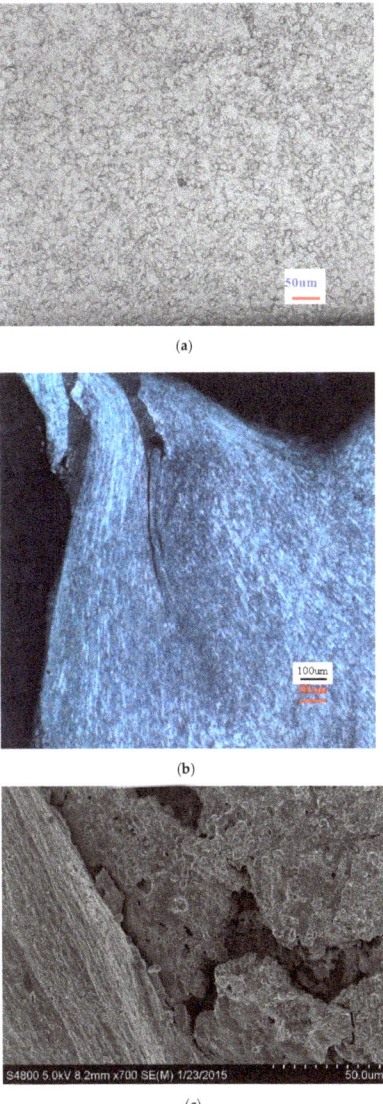

Figure 6. Comparison of microstructures in the original and recovered W–25Re alloy liner. (**a**) Original structure captured by metalloscope. (**b**) Microstructure of retrieved residual. (**c**) Fracture surface of retrieved residual.

Figure 7 illustrates schematically the major modes of failure and anticipated fracture surface in tungsten alloys. Little evidence of a matrix is normally seen, and tungsten cleavage and interface failure are regularly observed.

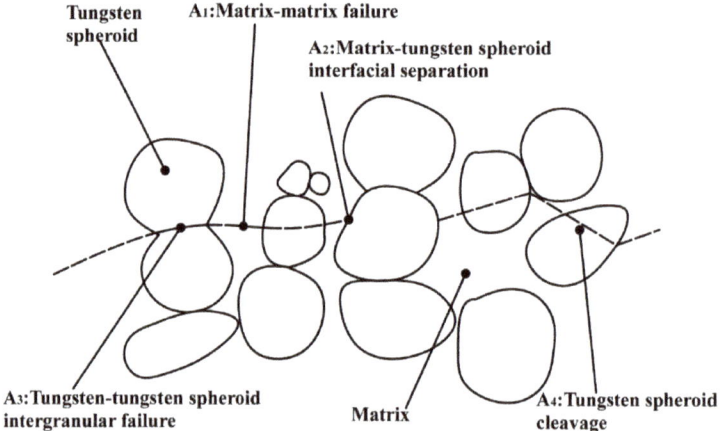

Figure 7. Schematic illustration of major failure modes in tungsten alloys. A_1: ductile transgranular fracture of matrix phase; A_2: intergranular fracture of matrix phase; A_3: transgranular cleavage of W-grains; A_4: intergranular fracture of W-grain network; A_5: tungsten side of W-matrix interface fracture; A_6: matrix side of W-matrix interface fracture.

The transgranular cleavage observed in fracture surfaces of the retrieved residual in Figures 5 and 6 demonstrates that the fracture mode belongs to brittle fracture, which falls into the type A_3 mode, as displayed in Figure 7. By comparing the microstructures of the original liner and retrieved residual of 90W–9Ni–Co and W–25Re alloys, it can be inferred: at high strain rates under explosive loading, both the W particles and matrix phase undergo tremendous deformation, and distortion of the W-grain network is more obvious, which could not satisfy the need of deformation in macroscopic scale. Micro-cracks occur in and around the W particles, which result in the transgranular cleavage of the W particles. Then, due to the severe stress concentration at the tip of the cracks, more cleavages of the W-grain arise, which further leads to the brittle fracture of the tungsten alloy under dynamic deformation [20–23].

In summary, the dynamic recrystallization and ductile fracture surface observed in the microstructure of copper explain the dynamic formation of a copper EFP under explosive loading, while the micro-cracks and cleavage observed in the 90W–9Ni–Co and W–25Re alloy which indicate the occurrence of brittle fracture are the predominant fracture mechanism and microstructure evolution of the two kinds of tungsten heavy alloy under explosive loading.

4. Numerical Simulation and Analysis

Due to the well-formed performance under explosive loading, copper EFP's forming characteristics are analyzed in the numerical simulation. Then, associated with the stress and strain conditions under explosive loading, the fracture phenomenon of tungsten heavy alloys can be analyzed in the fracture mechanism.

4.1. Numerical Model of Copper EFP

As shown in Figure 8a, all of the components of the EFP charge structure are modelled with the 2D Lagrange algorithm in LS-DYNA. Central point initiation is deployed to initiate the explosive. The elements are axisymmetric solid-area weighted shell, with mesh size of about 0.5 mm per grid, and a half model symmetric of the y axis is carried out. The mesh is shown in the grid model of Figure 8b.

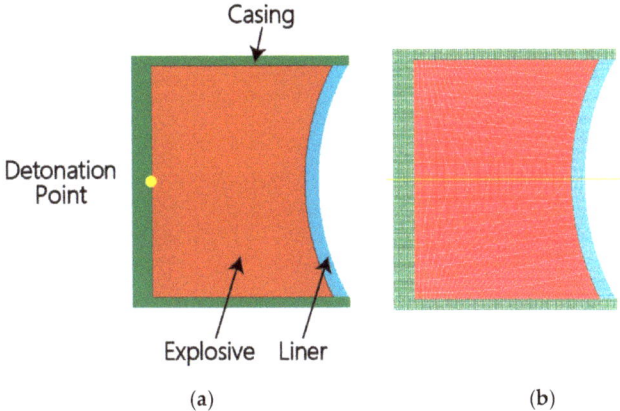

Figure 8. Numerical and grid model of EFP charge structure. (**a**) Diagram of numerical model. (**b**) Grid model.

The material models of charge, casing, and liner are listed in Table 2. The behavior of the high-explosive charge is characterized by the Jones–Wilkins–Lee (JWL) equation of state and high-explosive-burn constitutive model, which are widely used to describe the pressure–volume relationship of the explosive. The JWL equation of state defines the pressure as [24,25]:

$$p = A\left(1 - \frac{\omega}{R_1 V}\right)e^{-R_1 V} + B\left(1 - \frac{\omega}{R_2 V}\right)e^{-R_1 V} + \frac{\omega E}{V} \quad (1)$$

where A, B, R_1, R_2, and ω are constants to describe the relationship between the pressure and the relative volume of the charge. The EOS parameters of explosive 8701 are listed in Table 3.

Table 2. Material models used in numerical simulation.

Components	Material	Density ρ (g/cm³)	Equation of State	Constitutive Model
Charge	8701	1.71	Jones–Wilkins–Lee	High-Explosive-Burn
Casing	Steel #45	7.83	Grüneisen	Johnson–Cook
Liner	OFHC	8.96	Grüneisen	Steinberg–Guinan

Table 3. JWL EOS parameters of explosive 8701.

Explosive	D (m/s)	A (GPa)	B (GPa)	R_1	R_2	ω	E_0 (J·mm⁻³)
8701	8315	881.45	10.459	4.80	1.0	0.32	0.087

The selection of material model and setting of parameters of the liner are essential to predict the forming state of the copper EFP. In this paper, the Grüneisen equation of state is employed in conjunction with the Steinberg–Guinan constitutive model to simulate the forming of the copper EFP.

The Grüneisen EOS [26] can be used to describe how the materials interact with the shock wave and is based on Hugoniot's relation between the vs. and the v_p, as $v_s = c_0 + sv_p$, where vs. is the shock wave velocity, v_p is the material particle velocity, c_0 is the wave speed, and s is a material-related coefficient. The expression of equation of state of Grüneisen for compressed state is:

$$p = \frac{\rho_0 C^2 \mu \left[1 + \left(1 - \frac{\gamma_0}{2}\right)\mu - \frac{a}{2}\mu^2\right]}{\left[1 - (S_1 - 1)\mu - S_2 \frac{\mu^2}{\mu+1} - S_3 \frac{\mu^3}{(\mu+1)^2}\right]} + (\gamma_0 + a\mu)E. \quad (2)$$

In the expanded state,

$$p = \rho_0 C^2 \mu + (\gamma_0 + a\mu)E \quad (3)$$

where C is the intercept of velocity curve between shock wave and particle, S_1, S_2, and S_3 represent the slope of the v_s-v_p curve, γ_0 is the coefficient of Grüneisen, and a is one-order correction of γ_0. $\mu = \rho/\rho_0 - 1$ is a non-dimensional coefficient based on initial and instantaneous material densities. The parameters of equation of state are listed in Table 4.

Table 4. EOS parameters of the copper liner.

Material	Grüneisen Coefficient	C (m/s)	S_1	S_2	a
OFHC	2.02	3490	1.489	0	0.47

The Steinberg–Guinan model [27] is available for modelling materials at very high strain rate ($>10^5$ s^{-1}). The yield strength is a function of temperature and pressure. In the Steinberg–Guinan constitutive relation, the shear modulus, G, before the material melts, can be expressed as ...

$$G = G_0 \left[1 + bpV^{1/3} - h\left(\frac{E_i - E_c}{3R'} - 300\right)\right] e^{-fE_i/E_m - E_i} \quad (4)$$

where p is the pressure, V is the relative volume, and E_c is the cold compression energy:

$$E_c(x) = \int_0^x p\,dx - \frac{900R' \exp(ax)}{(1-x)^{2(\gamma_0 - a - 1/2)}}, \quad (5)$$

$$x = 1 - V, \quad (6)$$

and E_m is the melting energy:

$$E_m(x) = E_c(x) + 3R'T_m(x) \quad (7)$$

which is in terms of the melting temperature $T_m(x)$:

$$T_m(x) = \frac{T_{m0} \exp(2ax)}{V^{2(\gamma_0 - a - 1/3)}} \quad (8)$$

and the melting temperature at $\rho = \rho_0$, T_{m0}.

The yield stress σ_y is given by:

$$\sigma_y = \sigma_0' \left[1 + b'pV^{1/3} - h\left(\frac{E_i - E_c}{3R'} - 300\right)\right] e^{-fE_i/E_m - E_i} \quad (9)$$

when E_m exceeds E_i, here σ_0' is given by:

$$\sigma_0' = \sigma_0 [1 + \beta(\gamma_i + \bar{\varepsilon}^p)]^n \quad (10)$$

where σ_0 is the initial yield stress and γ_i is the initial plastic strain. If the work-hardened yield stress σ_0' exceeds σ_m, σ_0' is set to σ_m. After the melting point, σ_y and G are set to one half their initial value. the material parameters of the Steinberg–Guinan model for the copper liner are presented in Table 5.

Table 5. Material parameters of the Steinberg–Guinan model for the copper liner.

ρ (g/cm^3)	G_0 (MPa)	σ_0 (MPa)	β	n	γ_i	σ_m	b
8.96	47,700	120	36	0.45	0	640	36
b'	h	f	T_{m0}(K)	γ_0	a	PC	SPALL
0.45	0.000377	0.001	1790	2.02	1.5	−9	3

The steel casing adopts the Grüneisen EOS and Johnson–Cook constitutive model. The Johnson–Cook model [28,29] is a widely used constitutive model which incorporates the effect of strain rate dependent work hardening and thermal softening. The Johnson–Cook constitutive relation is given by:

$$\sigma = (\sigma_0 + B\varepsilon^n)\left(1 + C \ln \frac{\dot{\varepsilon}}{\dot{\varepsilon}_0}\right)(1 - T^{*m}) \quad (11)$$

where ε is the plastic strain and the temperature factor is expressed as:

$$T^* = \frac{T - T_r}{T_m - T_r} \quad (12)$$

where T_r is the room temperature, and T_m is the melt temperature of the material. σ_0, B, n, C, and n are material-related parameters. The material parameters of steel #45 for casing are presented in Table 6.

Table 6. Material parameters of steel #45 for casing.

Material	ρ(g/cm^3)	Grüneisen Coefficient	C(m/s)	S_1	S_2	a
	7.83	2.02	3490	1.489	0	0.47
Steel #45	σ_0 (MPa)	B (MPa)	n	C	m	T_m (K)
	175	380	0.32	0.006	1.0	1793

4.2. Numerical Results of Copper EFP

Figure 9 shows the shape and effective stress of the copper EFP at typical time in the forming stage. In the first 30 μs after the detonation of the charge, the detonation wave is transmitted to the top of the liner first. Thus, the top part of the liner accelerates and deforms in axial direction. As it interacts with the detonation wave, other parts of the liner deform and accelerate in sequence, with the bottom of the liner deforming last, which can be seen around 50 μs. At the same time, the liner is driven by the detonation wave to move forward along the axial direction. Then, due to the velocity gradient in the head and tail of the liner, the liner flips, and the inner surface of the liner will squeeze or even collide, which can be observed from 50 μs to 80 μs. The extrusion of the liner makes the inner wall of the liner close to the axis to form a rod-shaped projectile. After 80 μs, the shape of the EFP is basically stable.

As shown in Table 7, the maximum von Mises stress can be as much as 604 MPa at 68 μs, and the maximum shear stress can reach 341 MPa. The maximum plastic strain reaches 2.71 after 100 μs. So, it can be concluded that in the forming stage of the copper EFP, the liner undergoes maximum shear stress and maximum effective stress in the early 70 μs, and the maximum plastic strain can be as much as 3.0.

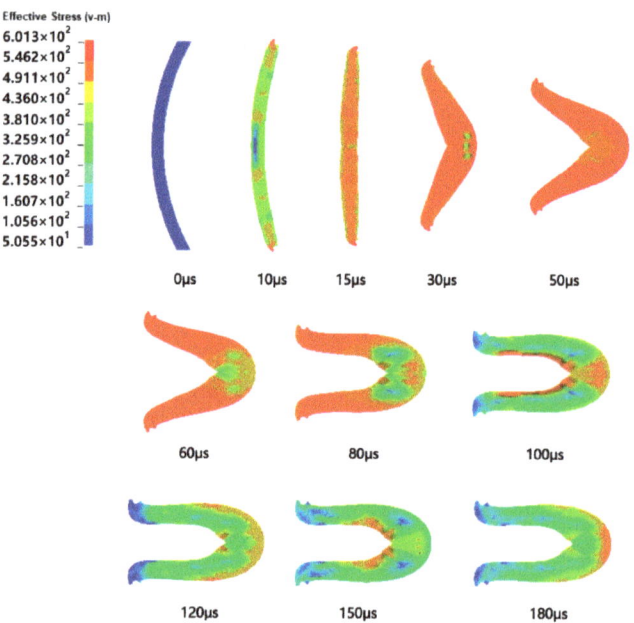

Figure 9. Effective stress of the copper EFP at typical time after detonation.

Table 7. Maximum stress and strain conditions in the forming of the Copper EFP.

Characteristics	Time (μs)	Forming Shape	Maximum Value
von Mises Stress	68		604 MPa
Shear Stress	32		341 MPa
Plastic Strain	≥100		2.74

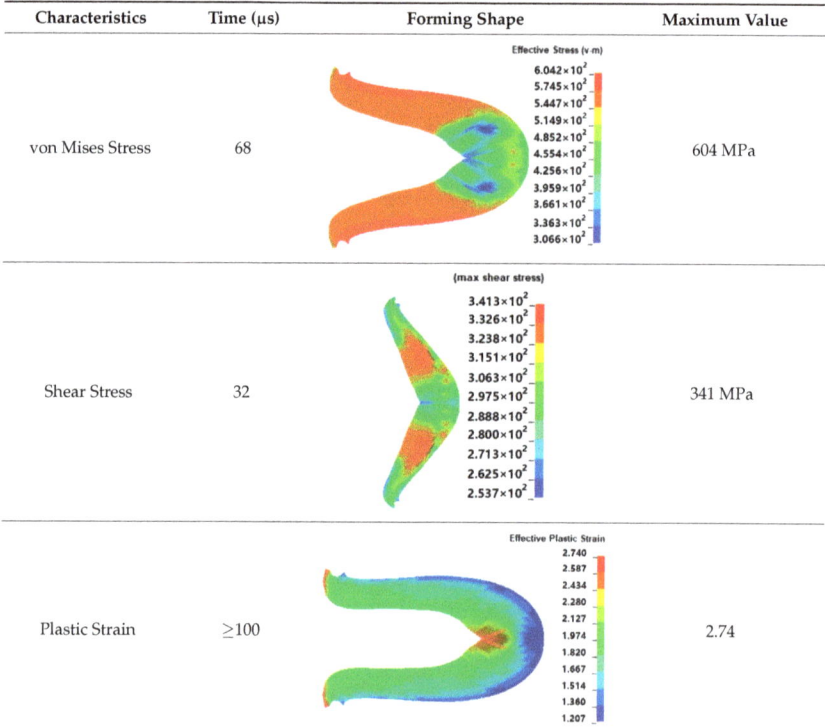

Table 8 makes a comparison of numerical simulation and experiment results. L and D are the length and diameter of EFP, and L/D is the length to diameter ration of EFP. By the comparison of length, diameter, and length to diameter ratio, the numerical simulation results agree well with the experiment results, which verify the accuracy of the numerical simulation.

Table 8. Comparison of the copper EFP's forming states in numerical simulation and the experiment results.

	Forming Shape	L/CD	D/CD	L/D
Numerical Simulation		0.672	0.372	1.81
Experimental Result		0.617	0.329	1.88

In conclusion, a feasible EFP liner should sustain tremendous stress in the early stage of forming without breaking. The liner has to bear large plastic strain under explosive loading, and undergo severe plastic deformation in the forming. Associated with the maximum stress and strain conditions in forming of EFP under explosive loading, the fracture phenomenon of tungsten heavy alloys can be analyzed in the fracture mechanism.

4.3. Analysis of Fracture Mechanism

In fracture mechanics, there are three modes of loading relative to a crack, as shown in Figure 10. Mode I is also called the opening mode, where the principal load is applied normally to the crack plane, which tends to open the crack. Mode II tends to slide one crack face with respect to the other. Mode III is called the tearing mode, which refers to out of plane shear [30]. On the microscopic scale, according to the feature of the fracture surface, the cleavage pattern is mainly produced by the tensile stress and results in the brittle fracture by separation (cleavage) across well-defined habit crystallographic planes [31]. As Mode I is the most dangerous loading pattern among the three modes, the stress field ahead of a crack tip in the opening mode is used to analyze the fracture mechanism of tungsten heavy alloys.

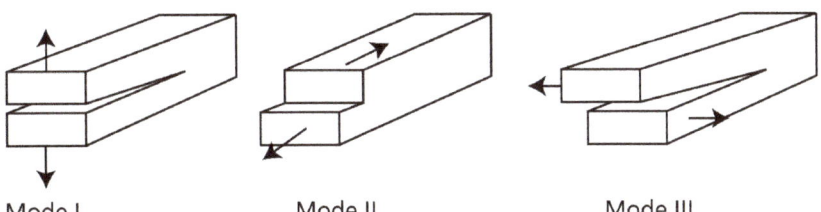

Figure 10. Three modes of loading relative to a crack: mode I (opening mode), mode II (shear or sliding mode), and mode III (tearing mode).

Figure 11 presents the definition of the coordinate axis ahead of a crack tip. σ_{ij} and τ_{ij} are the stress tensor. r and θ are the defined polar coordinate axis with the origin at the crack tip. K_I denotes the stress intensity factor, and ν is Poisson's ration. The stress fields ahead of a crack tip in an isotropic linear elastic material can be written in the form of (13) as an isotropic linear elastic material [32–34].

$$\begin{cases} \sigma_{xx} = \frac{K_I}{\sqrt{2\pi r}} \cos\frac{\theta}{2}\left(1 - \sin\frac{\theta}{2}\sin\frac{3\theta}{2}\right) \\ \sigma_{yy} = \frac{K_I}{\sqrt{2\pi r}} \cos\frac{\theta}{2}\left(1 + \sin\frac{\theta}{2}\sin\frac{3\theta}{2}\right) \\ \tau_{xy} = \frac{K_I}{\sqrt{2\pi r}} \sin\frac{\theta}{2}\cdot\cos\frac{\theta}{2}\cdot\cos\frac{3\theta}{2} \\ \sigma_{zz} = 0 \quad \text{(Plane stress)} \\ \sigma_{zz} = \nu(\sigma_{xx} + \sigma_{yy}) \text{(Plane strain)} \\ \tau_{xz} = \tau_{yz} = 0 \end{cases} \quad (13)$$

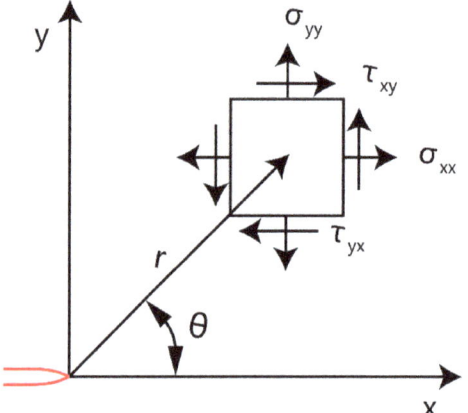

Figure 11. Definition of the coordinate axis ahead of a crack tip.

According to the von Mises criterion, yielding occurs when $\sigma_e = \sigma_{YS}$, the uniaxial yield strength. For plane stress or plane strain conditions, the principal stresses can be computed from the two-dimensional Mohr's relationship:

$$\sigma_1, \sigma_1 = \frac{\sigma_{xx} + \sigma_{yy}}{2} \pm \left[\left(\frac{\sigma_{xx} - \sigma_{yy}}{2}\right)^2 + \tau_{xy}^2\right] \quad (14)$$

For plane stress $\sigma_3 = 0$, and $\sigma_3 = \nu(\sigma_1 + \sigma_2)$ for plane strain. Then,

$$\begin{cases} \sigma_1 = \frac{K_I}{\sqrt{2\pi r}} \cos\frac{\theta}{2}\left(1 + \sin\frac{\theta}{2}\right) \\ \sigma_2 = \frac{K_I}{\sqrt{2\pi r}} \cos\frac{\theta}{2}\left(1 - \sin\frac{\theta}{2}\right) \\ \sigma_3 = 0 \quad \text{(Plane stress)} \\ \sigma_3 = \frac{K_I}{\sqrt{2\pi r}} \cos\frac{\theta}{2} \quad \text{(Plane strain)} \end{cases} \quad (15)$$

By substituting the equations into:

$$\sigma_e = \frac{1}{\sqrt{2}}\left[(\sigma_1 - \sigma_2)^2 + (\sigma_1 - \sigma_3)^2 + (\sigma_2 - \sigma_3)^2\right]^{1/2} \quad (16)$$

setting $\sigma_e = \sigma_{YS}$, and solving for r as a function of θ for plane stress:

$$r(\theta) = \frac{K_I^2}{2\pi\sigma_s^2} \cos^2\frac{\theta}{2}\left(1 + 3\sin^2\frac{\theta}{2}\right) \quad (17)$$

And for the plane strain,

$$r(\theta) = \frac{1}{2\pi}\left(\frac{K_I}{\sigma_s}\right)^2 \cos^2\frac{\theta}{2}\left((1-2\nu)^2 + 3\sin^2\frac{\theta}{2}\right) \qquad (18)$$

where set $\nu = 0.3$ for metal materials.

Fracture toughness K_{IC} and damage tolerance d_y of the three kinds of materials are listed in Table 9. The fracture toughness of copper is much higher than 90W–Ni–Co and W–25Re. The maximum fracture toughness of copper can reach 100 MPam$^{1/2}$, while the maximum value of tungsten alloy can just reach 60 MPam$^{1/2}$. The values of K_{IC} for 90W–Ni–Co and W–25Re almost stay the same. d_y, diameter of the process-zone at a crack tip, which indicates the damage tolerance of the plastic zone, are listed in the Table 9. d_y of the copper ranges from 1 mm to 1000 mm, while damage tolerance of 90W–9Ni–Co and W–25Re varies from 0.1 mm to 1 mm.

Table 9. Fracture toughness K_{IC} and damage tolerance d_y of the three kinds of materials.

Material	K_{IC} (MPam$^{1/2}$)	d_y (mm)	Fracture Mode	Reference
Copper	79–100	1–1000	Dimple	[35]
90W–9Ni–Co	55–60	0.1–1	Cleavage	[35]
W–25Re	54.2	0.1–1	Cleavage	[36,37]

The lower values of fracture toughness and damage tolerance of 90W–9Ni–Co and W–25Re can be a convincing evidence to explain the fracture phenomenon in the flash X-ray experiment, but more detailed discussion should be given. By substituting the maximum values of stress and the fracture toughness of the three kinds of materials in Equation (17) and Equation (18), the crack-tip plastic zone shapes estimated from the elastic solutions and the von Mises yield criterion for Mode I of loading can be obtained, as shown in Figure 12. The solid line is for the plane stress zone, while the dashed line is for the plane strain.

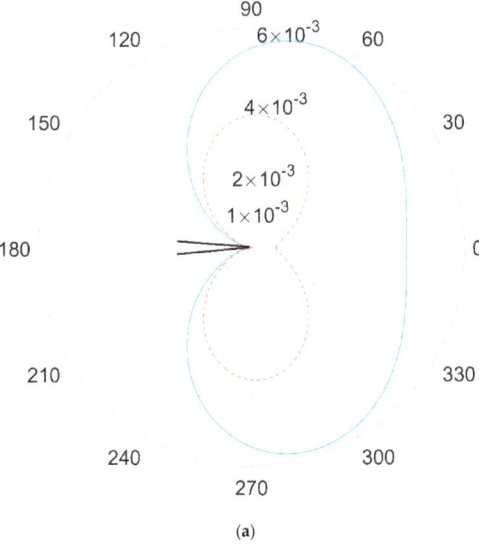

(a)

Figure 12. *Cont.*

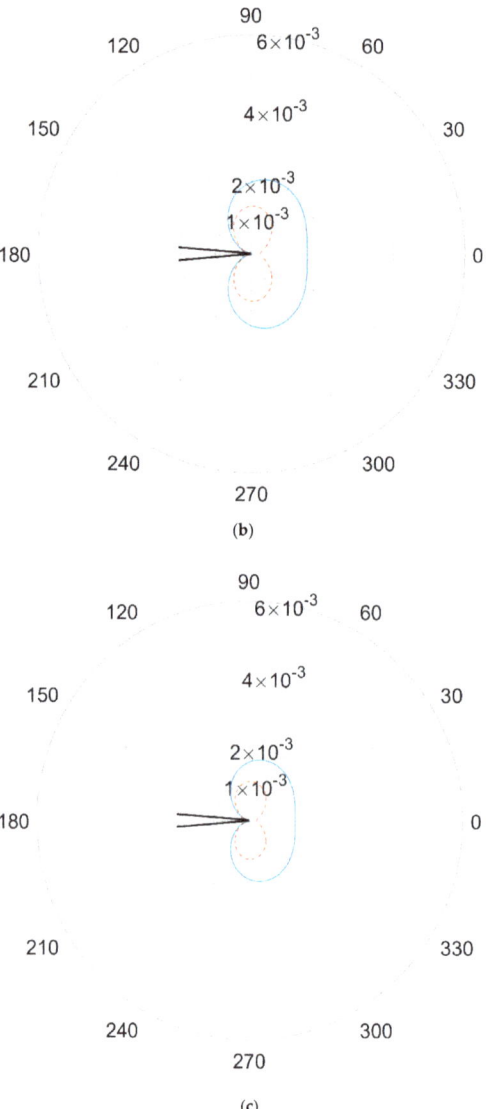

Figure 12. Crack-tip plastic zone shapes estimated. (a) Copper. (b) 90W-9Ni-Co. (c) W-25Re.

As presented in Figure 12, the plain strain condition suppresses yielding, resulting in a smaller plastic zone when compared with the plain stress. The maximums of $r(\theta)$ for the three kinds of materials are listed in Table 10. For the plane stress condition, the maximum $r(\theta)$ for plane stress can reach 58 mm for copper, while for the plane strain, the value could only reach 36 mm. For the tungsten alloy, the values of $r(\theta)$ are much lower than copper. The maximum $r(\theta)$ is about 17~21 mm, while for plane strain the values reduce to 1 mm. The maximum of $r(\theta)$ for the plane strain of tungsten alloys is consistent with the values of d_y, which indicates the damage tolerance of the plastic zone.

Table 10. The maximum of $r(\theta)$ for the three kinds of materials.

Material	Maximum of $r(\theta)$/mm	
	Plane Stress	Plane Strain
Copper	58	36
90W–9Ni–Co	21	1
W–25Re	17	1

In conclusion, the crack-tip plastic zones of 90W–9Ni–Co and W–25Re are much smaller than copper. As the plain strain is the most dangerous condition in the fracture mode, it has more reference significance to understand the fracture mechanism of tungsten alloys. Under explosive loading with severe stress and strain conditions, cracks may occur inside the material. Then, due to the severe stress concentration at the tip of the cracks, it is easy for the cracks to propagate and trigger the cleavage in tungsten alloys. In Figures 5 and 6, cracks and pores are easily observed, which are consistent with the prediction of the crack-tip plastic zone. While for copper with excellent ductility, it has the ability to bear a considerable amount of plastic deformation. In addition, it will not fracture even with cracks due to its big enough crack-tip plastic zone, as shown in Figure 12a.

In addition, the material selection criteria of the EFP liner can be further enriched and specified based on the value of $r(\theta)$ for the plane stress. The material selection criteria of the EFP liner can be summarized as below: (i) the most potential candidate of material should have fracture toughness K_{IC} be of 70–150 MPam$^{1/2}$, diameter of the process zone d_y be of 10–1000 mm. (ii) Impact toughness α_k can be an alternative guideline, which should be in the range of 1500–2000 KJm^{-2}. (iii) Fracture surface appearance of microvoid coalescence both in the quasi-static and dynamic failure is preferred [19]. (iv) $r(\theta)$ of the crack-tip plastic zone can also be an alternative guideline to explain the fracture phenomenon in explosive loading. The potential candidate of an EFP liner should have a crack-tip plastic zone $r(\theta)$ as much as 58 mm for plane stress, and reach 36 mm for the plane strain condition. Then, the material could have the ability to sustain tremendous plastic deformation in the forming under explosive loading and form an EFP like copper.

5. Conclusions

(1) Compared with Cu, 90W_9Ni_Co and W_25Re as two kinds of typical tungsten heavy alloy, are tested in the flash X-ray experiment to study their dynamic response under explosive loading. However, there are tremendous differences among the forming state and the penetration results of the three metal materials when used as an EFP liner.

(2) The copper liner could form a coherent EFP with a moderate penetration depth in the steel target, while the 90W–9Ni–Co liner broke into discrete fragments and the W–25Re liner turned over at first and fractured into three parts later. As the fracture states in the forming phase, the penetration depth was much lower than the copper liner.

(3) The microscopic features were examined to analyze the microstructure evolution of the dynamic response of the three metal materials: (i) in the copper liner, dynamic recrystallization occurs and the microstructure has experienced tremendous plastic deformation, which means copper as a validated EFP liner has the ability to bear extensive plastic deformation without break under explosive loading. (ii) In the 90W–9Ni–Co and W–25Re alloys liner, the cleavage observed in the fracture surface indicates the brittle fracture occurrence in the forming, belonging to type A_3 mode, which means it is easy for the cracks to propagate and trigger the cleavage of W particles in the tungsten heavy alloys.

(4) Copper EFP's forming characteristics are analyzed in the numerical simulation to obtain the stress and strain conditions in forming of EFP under explosive loading. The maximum von Mises stress can be as much as 604 MPa, while the maximum

shear stress can reach 341 MPa. In addition, the maximum plastic strain reaches 2.71. Associated with the maximum stress and strain conditions, the fracture mechanism of tungsten heavy alloys under explosive loading can be obtained.

(5) In the plane stress state, the crack-tip plastic zones of 90W–9Ni–Co and W–25Re reach 21 mm and 17 mm, respectively. While in the plane strain state, the value of $r(\theta)$ of the crack-tip plastic zone of 90W–9Ni–Co and W–25Re are around 1mm. By comparison, the crack-tip plastic zones of 90W–9Ni–Co and W–25Re are much smaller than copper, under explosive loading with severe stress and strain conditions, cracks may occur inside the material. Then, due to the severe stress concentration at the tip of the cracks, it is easy for the cracks to propagate and trigger the cleavage in tungsten alloys.

(6) The value of $r(\theta)$ of the crack-tip plastic zone can be used to explain the fracture phenomenon in explosive loading, which can be an alternative guideline for material selection criteria of the EFP liner. The potential candidate of an EFP liner should have a crack-tip plastic zone $r(\theta)$ as much as copper. For a feasible EFP liner, the crack-tip plastic zone $r(\theta)$ should be as much as 58 mm for plane stress, and reach 36 mm for plane strain condition.

The research work not only makes an attempt to study the dynamic response of typical tungsten heavy alloys under explosive loading, but to also provide an identification method associated to microcosmic scale with fracture mechanics to determine whether or not the alloy materials are capable of being used to form a coherent EFP. The research results are significant in understanding the dynamic forming, microstructure evolution, and fracture mechanism of tungsten heavy alloys.

Author Contributions: Methodology, validation, writing—original draft preparation, funding acquisition, L.D.; writing—review and editing, supervision, project administration, P.S.; data curation, resources, L.J. All authors have read and agreed to the published version of the manuscript.

Funding: This work was supported by the National Natural Science Foundation of China (Grant No. 11802142) and a project of the State Key Laboratory of Explosion Science and Technology (Grant No. KFJJ20-08M).

Institutional Review Board Statement: Not applicable.

Informed Consent Statement: Not applicable.

Data Availability Statement: The raw and processed data generated during this study will be made available upon reasonable request.

Acknowledgments: The authors thank the National Natural Science Foundation of China (Grant No. 11802142) and project of the State Key Laboratory of Explosion Science and Technology (Grant No. KFJJ20-08M).

Conflicts of Interest: The authors declare no conflict of interest.

References

1. Walters, W.P.; Zukas, J.A. *Fundamentals of Shaped Charges*; John Wiley: Hoboken, NJ, USA, 1989.
2. Sui, S.; Wang, S. *Terminal Effects*; National Defense Industry Press: Beijing, China, 2000.
3. Wang, S. *Terminal Effects*, 2nd ed.; China Science Publishing & Media Ltd.: Beijing, China, 2019.
4. Pappu, S.; Murr, L.E. Hydrocode and microstructural analysis of explosively formed penetrators. *J. Mater. Sci.* **2002**, *37*, 233–248. [CrossRef]
5. Held, M. Liners for shaped charges. *J. Battlef. Technol.* **2001**, *4*, 1–6.
6. Lassner, E.; Schubert, W.D. *Tungsten: Properties, Chemistry, Technology of the Element, Alloys, and Chemical Compounds*; Springer: Berlin/Heidelberg, Germany, 1999.
7. Patrik, A.; Patrik, L. Penetration performance of a calibre 5.56 mm Tungsten-Rhenium AP-Projectile into steel and ceramic targets. In Proceedings of the 29th International Symposium on Ballistics, Edinburgh, UK, 9–13 May 2016.
8. Prasad, B.L.; Annamalai, A.R. Effect of Rhenium addition on tungsten heavy alloys processed through spark plasma sintering. *Ain Shams Eng. J.* **2021**, *12*, 2957–2963. [CrossRef]
9. Guo, W.; Li, S.K.; Wang, F.C.; Wang, M. Dynamic recrystallization of tungsten in a shaped charge liner. *Scr. Mater.* **2009**, *60*, 329–332. [CrossRef]

10. Mutoh, Y.; Ichikawa, K.; Nagata, K.; Takeuchi, M. Effect of rhenium addition on fracture toughness of tungsten at elevated temperatures. *J. Mater. Sci.* **1995**, *30*, 770–775. [CrossRef]
11. Sadat, T.; Hocini, A.; Lilensten, L.; Faurie, D.; Tingaud, D.; Dirras, G. Data on the impact of increasing the W amount on the mass density and compressive properties of Ni–W alloys processed by spark plasma sintering. *Data Brief* **2016**, *7*, 1405–1408. [CrossRef] [PubMed]
12. Stawovy, M.T. Single Phase Tungsten Alloy. 2005. Available online: https://www.freepatentsonline.com/7360488.html (accessed on 17 December 2021).
13. Stawovy, M.T. Single Phase Tungsten Alloy for Shaped Charge Liner. 2005. Available online: https://www.freepatentsonline.com/y2005/0241522.html (accessed on 17 December 2021).
14. Rolc, S.; Buchar, J.; Akstein, Z. Computer simulation of Explosively formed projectiles (EFP). In Proceedings of the 23rd International Symposium on Ballistics, Tarragona, Spain, 16–20 April 2007; pp. 185–192.
15. Bingol, C.G.; Yildirim, R.O. Investigation of the effect of thickness and radius of curvature on explosively formed penetrators(EFP) formation. In Proceedings of the 25th International Symposium on Ballistics, Beijing, China, 17–21 May 2010; pp. 821–828.
16. Li, W.B.; Wang, X.M.; Zhou, H. Effect of the liner material on the shape of dual mode penetrators. *Combust. Explos. Shock. Waves* **2015**, *51*, 387–394. [CrossRef]
17. Wang, Y.J.; Li, W.B.; Li, W.B.; Wang, X.M.; Xu, H.Y. Influence of structural characteristics on EFP formation under different liner materials. In Proceedings of the 31th International Symposium on Ballistics, Hyderabad, India, 4–8 November 2019.
18. Koch, R.P.; Snyder, J.; Fong, R.; Redner, P.; Sadangi, R.; Kapoor, D. Using Nickel-Tungsten Alloys to Produce Multiple Explosively Formed Penetrator (MEFP) Warheads. In Proceedings of the 30th International Symposium on Ballistics, Long Beach, CA, USA, 11–15 September 2017.
19. Ding, L.; Jiang, J.; Men, J.; Wang, S.; Li, M. Research on Feasibility of Several High Density Materials for EFP Liner and Material Selection Criteria. *Propellants Explos. Pyrotech.* **2017**, *42*, 360–369. [CrossRef]
20. Lea, C.; Muddle, B.C.; Edmonds, D.V. Segregation to interphase boundaries in liquid-phase sintered tungsten alloys. *Metall. Trans. A* **1983**, *14*, 667–677. [CrossRef]
21. Edmonds, D.V. Structure/property relationships in sintered heavy alloys. *Int. J. Refract. Met. Hard Mater.* **1991**, *10*, 15–26. [CrossRef]
22. Kim, D.K.; Lee, S.; Noh, J.W. Dynamic and quasi-static torsional behavior of tungsten heavy alloy specimens fabricated through sintering, heat-treatment, swaging and aging. *Mater. Ence Eng. A* **1998**, *247*, 285–294. [CrossRef]
23. Fan, J. *Wu He Jin Ji Qi Zhi Bei Xin Ji Shu*; Metallurgical Industry Press: Beijing, China, 2006.
24. Hallquist, J. LS-DYNA Keyword User's Manual, Version 970. 2003. Available online: https://www.dynasupport.com/manuals/ls-dyna-manuals/ls-dyna-970-manual-for-keyword-format (accessed on 17 December 2021).
25. Hallquist, J.O. LS-Dyna Theory Manual. 2006. Available online: https://www.dynasupport.com/manuals/additional/ls-dyna-theory-manual-2005-beta/view (accessed on 17 December 2021).
26. Andre, M.M. *Dynamic Behavior of Materials*; John Wiley & Sons: Hoboken, NJ, USA, 1994.
27. Steinberg, D.J. *Equation of State and Strength Properties of Selected Materials*; Lawrence Livermore National Laboratory: Livermore, CA, USA, 1991.
28. Johnson, G.R. A constitutive model and data for metals subjected to large strains, high strain-rates and high temperatures. In Proceedings of the Seventh International Symposium on Ballistics, Hague, The Netherlands, 19–21 April 1983; p. 541.
29. Johnson, G.R.; Cook, W.H. Fracture characteristics of three metals subjected to various strains, strain rates, temperatures, and pressres. *Eng. Fract. Mech.* **1985**, *21*, 31–48. [CrossRef]
30. Kundu, T. *Fundamentals of Fracture Mechanics*; Taylor & Francis Group: Abingdon, UK, 2008.
31. Chen, J.H.; Cao, R. *Micromechanism of Cleavage Fracture of Metals: A Comprehensive Microphysical Model for Cleavage Cracking in Metals*; Butterworth Heinemann: Oxford, UK, 2014.
32. Anderson, T.L. *Fracture Mechanics: Fundamentals and Applications*; Taylor & Francis Group: Abingdon, UK, 1991.
33. Jin, C.; Shushan, Z. *Fracture Mechanics*; Science Press: Beijing, China, 2006.
34. Perez, N. *Fracture Mechanics*; Kluwer Academic Publishers: Dordrecht, The Netherlands, 2004.
35. Ashby, M.F. Materials Selection in Mechanical Design Third Edition. *MRS Bull.* **2005**, *30*, 995.
36. Gludovatz, B.; Faleschini, M.; Wurster, S.; Hoffmann, A.; Pippan, R. Influence of microstructure on the fracture toughness of tungsten alloys. In Proceedings of the Minerals, Metals & Materials Society (TMS) Annual Meeting, New Orleans, LA, USA, 9–13 March 2008.
37. Gludovatz, B.; Wurster, S.; Hoffmann, A.; Pippan, R. Fracture toughness of polycrystalline tungsten alloys. *Int. J. Refract. Met. Hard Mater.* **2010**, *28*, 674–678. [CrossRef]

Article

Shock Properties of One Unsaturated Clay and Its Equation of State Up to 30 GPa

Xianwen Ran [1,*], Xuan Zou, Jingyuan Zhou and Wenhui Tang

College of Liberal Arts and Sciences, National University of Defense Technology, Changsha 410073, China; zouxuan14@nudt.edu.cn (X.Z.); zhoujingyuan19@nudt.edu.cn (J.Z.); tangwenhui@nudt.edu.cn (W.T.)
* Correspondence: ranxianwen@nudt.edu.cn; Tel.: +86-135-7415-1246

Abstract: The complicated composition of unsaturated clay, e.g., solid mineral particles, water, and air, makes it difficult to get its precise equation of state (EOS) over a wide pressure range. In this paper, the high-pressure EOS of unsaturated clay was discussed at the mesoscale. With the original clay extracted from the southern suburbs of Luoyang city, China, three unsaturated clays with moisture contents of 0%, 8%, and 15%, respectively, were remolded. Their Hugoniot parameters in the pressure range of 0–30 GPa were measured using a one-stage or two-stage light gas gun. With the measured Hugoniot parameters, a high-pressure EOS of the unsaturated clay up to 30 GPa was developed and it is in good agreement with the experimental data.

Keywords: unsaturated clay; moisture content; Hugoniot parameters; high-pressure EOS

1. Introduction

Unsaturated clay is generally considered a three-constituent mixture composed of soil skeleton, water, and air, where the soil skeleton is formed by many solid mineral particles. Under external load, unsaturated clay will be compressed, and the three constituents inside will also deform correspondingly. During the initial loading period, the soil skeleton of unsaturated clay is generally the main bearing component and will deform from elastic to plastic with the increase of the external load. At the same time, most of the water and air inside can flow freely in the connected space among the solid particles due to the existence of the voids. When the external load increases to a threshold, the solid particles will be closely compacted and form many discrete enclosed rooms, in which nearly all the water and air are locked. At this stage, the soil skeleton, water, and air bear the same load. However, since the air has a greater compressibility than the water and the soil skeleton, it can be compressed to such an extent that its volume can be nearly ignored. Consequently, the force that the air bears is very small, compared with the force that the water and the soil skeleton bear, and it can also be neglected.

According to the Henrych theory [1], the deformation of an unsaturated clay under external load can be divided into two stages by the compacted pressure p_c: the low-pressure stage and the high-pressure stage. As shown in Figure 1a, the low-pressure stage can be further separated into an elastic-deformed part and a plastic-deformed part by the elastic limit p_e. In the elastic-deformed part, the soil skeleton deforms elastically, resulting in the elastic deformation of both the solid particles and the void content. When the external load is higher than the elastic limit p_e, the shear force among some solid particles will exceed the bond strength, which leads to the fracture of some solid-particle combinations and the displacement among some solid particles. This deformation cannot be recovered once the external load has been released, indicating that the plastic deformation of the unsaturated clay has happened. During the external load increases from the elastic limit p_e to the compacted pressure p_c, numerous solid particles will be sheared to slip and rearrange to a new position to accommodate the external load, and form more and more enclosed rooms in which air and water are locked. At the compacted pressure p_c, nearly all the solid particles are closely compacted, all the air and water are locked in, and there will be no void existing in the unsaturated clay. Therefore, in the low-pressure stage, the *p-alpha* equation of state (EOS) can be used to describe the state changes of the unsaturated clay. In the high-pressure stage, all of the soil skeleton, water, and air have been greatly compressed and the elastic–plastic effect is no longer the dominant factor in the deformation, as shown in Figure 1b. With the increase of the external load, the volume ratio of air in the unsaturated clay decreases because of the pressure balance among the solid particles, water, and air, and the relatively great compressibility of the air, which indicates that the effect of air on the state of the unsaturated clay in the high-pressure stage can be neglected. Hence, in the high-pressure state, it is adequate to consider only the contributions of the water and the solid particles in EOS of the unsaturated clay. In addition, mineral components and water usually have a relatively high thermal capacity and there will be a relatively small temperature-rise in the dynamic shock process for the unsaturated clay with low porosity. Therefore, when the shock pressure is not high enough, the thermal contribution on shock pressure is small compared with cold contribution, and the EOS of unsaturated clay can be simplified to the form $p = f(\rho)$ from the form $p = f(\rho, T)$.

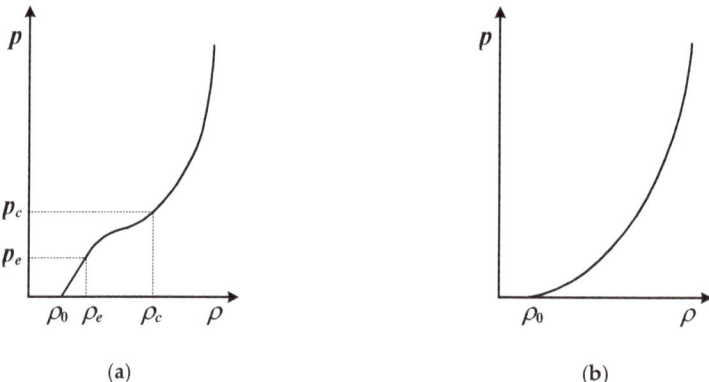

Figure 1. Schematic graph of the relation of p and ρ: (**a**) low pressure part; (**b**) high pressure part.

To study the mechanical and physical behaviors of the soil, many works have been conducted. Schofield [2] studied the mechanical behavior of saturated remolded soil, based on the critical state concept. Thiel [3], Kalashnikov [4], and Trunin [5] studied the shock wave data of the porous or hydrated earth materials and the sand–water mixture at several saturation levels. Tsembelis [6], Chapman [7], and Brown [8] completed a series of shock compressed experiments using a one-stage light gas gun to obtain the shock compressed behaviors of the dry sand. Resnyansky [9] theoretically put forward a two-substance EOS by regarding dry soil as the mixture of solid particles and gas. Using a three-substance EOS, Wang [10,11] carried out several numerical analyses on soil to obtain the dynamic response. These research results were important references for predicting the dynamic response of geotechnical material, but they could not be directly applied to unsaturated clay because moisture content was a significant factor affecting dynamic behavior. The shock loadings are the impact loads in terms of the wave dynamics.

To assess the effects of the underground explosion that happened in the southern suburbs of Luoyang city, China, in this study we conducted many experiments on the unsaturated clay with moisture contents of 0%, 8%, and 15%, respectively. Their Hugoniot parameters are obtained using a one- or two-stage light gas gun, based on which the high-pressure EOS up to 30 GPa is built. The developed EOS in this study can be applied for the numerical simulation of an underground explosion.

2. Experimental Measurement of the Hugoniot Parameters

2.1. Subsection

The original clay was extracted from the southern suburbs of Luoyang City in China. The detailed clay minerals of the clay are presented in Table 1. The density, moisture content, and solid particles of the original clay varied with the place and depth. Therefore, it was not possible to obtain the original clay samples with the same density and moisture content applied for experiments. All the samples used in this study were remolded according to the experimental needs.

Table 1. The mineral components of the dry original clay.

Components	Quartz	Calcite	Chlorite	Montmorillonite	Illite	Feldspar	Kaolinite	Amphibole	Hematite
Mass (%)	25	7	15	10	15	18	5	3	2

To study the influence of moisture content on the dynamical behaviors of clay, three unsaturated clay samples were prepared with a dry density of 1.70 g/cm^3 and initial moisture contents of 0%, 8%, and 15%, respectively. The clay samples were fabricated as follows: first, dry the dispersed clay in the oven; second, take the appropriate weight of clay with electronic scales; third, weigh the water with the measuring cylinder or injector and add it into the dried clay to the wanted moisture content; once the water diffuses in the clay uniformly, the clay sample should be put into the designed mold made of 2024 aluminum. As shown in Figure 2, the mold consists of a sample supporter, a reinforcement cylinder, and a compaction piston. The sample supporter had a circular indentation of 16 mm in diameter and 3 mm in depth. The reinforcement cylinder had a circular center hole of 16 mm in diameter and its depth was almost the same as the height of the compaction piston, which made the two parts connected closely and made the thickness of each sample almost the same. Before compressing the clay into the sample, an oil film should be evenly formed on the inner wall of the center bore of the reinforcement cylinder in order to reduce friction between the reinforcement cylinder and the piston. With the jack compressing the piston slowly, the clay in the mold can be compressed into the same size of $\varphi 16 \times 3$ mm, as shown in Figure 3. Table 2 gives the physical parameters of the final compressed clay samples of three moisture contents.

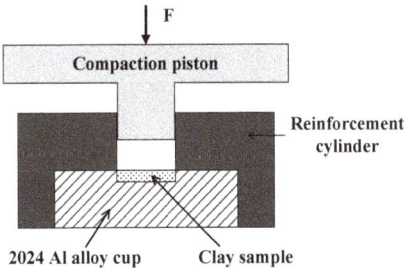

Figure 2. Schematic of sample mold.

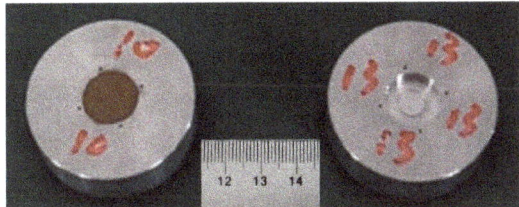

Figure 3. Schematic of experimental sample preparation.

Table 2. Physical parameters of clay in plate impact test.

Sample Number	1	2	3
initial dry density ρ_{d0} (g/cm^3)	1.70	1.70	1.70
initial wet density ρ_0 (g/cm^3)	1.70	1.84	1.96
sample weight m (g)	1.03	1.11	1.18
initial moisture content w (%)	0	8	15
initial saturation Sr_0 (%)	0	35.8	67.4

After the preparation of the experimental sample, the sample-holder was placed behind the target-stand to assemble a whole target, as shown in Figure 4a. After assembly, the target was installed in the terminal end of the light gas gun tube, as shown in Figure 4b.

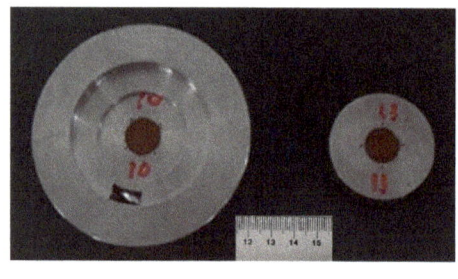
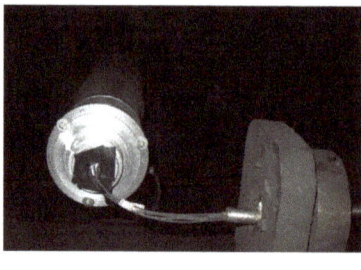

(a)   (b)

Figure 4. The target for plate-impact experiment graph: (**a**) the sample-holder and the target-stand; (**b**) back view of the assembled target.

2.2. Experimental Method

To obtain a wider range of loading pressure, the flyers made of 2024 aluminum alloy, copper, and tantalum were used in the plate-impact experiments. The flyers are discs of $\varphi 24 \times 3$ mm and their Hugoniot parameters [12] are shown in Table 3. Impacting velocities of flyers were measured by a velocity magnetic-measuring device [13].

Table 3. The parameters [11] of flyer materials and Hugoniot.

Material	ρ_0 (g/cm^3)	c_0 (km/s)	s
2024 aluminum alloy	2.785	5.328	1.338
copper	8.93	3.94	1.489
tantalum	16.656	3.437	1.19

Shock wave velocities were measured by the optical probes made from quartz fiber, which were calibrated in detonation experiments and had an uncertainty of about 1.8% [14]. When the shock wave propagated to the end of the probe, the quartz fiber would radiate optically due to shock wave arrival and the moment of shock wave arrival would be recorded by the digital oscillograph. The structure of the optical probe used in the experiments, as shown in Figure 5, consisted of one quartz fiber with a core diameter of 60 μm and an outside diameter of 175 μm, and one metal capillary with the internal diameters of 0.3 mm and 0.9 mm. The quartz fiber and metal capillary were cemented together with quick dry glue, in which the metal capillary plays an important role in enhancing the strength of the optical probe and keeping the optical probes in a fixed position perpendicularly. The surface of the end of the optical probe was coated with a 120~150 nm thick aluminum film which could prevent stray light from entering the quartz fiber.

The schematic diagram of one single-fiber probe system, as shown in Figure 6, consists of a photomultiplier tube and a digital oscillograph. The photomultiplier tube is the GDB-608 MCP and has an impulse response time of no more than 0.4 ns. The digital oscillograph used in the experiments has a sampling rate of 5 GS/s and an analog output bandwidth of 1 GHz.

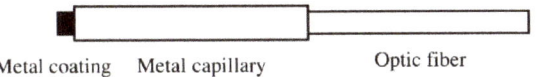

Figure 5. The schematic diagram of the optical probe of quartz fiber.

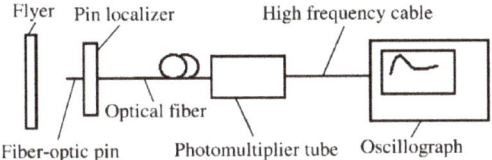

Figure 6. The schematic diagram of one single-fiber probe system.

In the experiments, there were a total of five fiber-optic pins (No. A-E) arranged for measuring the Hugoniot parameters of the sample. As shown in Figure 7, four fiber-optic pins (No. A-D) were arranged at the impacted surface and the outer edge of the sample to record the moment when the flyer just impacted the clay sample. Two diagonal fiber-optic pins formed a recording channel; there were a total of two channels that could amend the error arising from an oblique collision between the flyer and the clay sample. The last fiber-optic pin (No. E) was arranged in the center of the rear plane of the clay sample to record the moment of shock wave arrival.

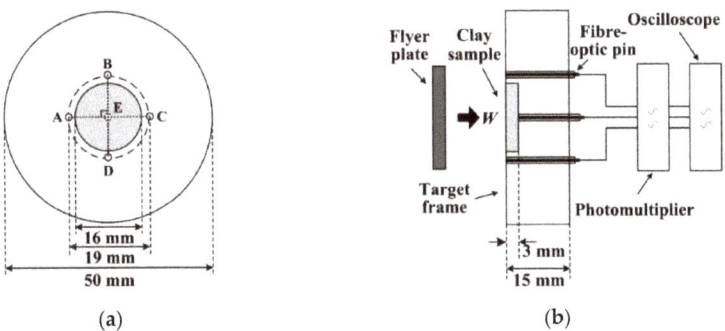

Figure 7. The schematic diagram for measuring Hugoniot parameters. (**a**) Arrangement position of fiber-optic pins. (**b**) Fiber-optic pins test system.

2.3. Data Processing Method

The typical signals in the experiments, as shown in Figure 8, were obtained from two measuring channels. The signal in Figure 8a was taken from the four fiber-optic pins (No. E, A, B, and C) and the signals from fiber-optic pins E, B, and C had almost the same wave-form. It can be seen from Figure 8 that the signals from fiber-optic pins A, B, C, and D had almost the same jumping moment, which indicated that the flyer had preferable planarity and a small deflection error when the flyer impacted the target. In all the signals, the inflection points showed the moments that the flyer or shock wave arrived at the corresponding fiber-optic pin. With these moments, the shock wave velocity U_s in the sample at this shocked state could be calculated from the following equation:

$$U_s = \Delta h / \Delta t \tag{1}$$

where Δh was the thickness of the sample and Δt was the interval that the shock wave took for traveling in the sample.

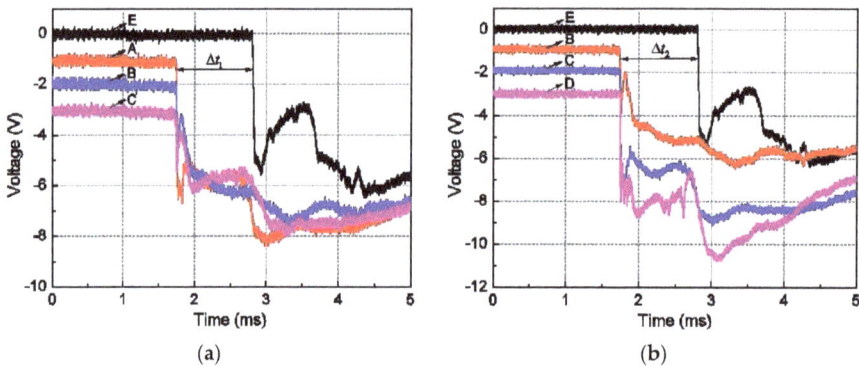

Figure 8. The typical signals of fiber optical pins in shock experiments: (**a**) optical pin signals from the first channel; (**b**) optical pin signals from the second channel.

For most materials, one linear relation between the shock wave velocity U_s and the particle velocity u_p existed as the following:

$$U_s - u_0 = c_0 + s(u_p - u_0) \tag{2}$$

where c_0 and s were the Hugoniot parameters and u_0 was the initial particle velocity. If a flyer impacted a sample plate with the impacting velocity W, according to the Rayleigh line and Equation (2), the shock pressure in the flyer should be written as:

$$p_f = \rho_{0f}(U_s - W)(u_p - W) = \rho_{0f}[-c_{0f} + s_f(u_p - W)](u_p - W) \tag{3}$$

According to Equation (3), for the given impacting velocity W and the measured Hugoniot parameters in Table 2, the shock pressure in the flyer could be expressed as a function of particle velocity u_p.

For the ith impact experiment with the velocity W_i and the measured shock wave velocity U_{si}, the particle velocity u_{pi} can be calculated according to the impedance match method [8] as shown in Figure 9:

$$u_{pi} = \frac{-B - \sqrt{B^2 - 4AC}}{2A} \tag{4}$$

where

$$A = \rho_{0f} s_f \tag{5}$$

$$B = -(\rho_{0f} c_{0f} + 2\rho_{0f} s_f W_i + \rho_{0s} U_{si}) \tag{6}$$

$$C = \rho_{0f} W_i (c_{0f} + s_f W_i) \tag{7}$$

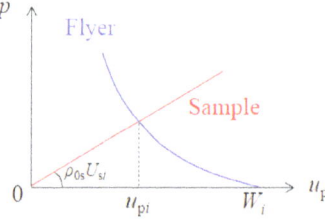

Figure 9. The impedance match method.

The subscript f and s meant the flyer and sample plates, and the subscript 0 meant the initial state.

2.4. Experimental Results and Analysis

According to the moisture content of the samples, the shock-impacted experiments were divided into three groups and each group had four samples with the same moisture content. Twelve effective experimental data were obtained in total. From these experimental data, the flyer plate impacting velocities, the shock wave velocities, and the particle velocities were derived with the experimental method in Section 2.2 and the experimental data processing method in Section 2.3. Moreover, the shock wave pressure in the samples was also obtained with the Hugoniot relation. All the experimental measured data and the processed data are shown in Table 4.

Table 4. Experimental results for unsaturated clay.

Clay Samples	No.	Flyer Materials	Flyer Velocity (km/s)	Shock Wave Arrival Time (ns)	Sample Thickness (mm)	Particle Velocity (km/s)	Shock Wave Velocity (km/s)	Pressure (GPa)
Dry Clay	1	Aluminum	0.561	1868	2.98	0.48	1.60	1.29
	2	Copper	1.042	1054	2.96	0.92	2.81	4.40
	3	Copper	2.340	658	2.94	1.97	4.47	14.95
	4	Tantalum	3.180	556	3.02	2.79	5.43	25.71
Wet clay (moisture content: 8%)	5	Aluminum	0.563	1564	2.94	0.46	1.88	1.58
	6	Copper	1.101	956	2.98	0.95	3.12	5.46
	7	Copper	2.290	640	3.00	1.89	4.69	16.26
	8	Tantalum	3.190	500	2.94	2.74	5.88	29.60
Wet clay (moisture content: 15%)	9	Aluminum	0.495	1192	3.00	0.37	2.52	1.84
	10	Copper	1.142	848	2.98	0.97	3.51	6.63
	11	Copper	2.370	538	2.88	1.89	5.35	19.81
	12	Tantalum	3.130	478	3.01	2.64	6.30	32.54

According to the shock wave velocities and the particle velocities (U_{si}, u_{pi}) for four samples of the same moisture content, the linear relation such as Equation (2) could be fitted with the least square method. Processed linear relations for the clay of the three different moisture contents are shown in Figure 10.

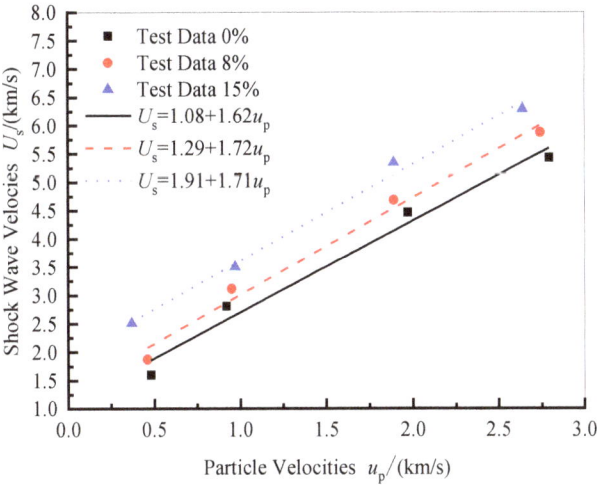

Figure 10. The linear relation U_s-u_p for the clay of three moisture contents.

With Figure 10, the material parameters in the Hugoniot linear relation for the clay of 0% moisture content, 8% moisture content, and 15% moisture content could be derived and they had the following expressions.

For the clay samples of 0% moisture content, the material parameters were:

$$c_0 = 1.08 \pm 0.30 \text{ km/s} \quad s = 1.62 \pm 0.17 \tag{8}$$

For the clay samples of 8% moisture content, the material parameters were:

$$c_0 = 1.29 \pm 0.24 \text{ km/s} \quad s = 1.72 \pm 0.14 \tag{9}$$

For the clay samples of 15% moisture content, the material parameters were:

$$c_0 = 1.91 \pm 0.18 \text{ km/s} \quad s = 1.71 \pm 0.10 \tag{10}$$

3. Equation of State of Unsaturated Clay

In Section 2, shock pressures in the experiments were not more than 30 GPa and the shock temperature rise can be neglected. Therefore, the EOS of the unsaturated clay can take the form $p = f(\rho)$. As shown in Figure 1, the equation of state of the unsaturated clay should include two deformation mechanisms when considering the critical pressure p_c. Our solution is to use the *p-alpha* compaction model [15] when the pressure $p < p_c$, and to use the EOS of a solid–liquid two-phase mixture when the pressure $p > p_c$.

3.1. p-alpha Compaction Model

The *p-alpha* compaction model was firstly presented by Herrmann [16] for porous materials. In this model, the stiffness of the skeleton material was neglected and the porous material could be thought of as an isotopic material. Therefore, the stress state of porous material can be described with hydrostatic pressure p, and the EOS for *p-alpha* compaction model has the following forms, containing the hydrostatic pressure p, specific volume v, and internal energy e:

$$p = f(v, e) \tag{11}$$

To differentiate the specific volume change of the skeleton material from the change of pore shape in the deformation process under external force, the porosity α was introduced and defined as:

$$\alpha = \frac{v}{v_s} = \frac{v(p, e)}{v_s(p, e)} \tag{12}$$

where v_s was the specific volume of skeleton material and v was the specific volume of corresponding porous material at the same state.

In the study on the state change of porous material, the surface energy of voids was usually neglected, thus the porous material had the same specific internal energy as the skeleton material. If the hydrostatic pressures of the porous and skeleton material were thought to be identical in any condition, it meant that only the specific volumes were different when the porous and skeleton material were in the same state. If the skeleton material was dense, the EOS of solid as the following could be used:

$$p = f(v_s, e) \tag{13}$$

After introducing porosity α, the following equation could be obtained:

$$p = f\left(\frac{v}{\alpha}, e\right) \tag{14}$$

In the low-pressure range $p < p_c$, the porosity α depended on the hydrostatic pressure p and the specific internal energy e. It was a key problem to determine the porosity α in the EOS of porous material under low-pressure. Herrmann thought that the porosity $\alpha(p, e)$ could be approximated by $\alpha(p)$ along the Hugoniot curve as shown in Figure 11 under the condition that the compressibility of the porous material was insensitive to the temperature.

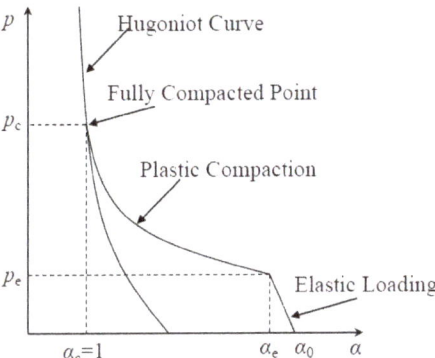

Figure 11. Schematic of porous material $\alpha(p)$.

The porosity α has little influence on the compaction during elastic loading, so the initial porosity α_0 can be approximated by α_e and then the porosity α can be specified as a function of pressure:

$$\alpha(p) = 1 + (\alpha_0 - 1)\left(\frac{p_s - p}{p_s - p_e}\right)^N \qquad (15)$$

where p_e was the elastic limit of porous material, when the pressure was larger than p_e, the voids began to collapse; p_s was the pressure at which the porous material was compacted into a completely solid state. N was one parameter and equals 2 in Herrmann's article, but later studies [17,18] showed that the parameter N could be determined according to the experimental results to get a better description of the compaction process of porous material.

In this study, the unsaturated clay was considered as a kind of porous material and $p_c = p_s$ was assumed. Therefore, p_c could be given by the Hugoniot equation:

$$p_c = \frac{\rho_c c_0^2 \eta_c}{(1 - s\eta_c)^2} \qquad (16)$$

where c_0 and s were Hugoniots and had been determined in experiments as shown in Section 2 and $\eta_c = 1 - \rho_0/\rho_c$, ρ_c was the density of the fully compacted clay.

After being fully compacted, the clay contained only water and solid particles, and the fully compacted density ρ_c was given by:

$$\rho_c = \frac{m}{v_c} = \frac{m_w + m_s}{v_{wc} + v_{sc}} \qquad (17)$$

where m was the total mass of clay; v_c was the volume of the fully compacted clay, corresponding to the pressure p_c; the subscript w and s represented water and solid particles, respectively.

3.2. p-alpha Compaction Model

Unsaturated clay was a three-phase media comprised of solid particles, water, and air. When the unsaturated clay had been fully compacted, the fully compacted pressure p_c (about 1 GPa) was so high that not much air was left in the clay and the compressibility of the skeleton could also be neglected. It meant that the high-pressure EOS of the unsaturated clay mainly came from the contribution of water and solid particles.

The relative volumes β_{wp} and β_{sp} for the water and the solid phase in the clay could be introduced as:

$$\beta_{wp} = \frac{v_{wp}}{v_c}, \quad \beta_{sp} = \frac{v_{sp}}{v_c} \qquad (18)$$

where v_{wp} (or v_{sp}) was the volume that the water (or the solid) phase had if the hydrostatic pressure in the clay was p. When the unsaturated clay was fully compacted, the relative volumes β_{wc} and β_{sc} were as follows:

$$\beta_{wc} = \frac{v_{wc}}{v_c}, \quad \beta_{sc} = \frac{v_{sc}}{v_c} \tag{19}$$

and satisfied the equation:

$$\beta_{wc} + \beta_{sc} = 1 \tag{20}$$

According to the mass conservation, the density ρ of the clay under the pressure p was given as:

$$\rho = \frac{m}{v} = \frac{\rho_c v_c}{v} = \frac{\rho_c}{\beta_{wp} + \beta_{sp}} \tag{21}$$

where $v = v_{wp} + v_{sp}$ was the volume of the clay under the pressure p. Here, the relative volume of water β_{wp} could be determined using the EOS of water [1]:

$$p = p_0 + \frac{\rho_{w0} \cdot c_{w0}^2}{k_w} \left[\left(\frac{\rho_w}{\rho_{w0}} \right)^{k_w} - 1 \right] \tag{22}$$

where $p_0 = 10^5$ Pa, $\rho_{w0} = 1.0 \times 10^3$ kg/m^3, $c_{w0} = 1415$ m/s, $k_w = 3$.

In the same way, the relative volume of solid particles β_{sp} could also be determined by the following EOS [1]:

$$p = p_c + \frac{\rho_{s0} \cdot c_{s0}^2}{k_s} \left[\left(\frac{\rho_s}{\rho_{s0}} \right)^{k_s} - 1 \right] \tag{23}$$

where $\rho_{s0} = 2.73 \times 10^3$ kg/m^3, $c_{s0} = 4500$ m/s, $k_s = 3$, p_c was the fully compacted pressure of clay determined by experiment results.

Substituting Equations (22) and (23) into Equation (21), the following equation could be obtained, by which the density of the clay under any pressure could be determined:

$$\rho = \rho(p) = \rho_c \left[\beta_{wc} \left(\frac{p - p_0}{\rho_{w0} c_{w0}^2} k_w + 1 \right)^{-1/k_w} + \beta_{sc} \left(\frac{p - p_c}{\rho_{s0} c_{s0}^2} k_s + 1 \right)^{-1/k_s} \right]^{-1} \tag{24}$$

Based on the aforementioned information and the experimental results, the physical parameters in the EOS of the unsaturated clay with the three moisture contents could be determined, and they are shown in Table 5.

Figure 12 shows the theoretical results from the EOS and the experimental results of the density variation of the clay samples with pressure. It is evident that the compressive strength of the clay sample increased with the increase of the moisture content. The water and the air trapped in the voids of the clay could not be removed under high strain rate and high loading; therefore, the shock compressible behavior of the unsaturated clay was dominated by the compressibility of the trapped water and air and the solid particles. Because water was a relatively incompressible material, the increase of water content led to an increase in the compressive strength of water-bearing clay. Al'tshuler and Pavlovskii [19] carried out shock compressed experiments on unsaturated clay samples with 4% and 20% moisture contents; the mineral components of their clay samples were mainly quartz and kaolin, which were similar to the samples used in this study. They also concluded the same results that the compressive strength of clay increases with the increase of the moisture content.

Table 5. The parameters of equation of state of unsaturated clay.

EOS	Parameters	Dry Clay	8% Clay	15% Clay
p-alpha model	Elastic yield strength, p_e (GPa)	0	0	0
	Plastic yield strength, p_s (GPa)	4.97	2.15	1.38
	Empirical fitting parameter, N	2	2	2
Solid–liquid two-phase model	Fully compacted density, ρ_c (g/cm^3)	2.73	2.42	2.23
	Water proportion under p_c, β_{wc}	0	0.18	0.29
	Solid proportion under p_c, β_{sc}	1	0.82	0.71
	Water phase	ρ_{w0} = 1.0 g/cm^3, c_{w0} = 1500 m/s, k_w = 7		
	Solid particle phase	ρ_{s0} = 2.73 g/cm^3, c_{s0} = 4500 m/s, k_s = 3		

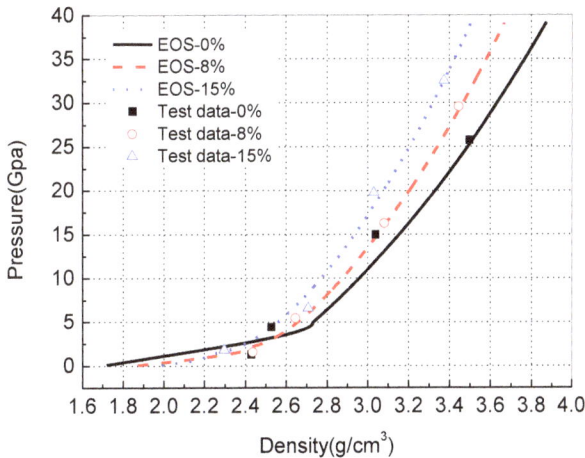

Figure 12. Comparison of theoretical and experimental data for the equation of state in terms of pressure and density.

As indicated by the agreement between the theoretical results and the experimental ones in Figure 12, the EOS of the unsaturated clay proposed in this study could give a better description of the relation between the pressure and the density of the clay, and it could also reflect the different shock compressed behaviors of the unsaturated clay resulting from the moisture content variation.

4. Discussion

(1) For the unsaturated clay, the EOS could be separated into low-pressure range and high-pressure range by the compacted pressure p_c. The *p-alpha* compacted model could be used in the low-pressure range and the solid-water two-constituent mixed EOS could be applied for the high-pressure range.

(2) With the original clay extracted from the southern suburbs of Luoyang city in China, three unsaturated clays with moisture content 0%, 8%, and 15%, respectively, were remolded. The Hugoniot parameters of three unsaturated clays were determined by means of the plate impact experiment on a one-stage and two-stage light gas gun. The results were: when the moisture content is 0%, ρ_0 = 1.70 g/cm^3, c_0 = 1.08 ± 0.30 km/s, s = 1.62 ± 0.17; when the moisture content is 8%, ρ_0 = 1.84 g/cm^3, c_0 = 1.29 ± 0.24 km/s, s = 1.72 ± 0.14; when the moisture content is 15%, ρ_0 = 1.96 g/cm^3, c_0 = 1.91 ± 0.18 km/s, s = 1.71 ± 0.10.

(3) With the Hugoniot parameters and the model of two-stage EOS presented here, the high-pressure EOS up to 30 GPa was developed for the unsaturated clays of three moisture contents and was consistent with the experimental results.

Author Contributions: Conceptualization, X.R.; methodology, X.R. and X.Z.; validation, X.R.; formal analysis, X.Z.; resources, W.T.; data curation, J.Z. and X.R.; writing—original draft preparation, X.R. and X.Z.; writing—review and editing, X.R. and J.Z.; supervision, W.T.; project administration, W.T. All authors have read and agreed to the published version of the manuscript.

Funding: This research received no external funding.

Institutional Review Board Statement: Not applicable.

Informed Consent Statement: Not applicable.

Data Availability Statement: The data can be requested from the corresponding authors.

Acknowledgments: This research was funded by the National Natural Science Foundation of China (Grant No. 11002162 and 11072262). The authors would like to thank Y.Y.Y. Cao for her warm help in revising the English writing. The authors express their sincere thanks to the anonymous reviewers and the editor for their invaluable help in revising this paper.

Conflicts of Interest: The authors declare no conflict of interest.

References

1. Henrych, J. *The Dynamics of Explosion and Its Use*; Elsevier Scientific Publishing Company: New York, NY, USA, 1979; pp. 73–84.
2. Schofield, A.; Wroth, P. *Critical State Soil Mechanics*; McGraw-Hill: New York, NY, USA, 1968.
3. Van Thiel, M.; Shaner, J.; Salinas, E. *Compendium of Shock Wave Data. Introduction. Section A1. Elements*; California Univ.: Oakland, CA, USA, 1977.
4. Kalashnikov, N.G.; Pavlovsky, M.N.; Simakov, G.V.; Trunin, R.F. Dynamic compressibility of calcite-group minerals. *Izv. Phys. Solid Earth* **1973**, *2*, 23–29.
5. Trunin, R.F. Rock compressibility in shock waves. *Izv. Earth Phys.* **1988**, *24*, 38–42.
6. Tsembelis, K.; Proud, W.; Vaughan, B. The behavior of sand under shock wave loading: Experiments and simulations. In Proceedings of the 14th DYMAT Technical Meeting on Behavior of Materials at High Strain Rates: Numerical Modeling, Sevilla, Spain, 1 January 2002.
7. Chapman, D.; Tsembelis, K.; Proud, W. The behaviour of dry sand under shock loading. In *Shock Compression of Condensed Matter*; American Institute of Physics: Baltimore, MD, USA, 2005.
8. Reinhart, W.D.; Thornhill, T.F., III; Chhabildas, L.C.; Vogler, T.J.; Brown, J.L. *Shock Response of Dry Sand*; Sandia National Laboratories: Albuquerque, NM, USA, 2007.
9. Resnyansky, A.; Bourne, N. *Shock Compression of Dry and Hydrated Sand*; AIP Conference Proceedings; American Institute of Physics: Baltimore, MD, USA, 2004; pp. 1474–1477.
10. Wang, Z.; Lu, Y. Numerical analysis on dynamic deformation mechanism of soils under blast loading. *Soil Dyn. Earthq. Eng.* **2003**, *23*, 705–714. [CrossRef]
11. Wang, Z.; Hao, H.; Lu, Y. A three-phase soil model for simulating stress wave propagation due to blast loading. *Int. J. Numer. Anal. Methods Geomech.* **2004**, *28*, 33–56. [CrossRef]
12. Marsh, S.P. *LASL Shock Hugoniot Data*; University of California Press: Berkeley, CA, USA, 1980; pp. 57–184.
13. Shi, S.; Chen, P.; Huang, Y. Velocity measurement of magnet induced system for projectile. *Chin. J. High Press. Phys.* **1991**, *5*, 205–214.
14. Zhao, S.W.; Wang, C.L.; Li, X. Experiment research of fiber probe applied in detonation velocity test. *J. Appl. Opt.* **2015**, *36*, 327–331.
15. Schmitt, D.R.; Ahrens, T.J. Shock temperatures in silica glass: Implications for modes of shock-induced deformation, phase transformation, and melting with pressure. *J. Geophys. Res. Solid Earth* **1989**, *94*, 5851–5871. [CrossRef]
16. Herrmann, W. Constitutive equation for the dynamic compaction of ductile porous materials. *J. Appl. Phys.* **1969**, *40*, 2490–2499. [CrossRef]
17. Borg, J.P.; Chapman, D.J.; Tsembelis, K. Dynamic compaction of porous silica powder. *J. Appl. Phys.* **2005**, *98*, 073509. [CrossRef]
18. Borg, J.P.; Cogar, J.R.; Lloyd, A. Computational simulations of the dynamic compaction of porous media. *Int. J. Impact Eng.* **2006**, *33*, 109–118. [CrossRef]
19. Al'tshuler, L.V.; Pavlovskii, M.N. Response of clay and clay shale to heavy dynamic loading. *J. Appl. Mech. Tech. Phys.* **1971**, *12*, 161–165. [CrossRef]

Article

Research on the Impact-Induced Deflagration Behavior by Aluminum/Teflon Projectile

Jianguang Xiao [1,2,*], Yanxin Wang [1], Dongmo Zhou [1], Chenglong He [1,2] and Xiangrong Li [3,*]

[1] College of Mechatronic Engineering, North University of China, Taiyuan 030051, China; s2001135@st.nuc.edu.cn (Y.W.); zhoudongmo@nuc.edu.cn (D.Z.); hechenglong@bit.edu.cn (C.H.)
[2] Science and Technology on Transient Impact Laboratory, No. 208 Research Institute of China Ordnance Industries, Beijing 102202, China
[3] Department of Weapons and Control, Academy of Armored Force Engineering, Beijing 100072, China
* Correspondence: xiaojg@nuc.edu.cn (J.X.); lixiangrong2022@gmail.com (X.L.)

Abstract: Although the ignition-and-growth model can simulate the ignition and detonation behavior of traditional energy materials well, it seems insufficient to simulate the impact-induced deflagration behavior of reactive materials (RMs) using current finite element codes due to their more complicated ignition threshold and lower reaction rates. Therefore, a simulation method for the impact-induced deflagration behavior of a reactive materials projectile (RMP) is developed by introducing tunable ignition threshold conditions for RMs, and a user-defined subroutine is formed by the secondary development on the equation of state (EOS). High-velocity impact experiments were performed to prove the validity of simulations. The results show that the user-defined subroutine for RMs is competent in simulating the ignition and deflagration behavior under impact conditions, because the reaction ratio, morphology and temperature distribution of RMP fragments are all well consistent with experiments, theory, and current reports from other researchers. In this way, the quantitative study on the deflagration reaction of RMs can be implemented and relevant mechanisms are revealed more clearly.

Keywords: reactive materials; impact-induced deflagration; simulation; ignition behavior; reaction ratio; temperature distribution

Citation: Xiao, J.; Wang, Y.; Zhou, D.; He, C.; Li, X. Research on the Impact-Induced Deflagration Behavior by Aluminum/Teflon Projectile. *Crystals* **2022**, *12*, 471. https://doi.org/10.3390/cryst12040471

Academic Editors: James L. Smialek, Yong He, Wenhui Tang, Shuhai Zhang, Yuanfeng Zheng and Chuanting Wang

Received: 9 February 2022
Accepted: 26 March 2022
Published: 28 March 2022

Publisher's Note: MDPI stays neutral with regard to jurisdictional claims in published maps and institutional affiliations.

Copyright: © 2022 by the authors. Licensee MDPI, Basel, Switzerland. This article is an open access article distributed under the terms and conditions of the Creative Commons Attribution (CC BY) license (https://creativecommons.org/licenses/by/4.0/).

1. Introduction

When perforating or penetrating the intended target, reactive materials (RMs) will chemically react due to the shock wave passing through them, thereby increasing the damaging effects from the combination of the kinetic energy (KE) and chemical energy (CE) of a reactive projectile [1–3]. RMs are a class of shock-induced energetic materials, including thermites, intermetallics, metal-polymer mixtures, metastable intermolecular composites (MICs), and so on. With the benefit of fine mechanical and chemical performance, polytetrafluoroethylene based RMs have been extensively researched recently. In order to investigate their lethality, ground tests have been conducted, and several physics-based models were established [4–8]. However, RMs are generally formulated to release appreciable CE under intense dynamic loads (such as high-velocity impact or detonation), so the activation time is extremely short, making the measurement of many physical quantities very difficult. Consequently, an appropriate analytical tool for RMs is required to predict target damage beyond that measured under experimental conditions [3]. Popular dynamic calculation codes, such as ANSYS-Autodyn or Ls-dyna, were employed to predict the response of energetic materials. In these codes, the simulation results are dependent on an appropriate material model, which includes the equation of state (EOS), strength model, failure model, erosion model, and so on. Researchers have paid close attention to the material model of RMs.

For the unreacted reactant of RMs, the shock EOS model was used to research the critical velocity of the reactive materials projectile (RMP) to initiate the covered explosive [9,10], and Instron compression tests and high-rate split Hopkinson bar experiments were carried out to determine the parameters of the Johnson–Cook strength model, which can be used to effectively simulate the deformation and penetration behavior of reactive materials [11,12]. For reaction product, the Jones–Wilkins–Lee (JWL) EOS was used to characterize the expansion behavior after the chemical reaction of RMs [13]. In fact, compared to the KE damage caused by traditional inert metal materials, it is mainly the CE that causes the remarkably high-efficiency damage during the impact events of RMs. The ignition time, reaction rate, reaction efficiency, and so on, play an important part in the energy release of CE. In particular, the initiation criterion, which is characterized by the values of the impact pressure P and its duration τ or by the values of an impacting projectile's velocity V and diameter d ($P^2\tau$ or V^2d criteria), significantly influences the damage event of RMs. The forest fire model was provided and developed to match pressure–time data obtained from gauges embedded in the energetic materials in a broader set of experiments [14]. Recently, the Naval Surface Warfare Center has estimated the impact velocity and pressure initiation threshold of reactive materials with different particle size with a gas gun experiment [15,16]. They found that the initiation reaction occurs earlier in reactive materials with smaller particle size; this is mainly induced by the shear band formed in the impact event, and empirical formulas ($t^a(\sigma - \sigma_{TS})^b = c$)) were proposed to characterize the ignition behavior of Al/PTFE reactive materials.

The above literature reviews show that the EOS and ignition model for RMs have been improved. However, an integrated, analytical method to reproduce the high-efficiency damage caused by RMs has not been presented, to the best of our knowledge. In past decades, relevant simulations often divided the damage event into two relatively independent phases. For example, the damaging effects on concrete targets produced by reactive material liner shaped charges were researched by dividing the physical process into an inert impact-penetration stage and an internal deflagration stage for RMs. The shock model was used to simulate the inert penetration behavior of RMs, and either the JWL or powder burn model was used to simulate the internal deflagration behavior of RMs [17–20]. Although the Lee–Tarver model embedded in Autodyn or Ls-dyna can characterize the detonation performance of high explosives, no interface is provided to adjust the initiation criterion that is crucial to simulate the damage event for RMs.

The purpose of this effort is to combine the divided stages into one by developing the EOS subroutine based on Autodyn code; then, quantitative research of real-time reaction can be conducted by using the tunable ignition criteria. A simulation method is proposed to investigate the impact-induced deflagration behavior of polytetrafluoroethylene based RMP. This work is of great value in the design of RMPs and understanding their damage mechanisms more clearly.

2. Experiment

2.1. Specimen Preparation

The Al/PTFE RMPs were prepared by the process of mechanical mixing–cold pressing–sintering. First, the raw Al and PTFE powders, with Al particle size of 5 μm and PTFE particle size of 34 μm (the purity of Al and PTFE are above 98% and 99%, respectively, according to the vendor's description), were poured into a container where they were mechanically stirred with a rotation speed of approximately 20,000 rpm. Then, the symmetrical distribution powders were poured into a cold-pressing mold where the pressure was loaded by a puncher pin on the powders with a linear speed of 5 mm/min. When the pressure reached approximately 1 MPa, the load speed of the puncher pin was controlled by pressure increments at approximately 1 MPa/s. The puncher pin did not stop until the pressure reached 100 Mpa; subsequently, the pressure was kept at this level for 1 min. In this way, the powders were pressed into a cylinder, then inserted into a vacuum sintering oven. The oven temperature rose to 375 °C at a rate of 60 °C/h and stayed at 375 °C for

0.5 h. After that, the oven temperature dropped to 327 °C at a rate of 40 °C/h and stayed at 327 °C for 1 h. Lastly, during the cooling process, the oven temperature dropped to ambient temperature at an average rate of approximately 50 °C/h. The density of the RMP fabricated by this method is close to the theoretical maximum density of 2.27 g/cm^3.

2.2. Experiment Setup

The experiment system was mainly composed of the one-stage gas gun, laser speed detector, chamber, double-spaced Al plates holder, and high-speed camera, as shown in Figure 1. After the projectiles were launched by the one-stage gas gun, the two laser beams (laser speed detector) recorded the speed of the projectiles. An on–off signal, produced by the veil effect of the projectiles on the laser beams, was used to trigger a high-speed camera; consequently, the images of the penetration and deflagration behavior of the RMPs on the double-spaced Al plates was recorded. Eight shots, with the impact velocity ranging from 293 to 652 m/s, were performed to investigate the penetration and deflagration behavior of RMs. Several square Al plates with a size of 400 × 400 mm^2 were used in this study and the distance between the front and rear plate was 200 mm. The size of the projectiles is listed in Figure 1. The penetration hole and bulges on the front plate (FP) and rear plate (RP) were used to characterize the penetration behaviors of RMPs and damage effects of the double-spaced Al plates, while the duration and size of the flame, resolved by high-speed camera, were used to characterize the deflagration behavior of RMPs.

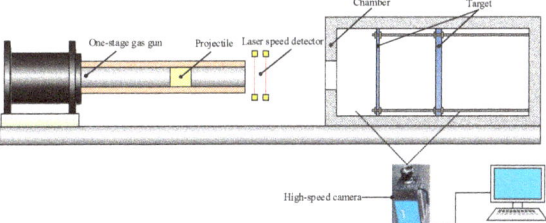

Figure 1. Schematic diagram of experimental setup.

2.3. The Damage Effects of Double-Spaced Al Plates

In the eight experiments with velocities ranging from 293 to 652 m/s, all front plates were perforated, while petalling or plugging damage formed on the backside of the front plates, depending on the various impact velocities. It is worth noting that there was a plastic zone (between the red and yellow dotted line in Figure 2) in the vicinity of the penetration hole in both petalling and plugging damage. On the rear plates, considerable soot was produced on the frontside because of the chemical reaction by residual RMPs, while various degrees of bulges were formed on the backside. The typical damage patterns and details on damage effects are shown in Figure 2 and Table 1, respectively.

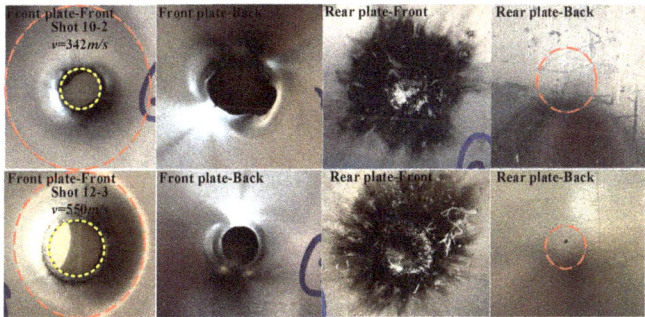

Figure 2. The damage patterns on the front and rear plate in shots 10-2 and 12-3.

Table 1. Experimental results.

Shot	Projectile Size ($\Phi D \times L$) (mm)	Velocity (m/s)	Penetration Hole Diameter on FP (mm)	Deflagration Time (μs)		
				Perforating FP	Impacting on RP	Fades Away
10-1	15.40 × 15.39	293	~15.47	210.24	210.24	1629.35
10-2	15.39 × 15.38	342	~18.01	420.48	998.63	5045.73
10-3	15.40 × 15.36	401	~24.53	367.92	2890.78	8672.34
10-4	15.41 × 15.21	397	~20.06	473.04	3521.50	9040.26
12-1	15.40 × 7.64	652	18.62	252.39	378.58	8539.10
12-2	15.50 × 7.76	570	18.73	294.45	3070.71	7697.81
12-3	15.50 × 7.80	550	17.33	210.32	252.39	4374.71
12-4	15.40 × 7.30	620	19.37	126.19	1724.65	8286.71

Notes: FP and RP represent the front and rear plate, respectively. The deflagration time means the duration of deflagration of RMs: (1) perforating FP means the whole deflagration time caused by the penetration to FP; (2) impacting RP means the time required for the reaction to increase to maximum from the impact on RP; and (3) fade away means the time from peak to end of deflagration reaction.

3. Numerical Simulation

3.1. Reaction Model of Energy Release Process by RMs

The impact-induced energy release process of RMs can be summarized as follows: (1) the fragmentation of RM samples; (2) the product of small gas molecules from the decomposition of the fluoropolymer matrix under impact loading (impact-induced hotspot or impact-induced fracture); (3) the exposure of reactive metal to small gas molecule atmospheres; and (4) the burning process of the fragmentized composite particle. For numerical simulations at the macro scale, the decomposition of the fluoropolymer matrix and the exposure of reactive metal to small gas molecule atmospheres are usually simplified, and attributed either to stress concentrated in a local point or temperature. When the stress or temperature achieve a threshold value, it is considered evidence that the decomposition of fluoropolymer matrix and exposure of reactive metal to small gas molecule atmospheres have occurred. In this way, the phenomenological research can be conducted based on any ignition criterion. However, no apparent interfaces are provided to adjust the ignition criterion in the ignition-and-growth model in Ls-dyna or Autodyn, which are commonly used to conduct simulations of the impact-induced burning process of energetic materials. Consequently, some modifications are required to simulate the impact-induced deflagration reaction of RMs.

In Autodyn code, the EOS subroutine is used to calculate pressure, energy, and sound speed as a function of density. When the density of the current timestep is updated and retrieved from the main program, the new pressure, energy, and sound speed can be updated by the EOS model, then returned to the main program to calculate other variables. In the energy release process of RMs, they undergo three material states, including solid reactant state, solid-gas mixing state, and gas state, in the same particle. Therefore, three different EOSs are generally needed to model the distinct states, which makes the solving process complex and expensive. To solve this issue, reaction ratio F is introduced to couple the equations of state (EOSs) for reactant and reaction product, based on the assumption of pressure and heat balance. After iterative computations on pressure and temperature, the reaction ratio, energy, and sound speed can be updated; this is discussed in detail in Section 3.1.5. In this way, a unitary EOS or reaction model was proposed to describe the mixture in the reaction zone [21].

3.1.1. EOS of Unreacted Reactant

For unreacted reactant, a shock EOS is employed to describe the pressure state under impact condition, which takes the following form [9]

$$P_u(V_u, T) = P_H(V_u) + \frac{\rho_0 \Gamma}{V_u}(E - E_H) = P_H(V_u)\left(1 - \frac{\Gamma}{2}\frac{1 - V_u}{V_u}\right) + \frac{\rho_0 \Gamma}{V_u} C_{v,u} T \quad (1)$$

where
$$\begin{cases} P_H(V_u) = \rho_0 c_0^2 (1-V_u)/[1-s(1-V_u)]^2 & V_u \le 1 \\ P_H(V_u) = \rho_0 c_0^2 (\eta_u - 1) & V_u > 1 \end{cases}$$

where subscript u denotes unreacted reactant, Γ is the Grüneisen coefficient, T is temperature, ρ_0 and c_0 are the initial density and elastic wave velocity of the reactant, respectively, $C_{v,u}$ is the specific heat capacity, and s is the slope of the shock velocity or particle velocity fit. E_H is the specific internal energy, which can be calculated as

$$\begin{cases} E_H(V_u) = \frac{1}{2}\{c_0(1-V_u)/[1-s(1-V_u)]\}^2 & V_u \le 1 \\ E_H(V_u) = \frac{1}{2}c_0^2 V_u \left(\frac{1}{V_u}-1\right)^2 & V_u > 1 \end{cases} \quad (2)$$

3.1.2. Ignition Criterion

The chemical reaction of RMs occurs when the particles obtain enough energy. Generally, mechanical, thermal, optical, electrical, chemical, and acoustic stimuli can cause the ignition of RMs. Under the impact-initiation scenario, ignition behavior often occurs in the shear band, so the peak stress has been employed to judge the appearance of the ignition event by the Naval Surface Warfare Center [15,16]. They point out that the RMs do not ignite at the beginning of the collision, but have an ignition delay time, which can also be interpreted as the time for the material to absorb energy under certain stress conditions. As described in Equation (3), when the stress exceeds the ignition threshold σ_{TS}, the term that represents the impulse on the left side of the equation accumulates with time. When it reaches a constant c, the local RMs may be ignited. In contrast, the ignition of RMs substantially starts with the decomposition reaction of fluorine polymer. When the instantaneous temperature reaches the critical decomposition temperature, the chemical reaction may also be stimulated. Therefore, the following expression is used to model the ignition behavior for RMs.

$$t^a (\sigma - \sigma_{TS})^b = c \quad or \quad T = T_d \quad (3)$$

where σ and σ_{TS} are real-time stress and ignition threshold stress, respectively; t is accumulated time in which the σ exceeds σ_{TS}; a, b, and c are constants; T is instantaneous temperature; and T_d is critical decomposition temperature of fluorine polymer contained in RMs.

3.1.3. EOS of Reaction Product

The JWL EOS is usually used to describe the state of the reaction product. The pressure and specific internal energy, respectively, can be calculated as follows [22]

$$P_p(V_p, T) = Ae^{-R_1 V_p} + Be^{-R_2 V_p} + \frac{\omega \rho_0}{V_p} C_{v,p} T \quad (4)$$

$$E_p(V_p, T) = \left(\frac{A}{R_1} e^{-R_1 V_p} + \frac{B}{R_2} e^{-R_2 V_p}\right)/\rho_0 + C_{v,p} T \quad (5)$$

where subscript p denotes reaction product and A, B, R_1, R_2, and ω are constants.

3.1.4. Reaction Rate

The ignition-and-growth model of explosive initiation is employed to characterize the impact-induced initiation behavior of RMs. The chemical reaction rate for the conversion from unreacted explosive to reaction product consists of three physically realistic terms: an ignition term of a small explosion occurs soon after the shock wave; a slow growth term of reaction speed as this initial reaction goes on; and a reaction completion term with a rapid reaction rate. The form of the reaction rate equation is [9]

$$\frac{dF}{dt} = I(1-F)^b \left(\frac{\rho}{\rho_0} - 1 - a\right)^x + G_1(1-F)^c F^d P^y + G_2(1-F)^e F^g P^z \quad (6)$$

where I, G_1, G_2, a, b, c, d, e, g, x, y and z are adjustable coefficients for energetic materials. $F = 0$ indicates the reactant did not react, and the variables are calculated by the equations of unreacted reactant. $F = 1$ indicates the chemical reaction is complete, and the variables are calculated by the equations of reaction product. If $0 < F < 1$, this indicates the chemical reaction is ongoing and the mixture is made of reactant and deflagration product, and the variables are calculated by introducing a mixing rule.

The three portions of the ignition-and-growth model exhibit distinguished chemical reaction rates during the three physically realistic terms discussed above. This model contains three more parameters: FMXIG, FMXGR, and FMNGR. The ignition rate is set equal to zero when $F \geq$ FMXIG, the growth rate is set equal to zero when $F \geq$ FMXGR, and the completion rate is set equal to zero when $F \leq$ FMNGR. Therefore, if no ignition criterion is employed, it is the constant "a" that dominates early ignition reactions, which could be considered the ignition criterion or threshold. When the compression of some particles increases to a value that is large than "a", the chemical reaction rate increases according to the ignition term. At this stage, the slow growth and completion term is approximate to zero as the "F", which is nearly zero in the early phase of ignition. The parameters contained in this model can be obtained through an explosive shock experiment with an iterative approach that adjusts the parameters until the simulation results agree with the test results [14].

3.1.5. Mixing Rule

An EOS subroutine in Autodyn is mainly required to update the specific internal energy and pressure. When $F = 0$ or $F = 1$, relevant calculations can be done by Equations (1), (2), (4) and (5). When $0 < F < 1$, the mixture is composed of unreacted reactant and reaction product. After the reactant is activated, the specific internal energy is defined as

$$E(V, T, F) = (1-F)E_u + FE_p + (1-F)Q \quad (7)$$

where Q is reaction heat released by reactive materials per unit of mass.

During an update process in a calculation circulation of an EOS subroutine, pressure and temperature equilibrium is assumed in the mixture. The temperature is first calculated based on the first law of thermodynamics. The differential form of specific internal energy is

$$dE = \left(\frac{\partial E}{\partial V}\right)_{T,F} dV + \left(\frac{\partial E}{\partial T}\right)_{V,F} dT + \left(\frac{\partial E}{\partial F}\right)_{V,T} dF$$
$$= JdV + C_v dT + HdF \quad (8)$$

According to the first law of thermodynamics, the internal energy absorbed by a small region of the material during a change from one state (or time) to another equals the heat (or energy) input into the region minus the work done by the region in the action of the internal pressure forces. This can be expressed in the relation $dE = -PdV$. Combining this with Equation (8), one can obtain the temperature increment

$$C_v dT = -(P+J)dv - HdF \quad (9)$$

In this paper, the heating effects produced by artificial viscosity and plastic work are included in the simulation, so Equation (9) becomes

$$C_v dT = -(P+J+q)dv - HdF + W_{plstic} \quad (10)$$

where q is artificial viscosity, C_v is constant-volume specific heat of energetic materials.

In this way, the temperature in a small region could be estimated based on the parameters of the former timestep. Then, the pressure of unreacted reactant and reaction product could be obtained by the (V, T) form of EOS, including Equations (1) and (4). The pressures in the unreacted and the gaseous phases of the RMs depend on the relative volume of the two phases, which is defined as follows

$$V = (1-F)V_u + FV_p \tag{11}$$

When the relative volume of the mixture is returned by main routine, the model iterates on the relative volume of the unreacted reactant until it meets pressure equilibrium with the reaction product, then the derived pressure P is returned. Meanwhile, the specific energies of unreacted and gaseous phases are obtained from Equations (2) and (5). The reaction rate can be calculated by Equation (6), and the specific energy of the mixture is calculated by Equation (7).

Finally, the local sound speed can be estimated as

$$\begin{aligned} c &= \sqrt{\frac{dp}{d\rho}} = \sqrt{\frac{dp}{d\eta} \cdot \frac{d\eta}{d\rho}} = \sqrt{\frac{1}{\rho_0} \cdot \frac{dp}{d\eta}} \\ &= \sqrt{\frac{1}{\rho_0} \cdot \left[(1-F)\left(\frac{\partial p_u}{\partial \eta_u} + \frac{\partial p_u}{\partial T}\frac{\partial T}{\partial E_u}\frac{\partial E_u}{\partial \eta_u}\right)\frac{d\eta_u}{d\eta} + F\left(\frac{\partial p_p}{\partial \eta_p} + \frac{\partial p_p}{\partial T}\frac{\partial T}{\partial E_p}\frac{\partial E_p}{\partial \eta_p}\right)\frac{d\eta_p}{d\eta}\right]} \end{aligned} \tag{12}$$

where $\eta = 1/V$ and subscripts u and p denote the unreacted reactant and reaction product, respectively.

3.1.6. Program Implementation in Autodyn User's Subroutines

Autodyn provides several subroutine interfaces for options such as material models, boundary conditions, and so on. The material model subroutines, such as EOSs, strength models, failure models, and erosion models, are commonly developed depending on user demand. In order to characterize the impact-induced initiation and energy release behavior of RMs, the EOS subroutine, including parameters initialization, parameters check, interfacial design, and solve loop, that dominates the pressure response and energy transformation, is developed in this work. The solve loop of the developed EOS subroutine is listed in Figure 3.

When called by the main program, the EOS subroutine checks whether the current cycle is the first cycle. If so, it initializes user-defined variables to zero, else it retrieves variable values from the previous timestep. Then, it estimates the temperature using Equation (10) so the pressure can be calculated with corresponding EOS models. Meanwhile, the local sound speed of a small region is estimated using Equation (12). Afterward, it updates ignition expression on the left side of Equation (3) using the pressure calculated in the current timestep. If this meets the ignition criterion, it updates the reaction fraction in Equation (6), else it implies the chemical reaction does not occur in this small region. Finally, it updates the specific energy using the new reaction ratio.

3.2. Constitutive Model

During the penetration process, the deformation and yield of materials on the impact interface between projectile and target dominates the damage effects on the target plate. In order to reproduce the deformation and yield behavior of the structures (RMP and Al plate), the Johnson–Cook strength model, which represents the strength behavior of materials subjected to large strains, high strain rates, and high temperatures, especially in problems of intense impulsive loading due to high-velocity impact, is employed in this study. This model defines the yield stress as

$$\sigma = \left(A + B\varepsilon_p^N\right)\left(1 + C\ln\dot{\varepsilon}_p^*\right)(1 - T_H^m) \tag{13}$$

where ε_p is the effective plastic strain, $\dot{\varepsilon}_p^*$ is the normalized effective plastic strain rate, T_H is the homologous temperature = $(T - T_{room})/(T_{melt} - T_{room})$, and A, B, C, N, and m are five material constants.

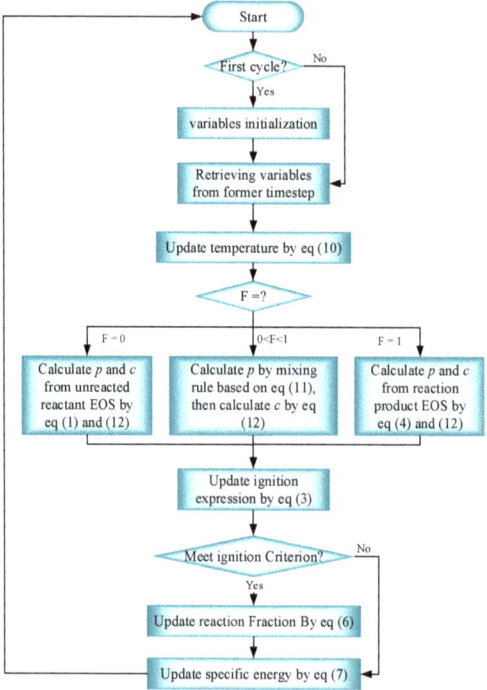

Figure 3. The solve loop of EOS subroutine.

When impacting the target with a high velocity, the projectile may suffer large deformation or even failure as a result of the high pressure caused by this event. The failure of materials means they can resist compressive but not tensile load, so the failure model is often employed to simulate the ejection behavior of fractured debris. Many failure models, such as minimum pressure, principal stress and strain, cumulative damage, and so on, are all permitted in Autodyn code. The Johnson–Cook failure model, which is commonly used to characterize the ductile failure of materials, is employed in this study. It consists of three independent terms that define the dynamic fracture strain as a function of pressure, strain rate, and temperature:

$$\varepsilon^f = \left(D_1 + D_2 e^{D_3 \sigma^*}\right)\left(1 + D_4 \ln \dot{\varepsilon}^*\right)(1 + D_5 T_H), \quad D = \sum \frac{\Delta \varepsilon}{\varepsilon^f} \qquad (14)$$

where $\Delta \varepsilon$ is an increment of effective plastic strain, ε_f is failure strain, σ^* is mean stress normalized by the effective stress, and $D_1, D_2, D_3, D_4,$ and D_5 are constants. The Johnson–Cook failure model essentially shows cumulative damage. The ratio of the incremental effective plastic strain to effective fracture strain for a considered particle is defined as the damage factor. The material is assumed to be intact until damage = 1.0. When failure occurs in some particle or element, the contained materials can no longer sustain tension.

3.3. Finite Element Model

The finite element model of RMP impacting double-spaced aluminum plates is shown in Figure 4. The impact and deflagration phenomenon of RMS is a highly nonlinear transient dynamic event; therefore, a meshless SPH method was employed to avoid the interruption of the simulation, which may be induced by severe deformation or distortion of the meshes used in the Lagrange or Euler methods. The purpose of an SPH method is to describe the continuous material with a group of interacting particles that bear various physical quantities, including mass, speed, etc. By solving the dynamic equation of the group of particles and tracking the movement of each particle track, the mechanical behavior of the whole system can be obtained. In this simulation, the separation distance between particles was set to 0.25 mm, and approximately 11,500 particles were used for the finite element model. The space between the two aluminum plates, one with a thickness of 2 mm and the other 4 mm, was set to 200 mm.

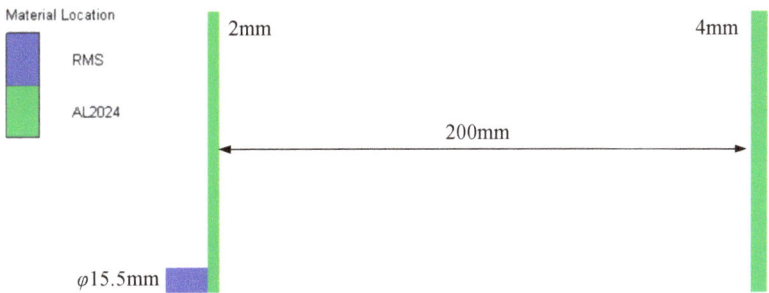

Figure 4. Scheme of simulation model.

4. Result and Discussion
4.1. Penetration Mechanism

The relatively heavier projectile, with a length–diameter ratio of approximately 1:1, produced irregular circular damage patterns on the front side of front plates, which was mainly caused by the unstable flight attitude of RMPs with relatively low kinetic energy. The lighter RMPs, with a length–diameter ratio of approximately 0.5:1, was fabricated to enhance the shot velocity. When the velocity of the RMPs reached above 550 m/s in this study, regular circular penetration holes were found on the front side of the front plates. As analyzed in Refs. [2–26], the RMPs impacting the thin plate with a low velocity were likely to produce petalling damage patterns, whereas high-velocity RMPs impacting thick plates were likely to produce plugging damage patterns. Hence, it is inferred that there is a minimum velocity that intensifies the stress suffered by the plate material to its shear strength, which could lead to the ultimate plugging damage. The relevant mechanisms may be considered as follows.

Generally, if the stress generated in the target plate material in contact with the cylindrical part of the projectile at low velocity is less than the shear strength of the target plate material in the early part of the impact, then the material does not fracture until it moves with the projectile for some distance, resulting in the petalling damage pattern on the plate. A higher velocity increases the stress to the shear strength of the target immediately after impact, resulting in the plugging damage pattern instead. In addition, when the aluminum plate is impacted at a velocity that produces plugging damage, two shockwaves propagate forward into the target and backward into the projectile. Because the considered impact velocity (<600 m/s) is much lower than the propagation velocity of shockwaves (>6000 m/s), the homogeneous stress in the Al plate is formed in a very short time compared to the penetration time, so it is reasonable to assume that every particle velocity in the plugger is the same. Furthermore, the plugger velocity, or the penetration velocity, reaches a steady level shortly after the initial impact time, so one can take the penetration velocity as a constant under the current impact conditions. The force equilibrium relationship on the plugger can be described as

$$\frac{\rho_p C_p \rho_t C_t (v_{0p} - v_{0t})}{\rho_p C_p + \rho_t C_t} A_s = \tau \cdot A_l \qquad (15)$$

where ρ and C are density and sound speed of related material, respectively; v_0 is the initial velocity of related material; A_s is the contact area on the impact interface; A_l and τ are the lateral area of the plugger and the shear stress implied on it, respectively; and the subscripts p and t denote the projectile and target, respectively.

The left side of Equation (16) represents the force that accelerates the plugger, whereas the right side is the resistance to the penetration. If the value of the left side is larger than that of the right side, or, for the given materials of projectile and target, if the velocity difference between them is high enough to produce a force exceeding the resistance force produced by the shear strength of the target plate, the plugger is formed immediately after impact; otherwise, the projectile will fly with the target in the vicinity of the impact interface for a longer distance before a penetration hole is formed. In the latter case, more remarkable bulges are found (as shown in Table 1), and petalling damage is produced. Taking τ as the value of the shear strength of the plate material, one can calculate the minimum velocity for plugging damage from Equation (15) for the known projectile and target materials, the diameter of projectile, and the target thickness. The contact area A_s is usually larger than the initial section area of projectile as a result of a mushrooming effect. In this study, we considered A_s the initial section area for simplification, $A_s = \pi D^2/4$, and the lateral area of the plugger was considered $A_l = \pi Dh$. Additionally, it is important to mention that the strain rate effect led to a higher shear strength of the considered materials, so a higher velocity was required for RMPs to produce plugging damage. Substituting the parameters into Equation (15), especially when $v_{0t} = 0$, one can obtain a simplified linear relationship between the minimum velocity for plugging damage and target plate thickness h for given projectile diameter D,

$$v_{0p} = 4\tau' h \frac{\rho_p C_p + \rho_t C_t}{\rho_p C_p \rho_t C_t D} \tag{16}$$

where τ' denotes the dynamic shear strength. Under current experiment conditions, we calculated v_{0p} to equal 496 m/s when taking τ' as 200 Mpa, which is reasonable because the minimum velocity for plugging damage in our experiments was between 401 and 550 m/s.

4.2. The Impact-Induced Deflagration Behavior

A high-speed camera was employed to investigate the impact-induced deflagration event produced by RMPs, the typical pictures are listed in Figures 5 and 6. As shown, a strong fire light is observed immediately after the impact event, indicating that the deflagration reaction is induced by the impact on the front plate. The fire light lasted several hundred microseconds, then gradually fades, and the residual fragments appear in Figures 5 and 6c. The residual fragments impact the rear plate with a relatively low velocity in the following moments, resulting in a more powerful deflagration reaction subsequently. At this time, the fire light lasts for an extended period, several milliseconds, implying more chemical energy was released after the second impact. Finally, the impact-induced deflagration reaction is finished, with the fire light fading away. The deflagration times for the impact on the front and rear plate are shown in Table 1.

The characteristics of fire light, such as appearance time, intensity, size, and so on, are closely related to the ignition and reaction ratio of RMPs. However, because of the shielding effect of fire light and insufficient testing methods at present, it is extremely difficult to study the ignition and relevant reaction mechanisms of RMPs in a high-velocity impact event. Based on this consideration, the numerical simulation method was developed to reproduce the impact-induced deflagration behavior using computer codes. An EOS with tunable ignition threshold was written in Autodyn codes for RMs, and simulations were carried out under the same conditions as the impact experiments, the results of which are shown in Figures 5 and 6. In the pictures, the scale represents the reaction ratio (ALPHA) of every particle included in the simulation model. In this way, the ignition behavior, time-resolved reaction ratio, and temperature distribution of RMPs were obtained. The energy and error time histories are illustrated in Figure 7, which indicates the energy balance at a good level, and the simulation results discussed in detail in the following part is reasonable.

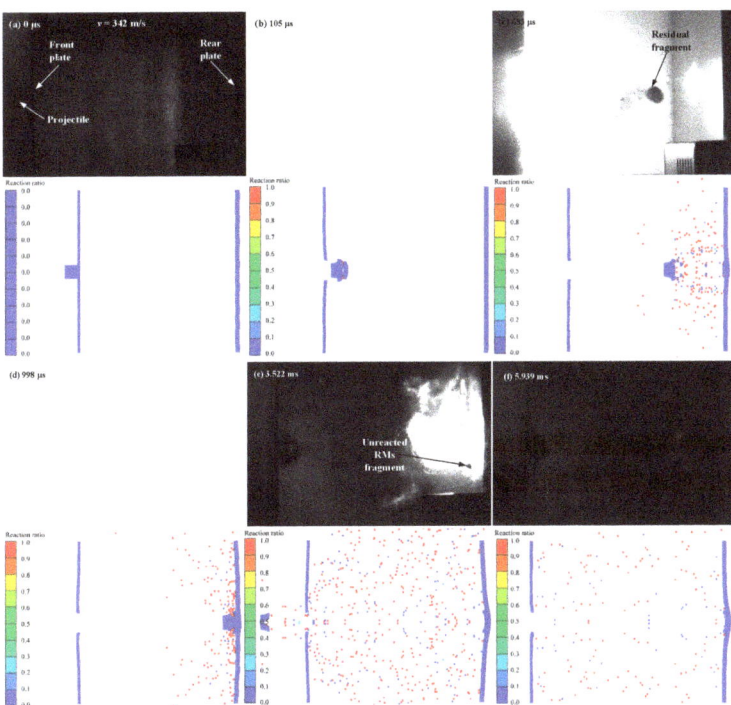

Figure 5. The impact-induced deflagration behavior of RMP with a velocity of 342 m/s.

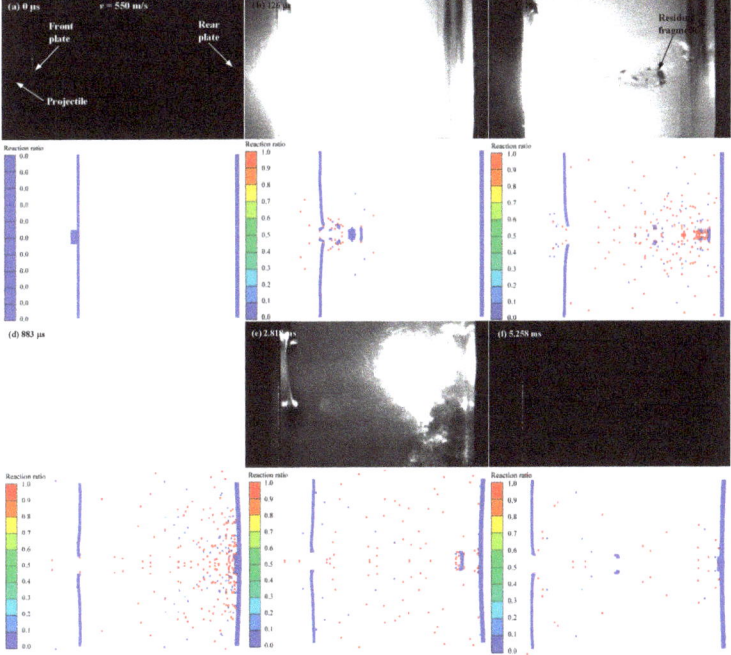

Figure 6. The impact-induced deflagration behavior of RMP with a velocity of 550 m/s.

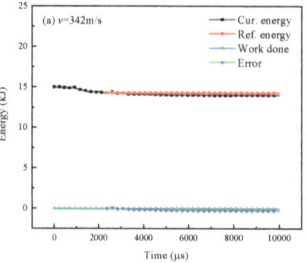 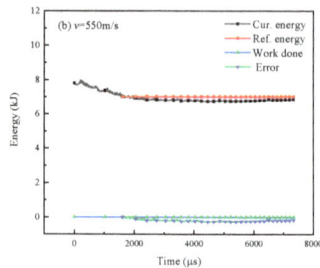

Figure 7. The energy and error versus time histories.

4.3. The Ignition Behavior

Generally, RMs are considered impact-initiated materials, so high-dynamic mechanical load is likely to be required to ignite RMs. In this impact-initiation scenario, one-dimensional impact stress is characterizing the ignition behavior [15,16]. However, quasi-static compression after a specific heat treatment procedure for RMs is also a resultful way to induce the rapid deflagration [26,27], though the chemical reaction may be suspended a short time after ignition because it is difficult to propagate the deflagration in high-density RMs, which may be attributed to the relatively lower deflagration pressure compared to high-energy explosives. Additionally, the ignition behavior of thermal, optical, and electrical stimuli can be found in Refs. [28–33]. In essence, the RMs could be successfully ignited when suffering enough energy stimuli. Based on the above analysis, we determined that the equivalent plastic stress and temperature are the main factors that dominate the ignition behavior of RMPs.

Figures 8 and 9 illustrate the typical ignition process of RMs at 160 μs after impact. When impacting the plate at 342 m/s, the ignition occurs at the strongest shear zone between the edge of the RMP and Al plate immediately after impact (Figure 8b). Then more RMs are activated to deflagration on the fracture surface (Figure 8d). It is interesting to note that the front side of the RMP fails to ignite, though it suffers the maximum pressure during the impact event, resulting in a cone of uninitiated RM fragments flying toward the rear plate. This phenomenon confirms that the ignition mechanisms of RMs are more likely to be attributed to shear instead of pressure, because the ignition result in this simulation is well consistent with the experiment result by Raftenberg [34]. When the impact velocity is enhanced to 550 m/s, more RMs are activated as a result of stronger shear effects between the edge of the RMP and plate (Figure 9d). Nevertheless, several RM fragments still fail to ignite (Figure 9e).

The relevant ignition mechanism can be analyzed based on the curves of ignition indicator, reaction ratio, pressure, and temperature versus time at the gauges set on the front side of the RMPs, as shown in Figures 8f and 9f. A high-pressure pulse is produced first after impact. However, we use equivalent plastic stress as the indicator of ignition. The equivalent plastic stress is divided into two parts in Autodyn. One part is hydrostatic pressure calculated by EOS; the other is deviator stress calculated by a constitutive model. Although the pressure is high, the integral indicator of ignition constant does not increase remarkably, indicating that the required stress threshold has not been reached at this time. After this, the forward-propagating compression wave is reflected by the free surface of the front plate, resulting in a rarefaction wave propagating backward to the RMP, which leads to a sharp decline in pressure on the materials. The rarefaction wave may also enhance the particle velocity change in the RMP, which causes high stress in RMs. Then, the indicator of ignition increases linearly to the value set by user (the value was set to one in this study). At this time the deflagration is successfully induced, and the pressure of the mixture of reactant and deflagration product is calculated based on the pressure and temperature equilibrium. As a result of the deflagration, massive chemical energy is released. The released energy exists in the form of pressure, potential energy, and internal energy. The pressure and temperature are remarkably enhanced by the increase of reaction ratio ALPHA. As shown in Figures 8f and 9f, when ALPHA increases to one, which indicates the deflagration stops at this time, the pressure and temperature of the particle present a declining trend.

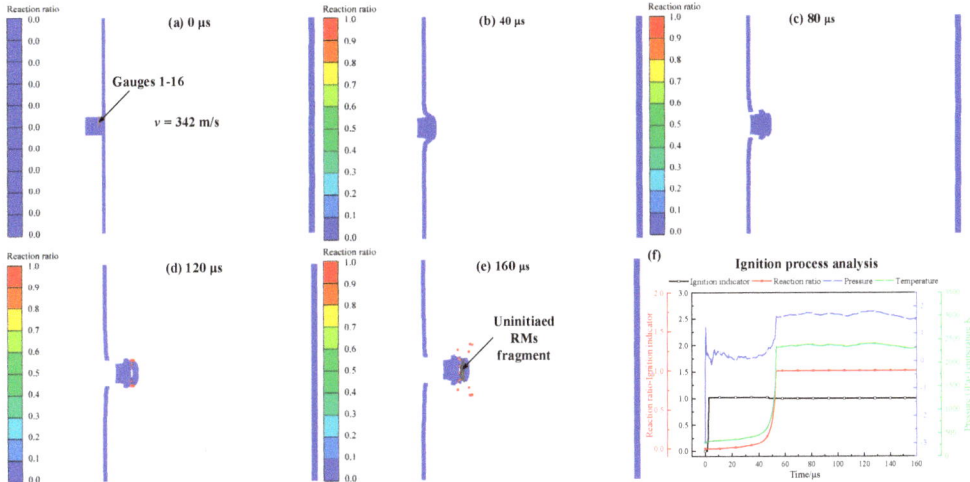

Figure 8. The ignition process of RMPs with a velocity of 342 m/s.

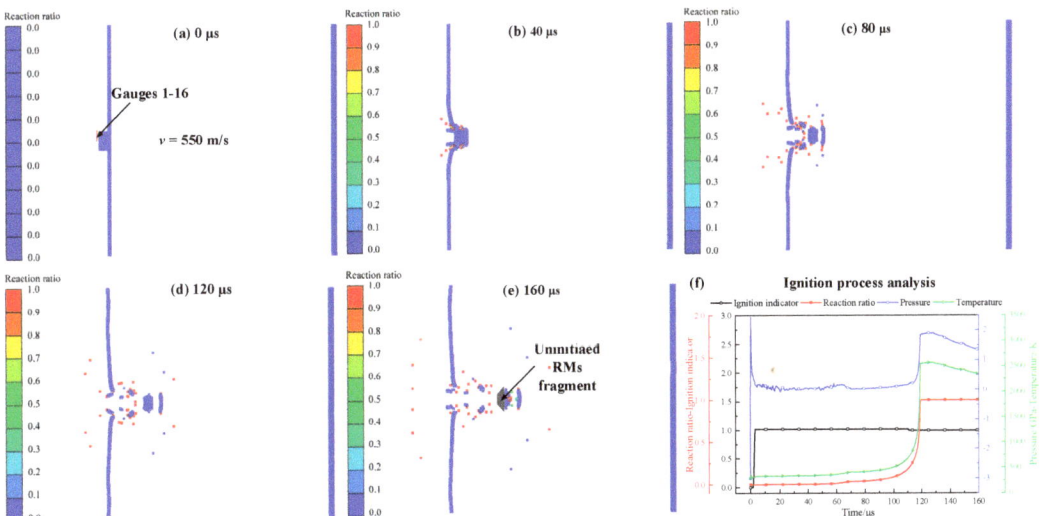

Figure 9. The typical ignition process of RMPs with a velocity of 550 m/s.

4.4. The Energy Release Behavior of RMPs

The energy release behavior of RMPs can be represented by a time-resolved reaction ratio. To get a time-resolved reaction ratio of an RMP for quantity research in the penetration and deflagration process, a subroutine executed at the end of every calculation circulation is compiled. The reaction ratio of an RMP is defined as

$$F = \frac{\sum m_i \alpha_i}{\sum m_i} \quad (17)$$

where m_i denotes the mass of every particle contained in the RMP and α_i corresponds to the reaction ratio of the particles.

The time-resolved reaction ratio of an RMP with an impact velocity of 342 and 550 m/s is shown in Figure 10. As analyzed above, the deflagration starts when the indicator of ignition constant reaches the preset value. After perforating the front plate, the reaction ratio of RMPs increases remarkably. Then, however, there is an approximate flat in the curve of the reaction ratio of RMPs, which implies the flameout of overall deflagration. This simulation result can be verified by the sharp decay of fire light in the experiments, and it does not greatly change until the residual RM fragments impact the rear plate. Then, partial RMs ignite at the second impact and the reaction ratio increases to a new level. It can be also seen from the pictures that the deflagration quantity after impact on the front and rear plate is dependent on impact velocity. When the impact velocity is 340 m/s, the deflagration quantity after impact on the front plate is less than 550 m/s (the reaction ratio rises to 9% for 340 m/s while 19% for 550 m/s). The indicator of ignition constants of uninitiated RMs accumulated during the first impact, so the impact on the rear plate induces more RMs to be deflagrated. The final reaction ratio increases with the impact velocity, but neither their final reaction ratio nor reaction efficiency reaches one, indicating that partial RMs in the RMP fail to ignite under current impact conditions.

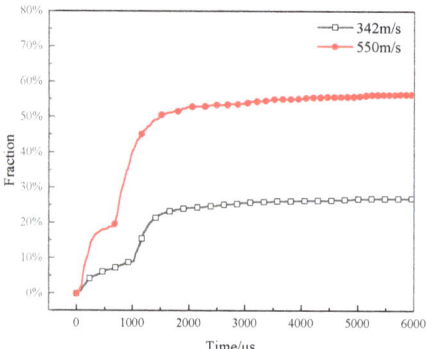

Figure 10. The time-resolved reaction ratio of RMP versus time.

The fragmentation of the RMs projectile is likely the prerequisite condition for ignition because of the non-self-sustaining reaction in RMs. The fragmentation is mainly caused by accumulation of damage related to the plastic deformation or plastic work. From Figure 10, one can conclude that the higher deflagration reaction degree includes two parts. First, after perforating the first plate, the plastic work absorbed by the cylindrical surface of the projectile is higher, inducing more RMs to react. Second, on the front of the rear plate, more reaction is induced. The reason may be that more damage has been produced inside the projectile after perforating the first plate with a higher velocity, and the higher residual velocity of RMs will also produce more serious fragmentation of RMs inside the residual penetration body after impact on the rear plate, resulting in higher reaction degree of the RMs projectile.

4.5. The Temperature Rise Effect

According to the first law of thermodynamics, the massive chemical energy released by the deflagration reaction will be transformed into pressure, potential energy, and internal energy of deflagration product. High pressure can cause the rapid expansion of product while internal energy heats the product to a high temperature. The early evolution of pressure and temperature is shown in Figures 8f and 9f. In Figure 11, the later temperature distribution of RMs at the impact on the rear Al plate is shown. The maximum steady temperatures are 2901 and 3154 K for the impact velocities 342 and 550 m/s, respectively.

This temperature is consistent with the results captured by the infrared framing camera and transient pyrometer in Ref. [8]. The chemical energy released during the impact process can be estimated as

$$4Al + 3(-C_2F_4-) \rightarrow 4AlF_3 + 6C + 2379.61 \text{ kJ} \tag{18}$$

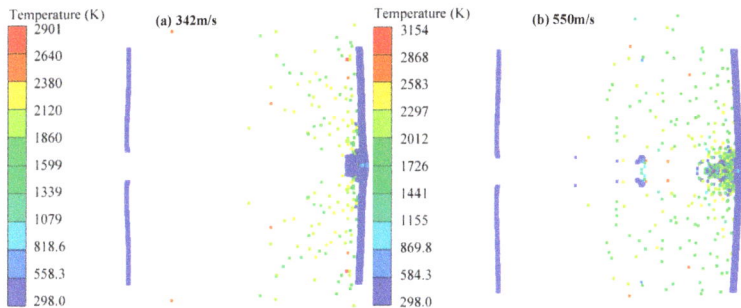

Figure 11. The temperature distribution at impact on the rear plate.

Assuming that the released chemical energy is absolutely transferred to internal energy of deflagration products, then the temperature could be estimated considering specific heat capacity of the deflagration products. Taking the values of the specific heat capacities of AlF_3 and C as 100.831 and 21.609 J/mol·K, respectively, the calculated temperature for the deflagration products is approximately 4464 K. The lower temperature in the simulation can be attributed to persistent expansion of the particles during or after impact, which leads to external work acting on surrounding particles, resulting in a cooling effect for relevant particles according to the first law of thermodynamics.

5. Conclusions

To understand the mechanisms of impact-induced deflagration reactions by RMPs more clearly, a simulation method is presented in this study. High-velocity impact experiments were performed to prove the validity of the simulations. The ignition indicator, reaction ratio, pressure, and temperature distribution are analyzed in detail. The results show that the equivalent plastic stress is more likely to dominate the ignition of RMs instead of the high pressure pulse produced immediately after impact, because no remarkable increase of the ignition indicator was observed when the pressure reached its maximum value during impact. The history curve of reaction ratio and temperature distribution of RMPs was obtained using a numerical simulation. The increasing trend of the reaction ratio is well consistent with the change of radiated fire light in experiments, and the calculated temperature is reasonable compared to experiment results reported in other studies. In summation, the presented simulation method is able to reproduce the impact-induced deflagration behavior of a reactive materials projectile.

Author Contributions: Conceptualization, J.X. and X.L.; methodology, C.H.; software, J.X.; formal analysis, D.Z.; data curation, Y.W.; writing—original draft preparation, J.X.; writing—review and editing, X.L.; funding acquisition, J.X. All authors have read and agreed to the published version of the manuscript.

Funding: The research was funded by National Natural Science Foundation of China (Grant No. 11702256), Natural Science Foundation of Shanxi Province (Grant No. 20210302124214), Scientific and Technological Innovation Programs of Higher Education Institutions in Shanxi (Grant No. 201802071), and Scientific and Technological Innovation Team Programs of North University of China (Grant No.TD201903).

Institutional Review Board Statement: Not applicable.

Informed Consent Statement: Not applicable.

Data Availability Statement: The data that support the findings of this study are available from the corresponding author upon reasonable request.

Conflicts of Interest: The authors declare no conflict of interest.

References

1. Ames, R.G. Vented chamber calorimetry for impact-initiated energetic materials. *43rd AIAA Aerosp. Sci. Meet. Exhib.-Meet. Pap.* **2005**, 15391–15403. [CrossRef]
2. Zheng, Y.; Zheng, Z.; Lu, G.; Wang, H.; Guo, H. Mesoscale study on explosion-induced formation and thermochemical response of PTFE/Al granular jet. *Def. Technol.* **2022**. [CrossRef]
3. Rosencrantz, S.D. Characterization and Modeling Methodology of Polytetrafluoroethylene Based Reactive Materials for the Development of Parametric Models. Ph.D. Thesis, Wright State University, Dayton, OH, USA, 2007.
4. Guo, H.; Zheng, Y.; Yu, Q.; Ge, C.; Wang, H. Penetration behavior of reactive liner shaped charge jet impacting thick steel plates. *Int. J. Impact Eng.* **2019**, *126*, 76–84. [CrossRef]
5. Liu, S.B.; Yuan, Y.; Zheng, Y.F.; Ge, C.; Wang, H.F. Enhanced ignition behavior of reactive material projectiles impacting fuel-filled tank. *Def. Technol.* **2019**, *15*, 533–540. [CrossRef]
6. Zheng, Y.; Su, C.; Guo, H.; Yu, Q.; Wang, H. Behind-Target Rupturing Effects of Sandwich-like Plates by Reactive Liner Shaped Charge Jet. *Propellants Explos. Pyrotech.* **2019**, *44*, 1400–1409. [CrossRef]
7. Xiao, J.; Zhang, X.; Guo, Z.; Wang, H. Enhanced Damage Effects of Multi-Layered Concrete Target Produced by Reactive Materials Liner. *Propellants Explos. Pyrotech.* **2018**, *43*, 955–961. [CrossRef]
8. Xiao, J.; Nie, Z.; Wang, Z.; Du, Y.; Tang, E. Energy release behavior of Al/PTFE reactive materials powder in a closed chamber. *J. Appl. Phys.* **2020**, *127*, 165106. [CrossRef]
9. Canonsburg, T.D. Application of CFD into an automotive torque converter. *Knowl. Creat. Diffus. Util.* **2012**, *15317*, 724–746.
10. Wang, H.F.; Zheng, Y.F.; Yu, Q.B.; Liu, Z.W.; Yu, W.M. Study on initiation mechanism of reactive fragment to covered explosive. *Beijing Ligong Daxue Xuebao/Trans. Beijing Inst. Technol.* **2012**, *32*, 786–823.
11. Raftenberg, M.N.; Mock, W.; Kirby, G.C. Modeling the impact deformation of rods of a pressed PTFE/Al composite mixture. *Int. J. Impact Eng.* **2008**, *35*, 1735–1744. [CrossRef]
12. Yang, S.-Q.; Xu, S.-L.; Zhang, T. Preparation and Performance of PTEF/Al Reactive Materials. *J. Natl. Univ. Def. Technol.* **2008**, *30*, 39–42.
13. Jiang, J.-W.; Wang, S.-Y.; Zhang, M.; Wei, Q. Modeling and simulation of JWL equation of state for reactive Al/PTFE mixture. *J. Beijing Inst. Technol.* **2012**, *21*, 150–156.
14. Lee, E.L.; Tarver, C.M. Phenomenological model of shock initiation in heterogeneous explosives. *Phys. Fluids* **1980**, *23*, 2362–2372. [CrossRef]
15. Mock, W.; Holt, W.H. Impact initiation of rods of pressed polytetrafluoroethylene (PTFE) and aluminum powders. *AIP Conf. Proc.* **2006**, *845*, 1097–1100.
16. Mock, W.; Drotar, J.T. Effect of aluminum particle size on the impact initiation of pressed PTFE/Al composite rods. *AIP Conf. Proc.* **2007**, *955*, 971–974.
17. Zhang, X.P.; Xiao, J.G.; Yu, Q.B.; Zheng, Y.F.; Wang, H.F. Demolition effect of reactive material liner shaped charge against concrete target. *Beijing Ligong Daxue Xuebao/Trans. Beijing Inst. Technol.* **2016**, *36*, 1211–1215.
18. Yi, J.; Wang, Z.; Yin, J.; Zhang, Z. Simulation study on expansive jet formation characteristics of polymer liner. *Materials* **2019**, *12*, 744. [CrossRef]
19. Zhang, X.F.; Shi, A.S.; Qiao, L.; Zhang, J.; Zhang, Y.G.; Guan, Z.W. Experimental study on impact-initiated characters of multifunctional energetic structural materials. *J. Appl. Phys.* **2013**, *113*, 083508. [CrossRef]
20. Su, C.; Wang, H.; Xie, J.; Ge, C.; Zheng, Y. Penetration and Damage Effects of Reactive Material Jet against Concrete Target. *Binggong Xuebao/Acta Armamentarii* **2019**, *40*, 1829–1835.
21. Zhang, Z.Y.; Tian, Z.D.; Chen, J.; Duan, Z.P. *Detonation Physics*; National University of Defense Technology Press: Changsha, China, 2016; p. 270.
22. Lee, E.L.; Hornig, H.C.; Kury, J.W. *Adiabatic Expansion of High Explosive Detonation Products*; Report UCRL-50422; Lawrence Radiation Labroratory of University of California: Livermore, CA, USA, 1968.
23. Xu, F.Y.; Zheng, Y.F.; Yu, Q.B.; Wang, Y.Z.; Wang, H.F. Experimental study on penetration behavior of reactive material projectile impacting aluminum plate. *Int. J. Impact Eng.* **2016**, *95*, 125–132. [CrossRef]
24. Xu, F.Y.; Zheng, Y.F.; Yu, Q.B.; Zhang, X.P.; Wang, H.F. Damage effects of aluminum plate by reactive material projectile impact. *Int. J. Impact Eng.* **2017**, *104*, 38–44. [CrossRef]
25. Xu, F.Y.; Yu, Q.B.; Zheng, Y.F.; Lei, M.A.; Wang, H.F. Damage effects of double-spaced aluminum plates by reactive material projectile impact. *Int. J. Impact Eng.* **2017**, *104*, 13–20. [CrossRef]
26. Feng, B.; Li, Y.; Wu, S.; Wang, H.; Tao, Z.; Fang, X. A crack-induced initiation mechanism of Al-PTFE under quasi-static compression and the investigation of influencing factors. *Mater. Des.* **2016**, *108*, 411–417. [CrossRef]

27. Feng, B.; Fang, X.; Li, Y.C.; Wang, H.X.; Mao, Y.M.; Wu, S.Z. An initiation phenomenon of Al-PTFE under quasi-static compression. *Chem. Phys. Lett.* **2015**, *637*, 38–41. [CrossRef]
28. Sippel, T.R.; Son, S.F.; Groven, L.J. Aluminum agglomeration reduction in a composite propellant using tailored Al/PTFE particles. *Combust. Flame* **2014**, *161*, 311–321. [CrossRef]
29. Osborne, D.T.; Pantoya, M.L. Effect of Al particle size on the thermal degradation of Al/Teflon mixtures. *Combust. Sci. Technol.* **2007**, *179*, 1467–1480. [CrossRef]
30. Mohan, S.; Trunov, M.A.; Dreizin, E.L. Heating and ignition of metallic particles by a CO_2 laser. *J. Propuls. Power* **2008**, *24*, 199–205. [CrossRef]
31. Wu, Y. *Study on Reaction Heat Measuring and Energy Release Characteristics of PTFE/Al Reactive Material*; National University of Defense Technology: Changsha, China, 2015.
32. Dolgoborodov, A.Y.; Makhov, M.N.; Kolbanev, I.V.; Streletskiĭ, A.N.; Fortov, V.E. Detonation in an aluminum-teflon mixture. *JETP Lett.* **2005**, *81*, 311–314. [CrossRef]
33. Dolgoborodov, A.Y.; Makhov, M.N.; Streletskii, A.N.; Kolbanev, I.V.; Gogulya, M.F.; Brazhnikov, M.A.; Fortov, V.E. Detonation-like phenomena in non-explosive oxidizer-metal mixtures. In Proceedings of the 31st International Pyrotechnics Seminar, Fort Collins, CO, USA, 11–16 July 2004; p. 569.
34. Raftenberg, M.N.; Scheidler, M.J.; Casem, D.A. *A Yield Strength Model and Thoughts on an Ignition Criterion for a Reactive PTFE-Aluminum Composite*; Report ARL-RP-219; Army Research Laboratory: Aberdeen Proving Ground, MD, USA, 2008.

Article

Penetration Failure Mechanism of Multi-Diameter Tungsten Fiber Reinforced Zr-Based Bulk Metallic Glasses Matrix Composite Rod

Chengxin Du [1,2], Huameng Fu [3], Zhengwang Zhu [3], Kehong Wang [1], Guangfa Gao [2,*], Feng Zhou [2], Lizhi Xu [2] and Zhonghua Du [2]

1. School of Materials Science and Engineering, Nanjing University of Science and Technology, Nanjing 210094, China; duchengxin4324@163.com (C.D.); wkh1602@126.com (K.W.)
2. School of Mechanical Engineering, Nanjing University of Science and Technology, Nanjing 210094, China; zhoufeng199191@163.com (F.Z.); xulznjust@163.com (L.X.); duzhonghua@aliyun.com (Z.D.)
3. Institute of Metal Research, Chinese Academy of Sciences, Shenyang 110016, China; hmfu@imr.ac.cn (H.F.); zwzhu@imr.ac.cn (Z.Z.)
* Correspondence: gfgao@ustc.edu.cn

Citation: Du, C.; Fu, H.; Zhu, Z.; Wang, K.; Gao, G.; Zhou, F.; Xu, L.; Du, Z. Penetration Failure Mechanism of Multi-Diameter Tungsten Fiber Reinforced Zr-Based Bulk Metallic Glasses Matrix Composite Rod. *Crystals* 2022, 12, 124. https://doi.org/10.3390/cryst12020124

Academic Editors: Pavel Lukáč and Alessandro Chiasera

Received: 6 December 2021
Accepted: 12 January 2022
Published: 18 January 2022

Publisher's Note: MDPI stays neutral with regard to jurisdictional claims in published maps and institutional affiliations.

Copyright: © 2022 by the authors. Licensee MDPI, Basel, Switzerland. This article is an open access article distributed under the terms and conditions of the Creative Commons Attribution (CC BY) license (https://creativecommons.org/licenses/by/4.0/).

Abstract: In order to increase the penetration ability of tungsten fiber-reinforced Zr-based bulk metallic glasses matrix composite rod, two multi-diameter tungsten fiber-reinforced Zr-based bulk metallic glasses matrix composites (MD-WF/Zr-MG) are designed. In MD-WF/Zr-MG-I, the diameters of tungsten fiber (WF) increase gradually from the inside to outside, which is the opposite in MD-WF/Zr-MG-II. Penetration experiment of two kinds of MD-WF/Zr-MG rods into rolled homogeneous armor (RHA) steel target from 1470 m/s to 1630 m/s is conducted. The average penetration depth of the MD-WF/Zr-MG-II rod is higher than that of the MD-WF/Zr-MG-I rod. Penetration failure modes of MD-WF/Zr-MG-I and MD-WF/Zr-MG-II rods are bending, backflow of WFs and shear failure respectively. The failure mode of MD-WF/Zr-MG is affected by the bend spaces and the ultimate bending diameters of WFs. If the bend spaces of all WFs are equal or larger than their ultimate bending diameters, the penetration failure mode is the bending and backflow of WFs, oppositely the penetration failure mode is the shear failure. The MD-WF/Zr-MG rod with shear failure exhibits high penetration ability because of low penetration resistance and little residual material in the crater. When designing MD-WF/Zr-MG, bend spaces of a part of WFs should be smaller than their ultimate bending diameter to cause shear failure.

Keywords: multi-diameter tungsten fiber-reinforced Zr-based bulk metallic glasses matrix composite; tungsten fiber; penetration failure mode; bend space; ultimate bending diameter

1. Introduction

Tungsten fiber-reinforced Zr-based bulk metallic glasses matrix composite (WF/Zr-MG) is a kind of unidirectional fiber-reinforced composite with tungsten fiber (WF) as the reinforcement phase and Zr-based bulk metallic glasses (Zr-MG) as the matrix [1,2]. WF/Zr-MG was firstly prepared by Dandliker et al. in 1998 [3]. Then researchers found that WF/Zr-MG was a potential penetrator material owing to advantageous properties. Firstly, there is little difference between the density of WF/Zr-MG and tungsten heavy alloy (WHA). When the volume fraction of WF in WF/Zr-MG is 80%, the density of WF/Zr-MG is more than 17.1 g/cm^3 [4,5], and comparatively that of WHA is 17.6 g/cm^3. Secondly, the compression strength of WF/Zr-MG is much higher than WHA. Normally, the quasi-static and dynamic compression strength of WF/Zr-MG is more than 1.8 GPa and 2.5 GPa, respectively [3–15]. In the research of Chen et al., the dynamic compressive strength of WF/Zr-MG even reached 3.5 GPa [15]. In contrast, the compression strength of WHA is about 1.2 GPa [16]. Moreover, WF/Zr-MG rods exhibited self-sharpening feature during the penetration experiments, which is similar with depleted uranium alloys. The diameter of the crater penetrated by WF/Zr-MG rods is about 15% lower than that of WHA rods [5,17]. Based on the advantageous properties of WF/Zr-MG, it exhibits high penetration ability. The penetration performance of WF/Zr-MG rods was about 10–20% higher than that of WHA rods under ideal penetration conditions [5,17–21]. However, WF/Zr-MG cannot be directly applied in an armor piercing projectile because its penetration performance is unstable. Chen et al. found that though the WF/Zr-MG rod can achieve effective penetration in the Q235 steel target, the WF/Zr-MG rod has no distinct advantage in the penetration performance compared with the WHA rod in 800~1000 m/s because of ballistic trajectory deflection [19]. Our previous research found that longitudinal splitting occurred in WF/Zr-MG rods with L/D of 3.75 during the penetration process, which resulted in a much lower penetration efficiency than that of WHA rods [17]. Therefore, research on covered material processing [6,22,23], experimental study [17–21], and numerical simulation [24–26] are conducted to analyze the causes and influencing factors of ballistic deflection and longitudinal splitting of WF/Zr-MG rods in order to improve its penetration stability.

The research results show that WF has a great influence on the strength, failure strain, and penetration behavior of WF/Zr-MG [8,9,27–30], especially the diameter of WF. Zhang et al. conducted systematic research on the influence of the diameter of WF on the compressive strength of composite materials [30]. The results showed that the diameter of WF has different effects on quasi-static compression and dynamic compression of composites. The compressive strength and fracture strain of WF/Zr-MG decrease with the increase of the diameter of WF. However, there is no linear relationship between the compressive strength of WF/Zr-MG and the diameter of WF during dynamic compression. Our previous research found that the diameter of tungsten fiber also has a great influence on the penetration behavior of composites. The penetration experiments of WF/Zr-MG reinforced by Ø0.3 mm, Ø0.5 mm, Ø0.7 mm, and Ø1.0 mm WFs were conducted respectively. The WF/Zr-MG rod reinforced by Ø0.3 mm WFs had the most stable penetration performance, and the WF/Zr-MG rod reinforced by Ø0.7 mm WFs had the deepest penetration depth [31]. As is seen, the diameter of WF has different effects on the quasi-static compression, dynamic compression, and penetration behavior of WF/Zr-MG, and WFs with different diameters have different advantages and disadvantages. Therefore, it is possible to make use of the advantages and avoid the disadvantages of WFs when WF/Zr-MG is reinforced by multiple diameter WFs. However, the research has not been studied up to now.

In this paper, two MD-WF/Zr-MG are designed and prepared. The penetration experiments of the two MD-WF/Zr-MG rods are conducted. According to the analysis of experimental results, the penetration ability and failure mode of the two WF/Zr-MG rods are revealed. A design method of MD-WF/Zr-MG with high penetration ability is also obtained according to the experimental results.

2. Design and Preparation of MD-WF/Zr-MG

Compared with homogeneous materials, such as WHA and depleted uranium alloys, WF/Zr-MG has the advantage of designability. The volume fraction and diameter of WFs in WF/Zr-MG can be changed according to the requirements. In our previous penetration experiments, WF/Zr-MG reinforced by Ø0.3 mm, Ø0.5 mm, Ø0.7 mm, and Ø1.0 mm WFs rods exhibit different advantages and disadvantages in penetration. Figure 1 shows the experimental results of our previous research [31]. Compared with the WHA rod, the WF/Zr-MG rod reinforced by Ø0.3 mm WFs had the most stable penetration performance, and the WF/Zr-MG rod reinforced by Ø0.7 mm WFs had the deepest penetration depth. In order to make full use of the advantages and avoid the disadvantages of these WF/Zr-MG rods, two multi-diameter tungsten fiber reinforced Zr-based bulk metallic glasses matrix composite (MD-WF/Zr-MG) are designed. In MD-WF/Zr-MG-I, the diameters of WF increase gradually from the inside to outside. On the contrary, the diameters of WF in MD-WF/Zr-MG-II decrease gradually from the inside to outside. The cross section of MD-WF/Zr-MG schematic is shown in Figure 2.

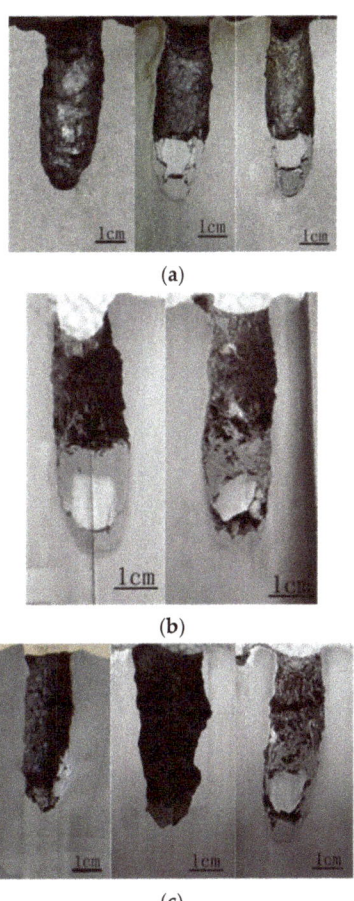

(a)

(b)

(c)

Figure 1. Longitudinal section of the craters penetrated by (**a**) Tungsten heavy alloy (WHA) rods, (**b**) WF/Zr-MG rods reinforced by Ø0.3 mm WFs, and (**c**) WF/Zr-MG rods reinforced by Ø0.7 mm WFs [31].

Ingots of $Zr_{41.25}Ti_{13.75}Ni_{10}Cu_{12.5}Be_{22.5}$ (in at. pct) were first prepared by combining the constitutive elements (purity 99.5 pct or higher) in an induction furnace under an argon atmosphere. WFs with a nominal diameter of 0.3 mm, 0.5 mm, and 1.0 mm were straightened. The fibers were cleaned in a bath of acetone through an ultrasonic method and then cleaned by ethanol. The WFs were placed in the sealed end of an evacuated quartz tube according to Figure 2. The ingots were then heated and melted in a resistive furnace, followed by pressure infiltration. The MD-WF/Zr-MG were prepared successfully using infiltration and rapid solidification. Figure 3 shows the cross sections of MD-WF/Zr-MG. The volume fractions of all the WFs and Ø1.0 mm WFs, Ø0.5 mm WFs, and Ø0.3 mm WFs in MD-WF/Zr-MG-I are 81.90 and 20.10%, and 40.22 and 21.57%, respectively. The volume fractions of all the WFs and Ø1.0 mm WFs, Ø0.5 mm WFs, and Ø0.3 mm WFs in MD-WF/Zr-MG-II are 81.16 and 46.14%, and 19.22 and 15.80%, respectively. The densities of MD-WF/Zr-MG-I and MD-WF/Zr-MG-II are 17.03 ± 0.05 g/cm^3 and 16.90 ± 0.05 g/cm^3, respectively.

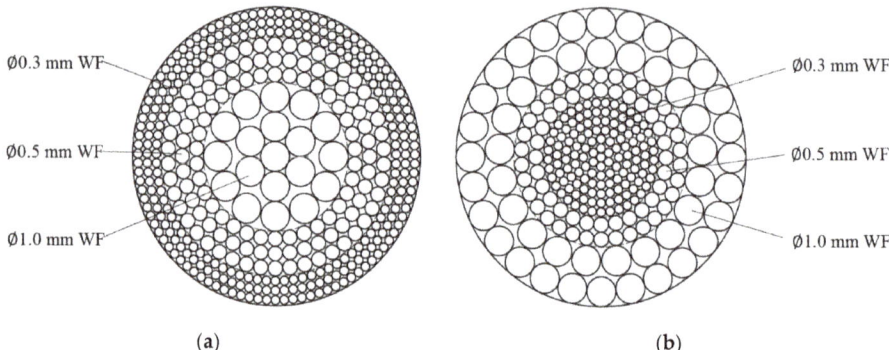

Figure 2. Cross section of MD-WF/Zr-MG schematic: (**a**) MD-WF/Zr-MG-I and (**b**) MD-WF/Zr-MG-II.

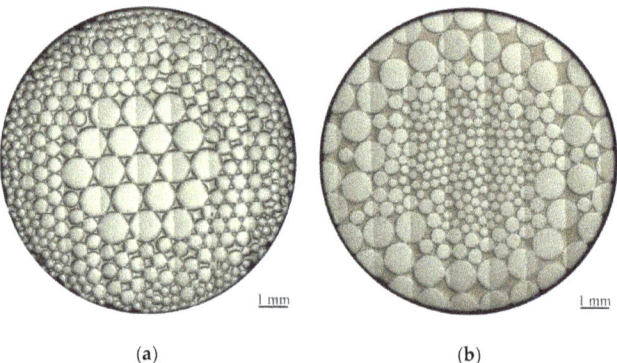

Figure 3. Cross section of MD-WF/Zr-MG: (**a**) MD-WF/Zr-MG-I and (**b**) MD-WF/Zr-MG-II.

3. Experiment of MD-WF/Zr-MG Rod

3.1. Setup of Experiment

Figure 4 shows the MD-WF/Zr-MG rod and the assembly drawing of the projectile. The largest diameter of WFs in MD-WF/Zr-MG is Ø1.0 mm, so the rod with a diameter of 10 mm is prepared. In order to compare with the previous experimental results in reference [17], the rod with the same length is selected, so the length of the MD-WF/Zr-MG rod is set to 54 mm. A tail and a nose are installed in the bottom and top of the rod respectively to enhance the flight stability of the rod, as is seen in Figure 4a. Both the tail and nose are made of aluminum alloy to reduce their interference with penetration results. Since the diameter of the rod is less than that of the Ø25 mm smooth bore gun, a sub-calibre launching technique of a long rod is employed in the present paper. As shown in Figure 4b, the aluminum alloy sabots are used to fit the larger diameter of smooth bore for firing a smaller rod. A nylon block is used to push the projectile. A steel platen is placed between the rod and block to increase the strength of the block. Three rods of each material are manufactured. The mass of projectiles and rods are shown in Table 1. The target in the experiment is made of RHA with a thickness of Ø80 mm.

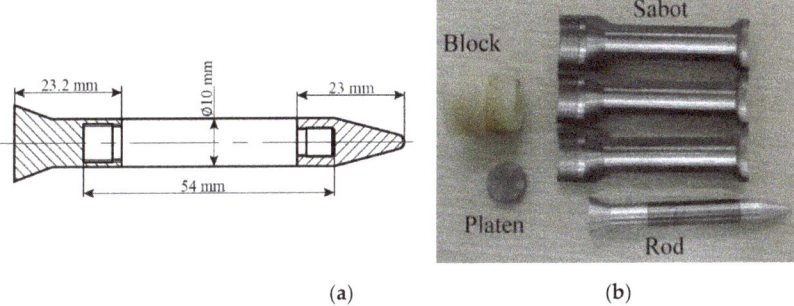

Figure 4. (a) Dimension of MD-WF/Zr-MG rod and (b) assembly drawing of the projectile.

Table 1. Mass of projectile and rod.

No.	Material	Mass of Projectile	Mass of Rod
I-1		108.1 ± 0.05 g	61.2 ± 0.05 g
I-2	MD-WF/Zr-MG-I	107.8 ± 0.05 g	61.6 ± 0.05 g
I-3		107.9 ± 0.05 g	61.6 ± 0.05 g
II-1		105.3 ± 0.05 g	59.1 ± 0.05 g
II-2	MD-WF/Zr-MG-II	107.3 ± 0.05 g	61.0 ± 0.05 g
II-3		107.7 ± 0.05 g	60.7 ± 0.05 g

A Ø25 mm smooth bore artillery was used to launch the projectiles. The setup of the experiments is shown in Figure 5. The average impact velocities of rods between 2 and 4 m in front of the target are measured by a velocimeter. The speed range is also selected as in reference [17], which varies from 1470 to 1650 m/s.

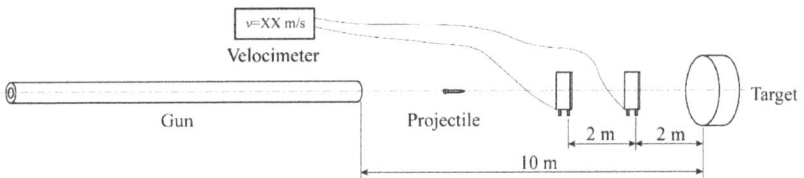

Figure 5. Setup of experiments.

3.2. Experimental Results and Analysis

Figure 6 shows experimental results of DOP vs the impact velocity of the MD-WF/Zr-MG-I and MD-WF/Zr-MG-II rods (the solid point in Figure 6). The dotted line shows the fitting curves of the experimental results. As shown in Figure 6, the average DOP of the MD-WF/Zr-MG-II rods is higher than that of the MD-WF/Zr-MG-I rods. According to the experimental results, the penetration ability of the MD-WF/Zr-MG-II rods is better than that of the MD-WF/Zr-MG-I rods.

In order to reveal the penetration mechanism and the reason for the difference of penetration ability between MD-WF/Zr-MG-I rods and MD-WF/Zr-MG-II rods, their penetration craters were dissected. Figure 7 is the longitudinal sections of craters penetrated by MD-WF/Zr-MG-I rods. Figure 7 shows that the craters are vertical with no trajectory deflection, which proves that the stiffness of the WF/Zr-MG rod increases with the addition of larger diameter WFs, as trajectory deflection often occurs during the penetration experiments of WF/Zr-MG rods with small aspect ratio [17]. According to the measurement, the diameters of the three craters are 16.4, 17.2, and 17.9 mm, respectively. The average diameters of craters are smaller than that penetrated by WHA rods (about 19~20 mm [31]), which proves that the MD-WF/Zr-MG-I rods exhibit a self-sharpening feature during the penetration process.

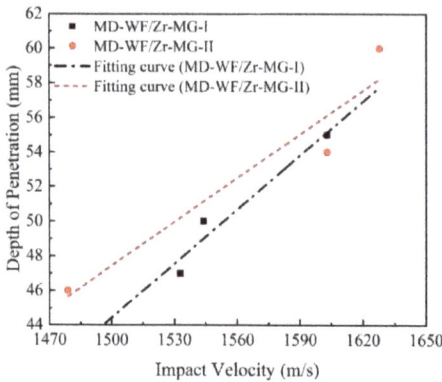

Figure 6. Experimental results of DOP of the MD-WF/Zr-MG-I and MD-WF/Zr-MG-II rods.

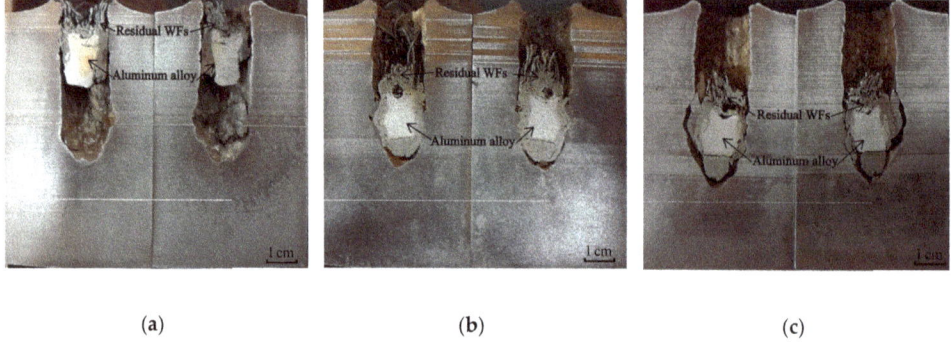

Figure 7. Longitudinal sections of craters penetrated by MD-WF/Zr-MG-I rods. (**a**) I-1 v = 1532.8 m/s. (**b**) I-2 v = 1544.0 m/s. (**c**) I-3 v = 1603.0 m/s.

Figure 8 shows the external view of the residual MD-WF/Zr-MG-I rods. Figure 8a is a front view of the residual rod in Figure 7a. As shown in Figure 8a, Ø1 mm WFs is on the outside of the residual rod, and Ø0.3 and Ø0.5 mm WFs are on the inside of the residual rod, which is opposite to the original MD-WF/Zr-MG-I rod. Figure 8b is a magnified view of the residual rod in Figure 7b. Figure 8b reveals the reason why the Ø1 mm WFs in Figure 8a is on the outside of the residual rod. Figure 8b shows that the WFs at the head of the rod turn out layer by layer, which results in the arrangement of WFs in the residual rod is opposite to that in the original rod. Figure 8 proves that the WFs at the head of the rod bending and backflow during the penetration process. This penetration failure mode of the MD-WF/Zr-MG-I rod is different from the adiabatic shear failure mode of depleted uranium alloys. The penetration failure mode of the MD-WF/Zr-MG-I rod leads to a hemispherical head, which is similar to that of WHA [16,32].

Figure 9 is the longitudinal sections of craters penetrated by MD-WF/Zr-MG-II rods. Similar to the craters penetrated by MD-WF/Zr-MG-I rods (Figure 7), the craters penetrated by MD-WF/Zr-MG-II rods are also vertical and exhibit self-sharpening feature during the penetration process (the average diameters of craters in Figure 9 are 17.3, 18.5, and 18.2 mm, respectively). But unlike the craters penetrated by the MD-WF/Zr-MG-I rods, there is no residual materials in the craters penetrated by MD-WF/Zr-MG-II rods, and the bottom shape of the craters is conical rather than hemispherical. The difference of the craters proves that the penetration failure mode of MD-WF/Zr-MG-II rods is different from that of MD-WF/Zr-MG-I rods. According to the longitudinal sections of craters in Figure 9, the penetration failure mode of MD-WF/Zr-MG-II rods should be the shear failure which is similar to that of depleted uranium alloys. During the penetration process of MD-WF/Zr-MG-II rods, the bending and breaking of tungsten fiber leads to shear failure at the head. When the shear failure occurs, the broken rod material is small and easy to flow out from the crater, so there is no residual rod in the crater. Then the head of the rod with shear failure is conical, so the bottom shape of the crater is also conical.

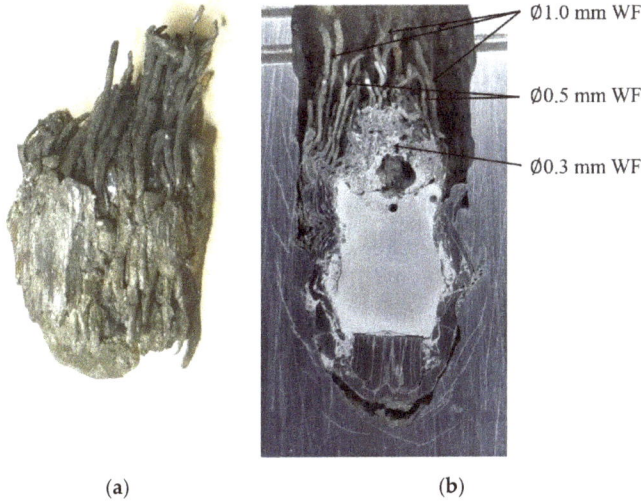

(a) (b)

Figure 8. (a) Front view of the residual rod in Figure 7a,b. Magnified view of the residual rod in Figure 7b.

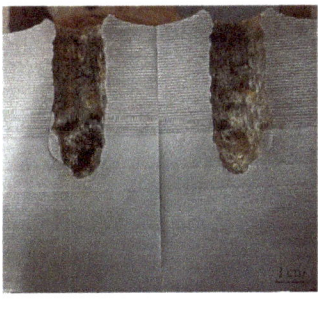

(a)

Figure 9. *Cont.*

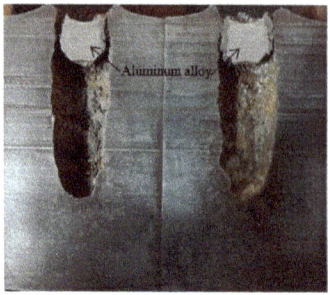

(b)

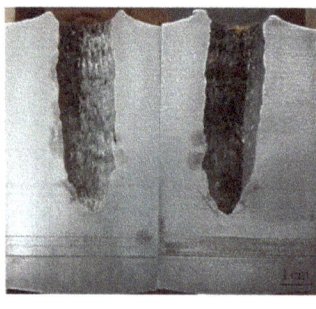

(c)

Figure 9. Longitudinal sections of craters penetrated by the MD-WF/Zr-MG-II rods. (**a**) II-1 v = 1478.8 m/s. (**b**) II-2 v = 1603.0 m/s. (**c**) II-3 v = 1627.9 m/s.

4. Penetration Failure Mechanism of the MD-WF/Zr-MG Rod

According to the experimental results in Figures 7 and 9, the penetration failure modes of the MD-WF/Zr-MG-I rod and MD-WF/Zr-MG-II are different. The WFs at the head of the MD-WF/Zr-MG-I rod show bending and backflow during the penetration process, but the WFs are bent and broken during the penetration process of MD-WF/Zr-MG-II rods. Therefore, the bending resistance of WF is an important factor affecting the penetration failure mechanism of the MD-WF/Zr-MG rod.

4.1. Ultimate Bending Diameter of WF

The experimental results show that the penetration failure mode of MD-WF/Zr-MG-I rods is the bending and backflow of WFs, and the penetration failure mode of the MD-WF/Zr-MG-II rods is shear failure. No bending behavior of WFs occurs during the penetration process of MD-WF/Zr-MG-II rods. Therefore, the bending deformation ability of WF is an important factor affecting the penetration failure mode of the MD-WF/Zr-MG rod.

WF is a kind of brittle material. The test results of Terentyev et al. show that the failure strain of tungsten fiber is about 1.5% at room temperature and 5% at 2100 °C [33–35]. Therefore, according to the mechanics of materials, the ultimate bending diameter of Ø1.0 mm WF is about 66 mm. However, in Figure 8b, Ø1.0 mm WFs are not broken when its bending diameter of Ø1.0 mm WF is far less than 66 mm, which proves that the ultimate bending diameter of WF cannot be directly calculated by mechanics of materials. Therefore, a bending test is conducted to measure the ultimate bending diameter of WF.

After 5 repeated tests, the bending test results of WFs are obtained. The test results of Ø1.0 mm WF is shown in Figure 10. Ø1.0 mm WF is not broken when its bending diameter is about 4 mm (Figure 10a), and Ø1.0 mm WF breaks when its bending diameter decreases to about 2 mm (Figure 10b). The test results show that the ultimate bending diameter of Ø1.0 mm WF is between 2 and 4 mm. Figure 10c,d show the test results of Ø0.5 mm and Ø0.3 mm WFs. Ø0.5 mm and Ø0.3 mm WFs are not broken during the bending test. According to the test results, the ultimate bending diameter of Ø0.5 mm WF is less than 0.6 mm, and the ultimate bending diameter of Ø0.3 mm WF is less than 0.4 mm.

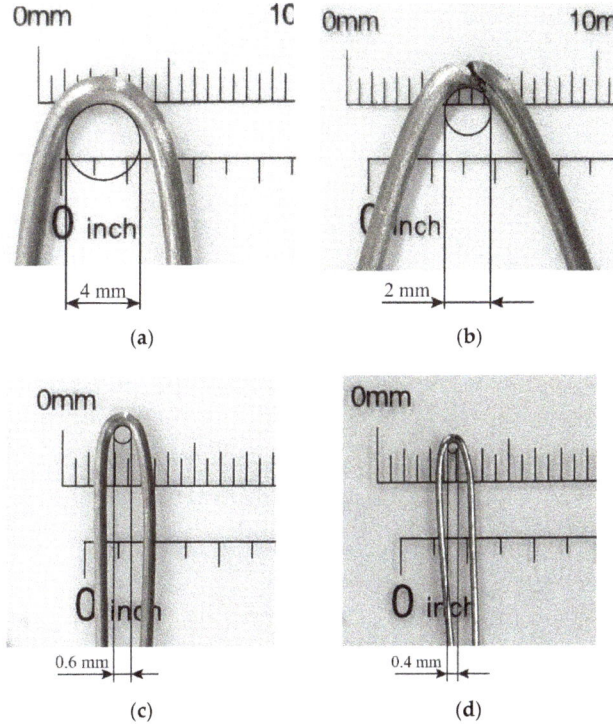

Figure 10. Bending test results of Ø1.0 mm WF Ø0.5 mm WF and Ø0.3 mm WF: (**a**) Bending diameter of Ø1.0 mm WF is about 4 mm, (**b**) bending diameter of Ø1.0 mm WF is about 2 mm, (**c**) bending diameter of Ø0.5 mm WF is about 0.6 mm, and (**d**) bending diameter of Ø0.3 mm WF is about 0.4 mm.

In addition, the temperature of WF/Zr-MG will rise during the penetration process and the high temperature will lead to the decrease of the ultimate bending diameter of WF. However, the temperature of tungsten increases slowly during the penetration process. The temperature rise of tungsten under the impact pressure of 60 GPa (about 2000 m/s impact velocity) is about 300 °C [36], which is about one tenth of the melting point of tungsten. Therefore, the impact temperature rise would have no great effect on the ultimate bending diameter of WF.

4.2. Penetration Failure Mode

The cross-section diagram of the MD-WF/Zr-MG rod penetrating the target is shown in Figure 11. The radius of the crater is R and the radius of the rod is r. One WF at point "A" in the MD-WF/Zr-MG, the distance between point "A" and "O" point (center of rod) is r_1. If the WF can bend and backflow during the penetration process, its position in the crater will be "A'". The distance between point "A'" and the wall of the crater is r_1'. Then the bending space of the WF is the distance x between point "A" and point "A'".

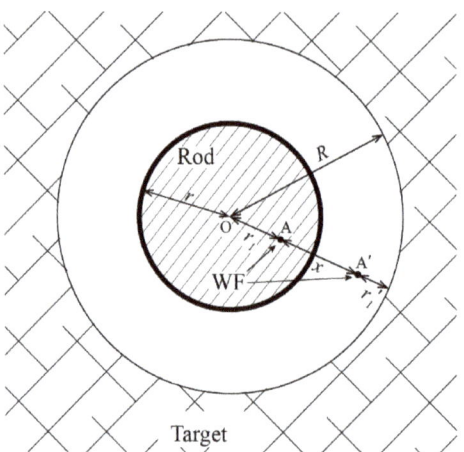

Figure 11. Cross-section diagram of the MD-WF/Zr-MG rod penetrating the target.

It is assumed that all WFs in the MD-WF/Zr-MG rod bend and backflow during the penetration process, and are evenly distributed in the space between the rod and crater. The distance between point "A'" and the wall of the crater is:

$$r_1' = \frac{r_1(R-r)}{r}. \tag{1}$$

The bending space x of the WF is:

$$x = R - r_1 - r_1' = \frac{R(r-r_1)}{r}. \tag{2}$$

If the bending space of WF is equal or larger than its ultimate bending diameter, the WF could bend and backflow during the penetration process. Instead, the WF will break during the penetration process.

The average diameter of the carters penetrated by MD-WF/Zr-MG-I rods is about Ø17 mm, then Figure 12a shows the distance between WFs and center of the rod during the penetration process of the MD-WF/Zr-MG-I rod. According to Equation (2) and the test results, the bending space and ultimate bending diameter of WFs in MD-WF/Zr-MG-I is shown in Table 2. Compared with the bending space and ultimate bending diameter of WFs in MD-WF/Zr-MG-I, only part of Ø0.3 mm WFs may not have enough bending space during the penetration process because the test result of Ø0.3 mm WFs shows its ultimate bending diameter is smaller than 0.4 mm. However, due to the volume fraction of the broken Ø0.3 mm WFs being very small, it has no effect on the penetration failure mode of the rod. Most of the WFs can bend and backflow, and a spherical head is formed, as shown in Figure 12b. As the WFs backflow layer by layer, Ø1.0 mm WFs become the outer layer, as shown in Figure 8. The Ø1.0 mm WFs become a stable protective layer for Ø0.3 mm and Ø0.5 mm WFs, so Ø0.3 mm and Ø0.5 mm WFs do not break continuously during the penetration process.

The average diameter of the carters penetrated by MD-WF/Zr-MG-II rods is also about Ø 17 mm; Figure 13a shows the distance between WFs and center of the rod during the penetration process of the MD-WF/Zr-MG-I rod. According to Equation (2) and the test results, the bending space and ultimate bending diameter of WFs in MD-WF/Zr-MG-II is shown in Table 3. The bend spaces of Ø0.3 mm WFs and Ø0.5 mm WFs in the MD-WF/Zr-MG-II rod are larger than their ultimate bending diameter, but the bend spaces of Ø1.0 mm WFs are smaller than their ultimate bending diameter. Therefore, during the penetration process of the MD-WF/Zr-MG-II rod, Ø1.0 mm WFs shear failure occurs firstly, rather than bending deformation. Although the bend spaces of Ø0.3 mm WFs and Ø0.5 mm WFs are enough to bend, they will be cut by the Ø1.0 mm WFs and the crater wall because of the tiny distance between the broken Ø1 mm WFs and crater wall. Therefore, a conical head similar to depleted uranium alloys is formed during the penetration process of the MD-WF/Zr-MG-II rod, as shown in Figure 13b. The penetration failure mode of the MD-WF/Zr-MG-II rod is the shear failure layer by layer. The failed rod material is small and easy to flow out from the crater, so there is no residual rod in the crater, as shown in Figure 9.

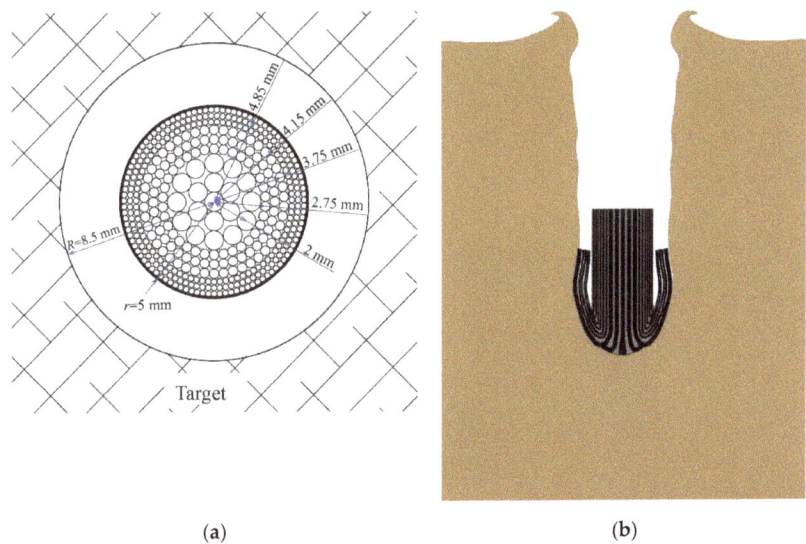

Figure 12. (a) Cross-section diagram of the MD-WF/Zr-MG-I rod penetrating the target and (b) schematic diagram of failure process of the MD-WF/Zr-MG-I rod.

Table 2. Bending space and ultimate bending diameter of WFs in MD-WF/Zr-MG-I.

Diameter of WF	Position (r_1)	Bending Space	Ultimate Bending Diameter
Ø0.3 mm	4.15 mm~4.85 mm	0.255 mm~1.445 mm	<0.4 mm
Ø0.5 mm	2.75 mm~3.75 mm	2.125 mm~3.825 mm	<0.6 mm
Ø1.0 mm	0~2 mm	5.1 mm~8.5 mm	2 mm~4 mm

Comparing with Figures 12 and 13b, it can be found that the shape of the residual rod head is hemispherical when the penetration failure mode of the rod is bending and there is backflow of WFs, while the shape of the residual rod head is conical when the penetration failure mode of the rod is shear failure. According to the penetration mechanics, the penetration resistance of conical head is less than that of the hemispherical head despite the diameters of the two craters being same, so the penetration ability of the MD-WF/Zr-MG-II rods is higher than that of the MD-WF/Zr-MG-I rods, as shown in Figure 6. Moreover, if the aspect ratio of the rod increases, the residual WFs in the crater penetrated by the MD-WF/Zr-MG-I rod will hinder the outflow of the rod and target residues, which will have a negative influence on the penetration ability of the rod. Therefore, in the case of the large aspect ratio rod penetration, the MD-WF/Zr-MG-II rod will have greater advantage than the MD-WF/Zr-MG-I rod.

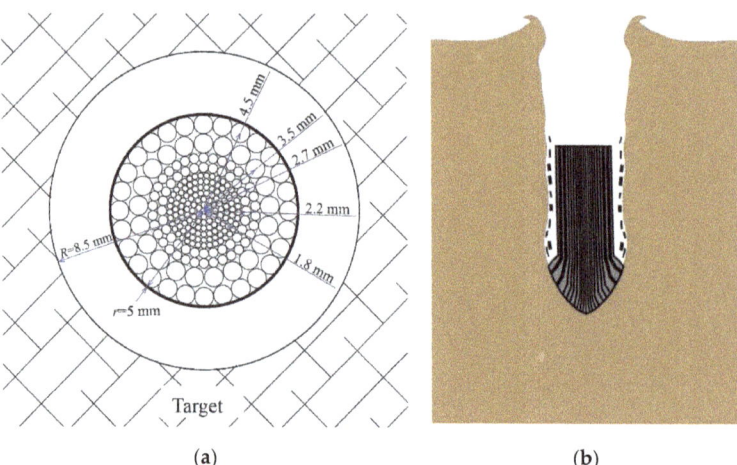

Figure 13. (a) Cross-section diagram of the MD-WF/Zr-MG-II rod penetrating the target and (b) the schematic diagram of the failure process of the MD-WF/Zr-MG-II rod.

Table 3. Bending space and ultimate bending diameter of WFs in MD-WF/Zr-MG-II.

Diameter of WF	Position (r_1)	Bending Space	Ultimate Bending Diameter
Ø0.3 mm	0~1.8 mm	5.44 mm~8.5 mm	<0.4 mm
Ø0.5 mm	2.2 mm~2.7 mm	3.91 mm~4.76 mm	<0.6 mm
Ø1.0 mm	3.5 mm~4.5 mm	0.85 mm~2.55 mm	2 mm~4 mm

5. Discussion

The above analysis shows that the relationship of the bend space and ultimate bending diameter of WFs in the MD-WF/Zr-MG rod is an important factor affecting the penetration failure mode of the rod. If the bend spaces of all WFs in a rod are larger than their ultimate bending diameter, the penetration failure mode of the rod is the bending and backflow of WFs. In Figure 10, the test results show that the ultimate bending diameter of Ø0.5 mm WF is less than 0.6 mm, and the ultimate bending diameter of the Ø0.3 mm WF is less than 0.4 mm. According to Equation (2), even Ø0.5 mm WFs and Ø0.3 mm WFs are arranged in the outer layer of the rod, their bending spaces being larger than their ultimate bending diameter, which leads to the penetration failure of the bending and backflow of WFs. Previous experiments have shown that the penetration failure mode of the WF(0.3)/Zr-MG rod is the bending and backflow of WFs when the rod diameter is Ø5.6 mm, as shown in Figure 14.

Figure 14. Longitudinal section of crater penetrated by the Ø5.6 mm WF(0.3)/Zr-MG rod.

However, if the bend spaces of part of WFs in a MD-WF/Zr-MG rod are smaller than their ultimate bending diameter, shear failure of the rod will occur. Therefore, the relationship of the bend space and ultimate bending diameter of the WFs should be calculated by the design of MD-WF/Zr-MG.

It is assumed that the diameter of WFs of the MD-WF/Zr-MG rod is d, the ultimate bending diameter of the outer layer WFs will be:

$$D_{bend} = f(d) \qquad (3)$$

with $f(d)$ as the function related to the properties of WF, it can be measured by bend tests.

It is assumed that the radius of MD-WF/Zr-MG rod is r, the radius of crater penetrated by MD-WF/Zr-MG rod will be:

$$R = g(r) \qquad (4)$$

with $g(r)$ as a function related to the properties of impact velocity, properties of rod materials, and target materials. Usually, the radius of the crater is proportional to the radius of the rod [37], then:

$$R = \beta r \qquad (5)$$

with β as the parameter related to impact velocity, properties of rod materials, and target materials.

Compared with the bend space and the ultimate bending diameter of the WFs,

$$\eta = \frac{\beta(r - r_1)}{f(d)}. \qquad (6)$$

According to bending test results of WFs, the ultimate bending diameter of Ø1.0 mm WF is $D_{bend(1.0)} = 2\text{mm} \sim 4\text{mm}$, the ultimate bending diameter of Ø0.5 mm WF is $D_{bend(0.5)} < 0.6\text{mm}$, and the ultimate bending diameter of Ø0.3 mm is WF $D_{bend(0.3)} < 0.4\text{mm}$. According to Equations (3) to (6) and the arrangement of WFs Figure 2, the η value and volume fraction of WFs in the MD-WF/Zr-MG-I and MD-WF/Zr-MG-II rods during the penetration process is calculated and shown in Table 4. According to the calculation results in Table 4, only 8.05% of WFs in the MD-WF/Zr-MG-I rod do not have enough bending space during penetration, and fracture will occur. However, about 46.14% of WFs will break during MD-WF/Zr-MG-II rod penetration. Therefore, the penetration failure mode of the MD-WF/Zr-MG-I rod is bending and backflow of WFs, and the penetration failure mode of the MD-WF/Zr-MG-II rod is shear failure.

As the MD-WF/Zr-MG rod with shear failure mode exhibits stronger penetration capability than the MD-WF/Zr-MG rod with a failure mode of bending and backflow of WFs, when designing the MD-WF/Zr-MG, calculating the η of WFs during penetration is necessary. The η of as many WFs as possible should be less than 1.

Table 4. Value and the volume fraction of WFs in the MD-WF/Zr-MG-I rod and MD-WF/Zr-MG-II rod.

Materials	Diameter	η	Volume Fraction
MD-WF/Zr-MG-I	0.3	>0.6375	8.05%
		>2.125	7.08%
		>3.6125	6.43%
	0.5	>3.542	15.42%
		>4.958	13.41%
		>6.375	11.40%
	1	1.275~2.55	12.69%
		1.7~3.4	6.35%
		2.125~4.25	1.06%

Table 4. *Cont.*

Materials	Diameter	η	Volume Fraction
MD-WF/Zr-MG-II	0.3	>13.6	4.46%
		>14.875	3.74%
		>16.15	3.02%
		>17.425	2.29%
		>18.7	1.45%
		>19.975	0.60%
		>21.25	0.12%
	0.5	>6.517	10.54%
		>7.933	6.68%
	1	0.213~0.425	26.92%
		0.638~1.275	19.23%

6. Conclusions

In this paper, two kinds of MD-WF/Zr-MG are designed. Both the two MD-WF/Zr-MG contain Ø0.3 mm WFs, Ø0.5 mm WFs, and Ø1.0 mm WFs. In MD-WF/Zr-MG-I, the diameters of WFs increase gradually from inside to outside. In MD-WF/Zr-MG-II, the diameters of WFs decrease gradually from inside to outside. Penetration experiments of two MD-WF/Zr-MG rods are conducted respectively. According to the analysis of the residual rods and craters, the following conclusions are obtained.

1. When multi-diameter WFs are added to WF/Zr-MG, the WF/Zr-MG rod retains the penetration self-sharpening feature and increases penetration stability. The WF/Zr-MG makes use of the advantages and avoids the disadvantages of WFs when WF/Zr-MG is reinforced by multiple diameter WFs. The penetration ability of MD-WF/Zr-MG-II rods is better than that of MD-WF/Zr-MG-I rods.
2. The arrangement of WFs in MD-WF/Zr-MG has an important influence on its penetration failure mode. When the diameters of WFs increase gradually from inside to outside (MD-WF/Zr-MG-I), the penetration failure mode of MD-WF/Zr-MG rods into RHA is the bending and backflow of WFs. When the diameters of WFs decrease gradually from inside to outside (MD-WF/Zr-MG-II), the penetration failure mode of MD-WF/Zr-MG rods into RHA is the shear failure.
3. The penetration failure mode of MD-WF/Zr-MG rods is related to the bend spaces and the ultimate bending diameters of WFs in it. If the bend spaces of all WFs are equal or larger than their ultimate bending diameters, the penetration failure mode of the MD-WF/Zr-MG-I rod is the bending and backflow of WFs, oppositely the penetration failure mode of the MD-WF/Zr-MG-II rod is the shear failure.
4. When designing MD-WF/Zr-MG, the bending space and ultimate bending diameter of WFs should be calculated to ensure that the bending space of most parts of WFs is larger than their ultimate bending diameter.

Author Contributions: Conceptualization, C.D. and H.F.; methodology, Z.Z. and Z.D.; validation, K.W. and G.G.; formal analysis, G.G.; investigation, C.D., F.Z. and Z.D.; resources, H.F., Z.Z. and K.W.; data curation, L.X.; writing—original draft preparation, C.D.; writing—review and editing, G.G. and L.X.; visualization, F.Z.; supervision, G.G.; project administration, G.G. and Z.D.; funding acquisition, C.D. and G.G. All authors have read and agreed to the published version of the manuscript.

Funding: This research was funded by National Natural Science Foundation of China, grant number 12102201, 11802141, and 11772160 and the China Scholarship Council, grant number 201806840024.

Institutional Review Board Statement: Not applicable.

Informed Consent Statement: Not applicable.

Data Availability Statement: Not applicable.

Acknowledgments: We wish to express our gratitude to the members of our research team, Zhaojun Pang, Xi Chen, Shuai Yue, Jiangbo Wang, and Xiaodong Wang.

Conflicts of Interest: The authors declare no conflict of interest.

References

1. Wang, W.H.; Dong, C.; Shek, C.H. Bulk metallic glasses. *Mater. Sci. Eng. R Rep.* **2004**, *44*, 45–89. [CrossRef]
2. Schroers, J. Bulk Metallic Glasses. *Phys. Today* **2013**, *66*, 32–37. [CrossRef]
3. Dandliker, R.B.; Conner, R.D.; Johnson, W.L. Melt infiltration casting of bulk metallic-glass matrix composites. *J. Mater. Res.* **1998**, *13*, 2896–2901. [CrossRef]
4. Choi-Yim, H.; Conner, R.D.; Szuecs, F.; Johnson, W.L. Quasistatic and dynamic deformation of tungsten reinforced $Zr_{57}Nb_5Al_{10}Cu_{15.4}Ni_{12.6}$ bulk metallic glass matrix composites. *Scr. Mater.* **2001**, *45*, 1039–1045. [CrossRef]
5. Conner, R.D.; Dandliker, R.B.; Scruggs, V.; Johnson, W.L. Dynamic deformation behavior of tungsten-fiber/metallic glass matrix composites. *Int. J. Impact Eng.* **2000**, *24*, 435–444. [CrossRef]
6. Zhang, H.F.; Li, H.; Wang, A.M.; Fu, H.M.; Ding, B.Z.; Hu, Z.Q. Synthesis and characteristics of 80 vol.% tungsten (W) fibre/Zr based metallic glass composite. *Intermetallics* **2009**, *17*, 1070–1077. [CrossRef]
7. Qiu, K.Q.; Wang, A.M.; Zhang, H.F.; Ding, B.Z.; Hu, Z.Q. Mechanical properties of tungsten fiber reinforced ZrAlNiCuSi metallic glass matrix composite. *Intermetallics* **2002**, *10*, 1283–1288. [CrossRef]
8. Zhang, B.; Fu, H.; Sha, P.; Zhu, Z.; Dong, C.; Zhang, H.; Hu, Z. Anisotropic compressive deformation behaviors of tungsten fiber reinforced Zr-based metallic glass composites. *Mater. Sci. Eng. A* **2013**, *566*, 16–21. [CrossRef]
9. Zhang, B.; Fu, H.; Li, Z.; Zhu, Z.; Zhang, H.; Hu, Z. Anisotropic tensile properties of tungsten fiber reinforced Zr based metallic glass composites. *Mater. Sci. Eng. A* **2014**, *619*, 165–170. [CrossRef]
10. Zhang, H.; Zhang, Z.F.; Wang, Z.G.; Zhang, H.F.; Zang, Q.S.; Qiu, K.Q. Effects of tungsten fiber on failure mode of zr-based bulk metallic glassy composite. *Metall. Mater. Trans. A* **2006**, *37*, 2459–2469. [CrossRef]
11. Choi-Yim, H.; Schroers, J.; Johnson, W.L. Microstructures and mechanical properties of tungsten wire/particle reinforced $Zr_{57}Nb_5Al_{10}Cu_{15.4}Ni_{12.6}$ metallic glass matrix composites. *Appl. Phys. Lett.* **2002**, *80*, 1906–1908. [CrossRef]
12. Chen, J.H.; Chen, Y.; Jiang, M.Q.; Chen, X.W.; Zhang, H.F.; Dai, L.H. On the compressive failure of tungsten fiber reinforced Zr-based bulk metallic glass composite. *Int. J. Solids Struct.* **2015**, *69*, 428–441. [CrossRef]
13. Ma, W.; Kou, H.; Chen, C.; LI, J.; Hu, R. Interfacial characteristics and dynamic mechanical properties of Wf/Zr-based metallic glass matrix composites. *T. Nonferr. Met. Soc.* **2008**, *18*, 77–81. [CrossRef]
14. Ma, W.; Kou, H.; Chen, C.; Li, J.; Chang, H.; Zhou, L.; Fu, H. Compressive deformation behaviors of tungsten fiber reinforced Zr-based metallic glass composites. *Mater. Sci. Eng. A* **2008**, *486*, 308–312. [CrossRef]
15. Chen, G.; Hao, Y.; Chen, X.; Hao, H. Compressive behaviour of tungsten fibre reinforced Zr-based metallic glass at different strain rates and temperatures. *Int. J. Impact Eng.* **2017**, *106*, 110–119. [CrossRef]
16. Zhu, F.L.; Chen, Y.; Zhu, G.L. Numerical simulation study on penetration performance of depleted Uranium (DU) alloy fragments. *Def. Technol.* **2021**, *17*, 50–55. [CrossRef]
17. Du, C.; Shu, D.; Du, Z.; Gao, G.; Wang, M.; Zhu, Z.; Gao, G.; Wang, M.; Zhu, Z.; Xu, L. Effect of L/D on penetration performance of tungsten fibre/Zr-based bulk metallic glass matrix composite rod. *Int. J. Refract. Met. Hard Mater.* **2019**, *85*, 105042. [CrossRef]
18. Chen, X.W.; Chen, G. Experimental research on the penetration of tungsten-fiber/metallic-glass matrix composite material bullet into steel target. *EPJ Web Conf.* **2012**, *26*, 1049. [CrossRef]
19. Chen, X.W.; Wei, L.M.; Li, J.C. Experimental research on the long rod penetration of tungsten-fiber/Zr-based metallic glass matrix composite into Q235 steel target. *Int. J. Impact Eng.* **2015**, *79*, 102–116. [CrossRef]
20. Rong, G.; Huang, D.W.; Yang, M.C. Penetrating behaviors of Zr-based metallic glass composite rods reinforced by tungsten fibers. *Theor. Appl. Fract Mec.* **2012**, *58*, 21–27. [CrossRef]
21. Guo, W.; Jiang, H.; Wang, S.; Wan, M.; Fang, X.; Tian, S. Self-sharpening ability enhanced by torque gradient in twisted tungsten-fiber-reinforced Cu-Zn matrix composite. *J. Alloys Compd.* **2019**, *794*, 396–401. [CrossRef]
22. Lee, K.; Lee, S.; Lee, S.; Lee, S. Compressive and Tensile Properties of Tungsten-Continuous-Fiber-Reinforced Zr-Based Amorphous Alloy Matrix Composite Fabricated by Liquid Pressing Process. *Metall. Mater. Trans. A* **2008**, *39*, 1319–1326. [CrossRef]
23. Lee, S.; Lee, S.; Lee, S.; Kim, N.J. Microstructure and Mechanical Properties of Two Continuous-Fiber-Reinforced Zr-Based Amorphous Alloy Composites Fabricated by Liquid Pressing Process. *Metall. Mater. Trans. A* **2008**, *39*, 763–771. [CrossRef]
24. Li, J.C.; Chen, X.W.; Huang, F.L. FEM analysis on the deformation and failure of fiber reinforced metallic glass matrix composite. *Mater. Sci. Eng. A* **2016**, *652*, 145–166. [CrossRef]
25. Li, J.C.; Chen, X.W.; Huang, F.L. FEM analysis on the "self-sharpening" behavior of tungsten fiber/metallic glass matrix composite long rod. *Int. J. Impact Eng.* **2015**, *86*, 67–83. [CrossRef]
26. Xue, Y.; Zhong, X.; Wang, L.; Fan, Q.; Zhu, L.; Fan, B.; Zhang, H.; Fu, H. Effect of W volume fraction on dynamic mechanical behaviors of W fiber/Zr-based bulk metallic glass composites. *Mater. Sci. Eng. A* **2015**, *639*, 417–424. [CrossRef]
27. Chen, S.; Zhang, L.; Fu, H.M.; Li, Z.K.; Zhu, Z.W.; Li, H.; Zhang, H.W.; Wang, A.M.; Wang, Y.D.; Zhang, H.F. Compressive mechanical properties and failure modes of Zr-based bulk metallic glass composites containing tungsten springs. *Mater. Des.* **2018**, *160*, 652–660. [CrossRef]

28. Chen, S.; Li, W.Q.; Zhang, L.; Fu, H.M.; Li, Z.K.; Zhu, Z.W.; Li, H.; Zhang, H.W.; Wang, A.M.; Wang, Y.D.; et al. Dynamic compressive mechanical properties of the spiral tungsten wire reinforced Zr-based bulk metallic glass composites. *Compos. Part B Eng.* **2020**, *199*, 108219. [CrossRef]
29. Zhang, Z.; Kong, J.; Liu, X.; Song, X.; Dong, K. Preparation, microstructure and mechanical properties of tungsten fiber reinforced LaAlCuNi metallic glass matrix composites. *Intermetallics* **2021**, *132*, 107139. [CrossRef]
30. Zhang, B.; Fu, H.M.; Zhu, Z.W.; Zhang, H.F.; Dong, C.; Hu, Z.Q. Effect of W fiber diameter on the compressive mechanical properties of the Zr-based metallic glass composites. *Acta Metall. Sin.* **2013**, *49*, 1191–1200. [CrossRef]
31. Du, C.X.; Du, Z.H.; Zhu, Z.W. Effect of impact velocity and diameter of tungsten fiber on penetration ability of Wf/Zr-based metallic glass composite penetrator. *Rare Met. Mater. Eng.* **2017**, *46*, 1632–1637.
32. Magness, L.S. High strain rate deformation behaviors of kinetic energy penetrator materials during ballistic impact. *Mech. Mater.* **1994**, *17*, 147–154. [CrossRef]
33. Terentyev, D.; Van Renterghem, W.; Tanure, L.; Dubinko, A.; Riesch, J.; Lebediev, S.; Khvan, T.; Verbeken, K.; Coenen, J.W.; Zhurkin, E.E. Correlation of microstructural and mechanical properties of K-doped tungsten fibers used as reinforcement of tungsten matrix for high temperature applications. *Int. J. Refract. Met. Hard Mater.* **2019**, *79*, 204–216. [CrossRef]
34. Terentyev, D.; Riesch, J.; Lebediev, S.; Bakaeva, A.; Coenen, J.W. Mechanical properties of as-fabricated and 2300 °C annealed tungsten wire tested up to 600 °C. *Int. J. Refract. Met. Hard Mater.* **2017**, *66*, 127–134. [CrossRef]
35. Terentyev, D.; Riesch, J.; Lebediev, S.; Khvan, T.; Dubinko, A.; Bakaeva, A. Strength and deformation mechanism of tungsten wires exposed to high temperature annealing: Impact of potassium doping. *Int. J. Refract. Met. Hard Mater.* **2018**, *76*, 226–233. [CrossRef]
36. Wilbeck, J.S.; Charles, E.; Anderson, J.; John, P.; Riegel, I.; Lankford, J.; Mullin, S.A.; Bodner, S.R. *A Short Course on Penetration Mechanics*; Southwest Research Institute: San Antonio, TX, USA, 1986.
37. Rosenberg, Z.; Kositski, R. The hole diameter in metallic plates impacted by hypervelocity projectiles. *Int. J. Impact Eng.* **2017**, *102*, 147–155. [CrossRef]

Article

Study on Anti-Penetration Performance of Semi-Cylindrical Ceramic Composite Armor against 12.7 mm API Projectile

Anbang Jiang, Yongqing Li, Dian Li and Hailiang Hou *

Department of Naval Architecture Engineering, Naval University of Engineering, Wuhan 430000, China
* Correspondence: hou9611104@163.com

Abstract: To explore the anti-penetration performance of the specially shaped ceramic/metal composite armor, such an armor is designed and fabricated using a semi-cylindrical projectile resistant ceramic and metal back plate, and its anti-penetration performance for the 12.7 mm armor-piercing incendiary (API) projectile (also known as the 0.50 caliber API projectile) is investigated experimentally and numerically. The results show that due to the significant attitude deflection during projectile penetration, the penetration into the designed ceramic composite armor is quite different from that into the conventional homogeneous ceramic/metal composite armor, which can be roughly divided into the following four stages: asymmetric erosion of the projectile, ceramic cone squeezing movement, back plate failure and projectile exit. The failure mode of the back plate is mainly dishing deformation and petaling failure. When obvious attitude deflection occurs to the projectile, the breaches in the back plate are elliptical in varying degrees, and the height and size of petals are apparently different. The area of the composite armor is divided into different zones according to its anti-penetration performance. The influence of the ratio of semi-cylindrical ceramic diameter to projectile core diameter ζ on the anti-penetration performance is studied under constant areal density. The results show that the deflection effect of the composite armor is small when the ratio ζ is less than 2, and the anti-penetration performance is the strongest when ζ is close to 2. With the increase in the initial velocity of the projectile, the deflection effect of the composite armor on the projectile gradually weakened, and the erosion effect gradually increased.

Keywords: ceramic composite armor; 12.7 mm API projectile; ballistic performance; deflection and yaw

Citation: Jiang, A.; Li, Y.; Li, D.; Hou, H. Study on Anti-Penetration Performance of Semi-Cylindrical Ceramic Composite Armor against 12.7 mm API Projectile. *Crystals* **2022**, *12*, 1343. https://doi.org/10.3390/cryst12101343

Academic Editors: Tomasz Sadowski and Vladislav V. Kharton

Received: 21 August 2022
Accepted: 17 September 2022
Published: 22 September 2022

Publisher's Note: MDPI stays neutral with regard to jurisdictional claims in published maps and institutional affiliations.

Copyright: © 2022 by the authors. Licensee MDPI, Basel, Switzerland. This article is an open access article distributed under the terms and conditions of the Creative Commons Attribution (CC BY) license (https://creativecommons.org/licenses/by/4.0/).

1. Introduction

A lightweight and high-performance armor protection structure has always been an important goal of research on modern protection technology, as its lightweight and high ballistic performance can help ensure the mobility of military weapons, while reaching the protection goal [1,2]. Ceramic materials have lower densities than metal materials, as well as high hardness and compressive strength and are widely used in lightweight and high-performance armor [3–5]. On the other hand, because of the brittleness and low tensile strength of ceramics, they cannot absorb a large amount of energy in the penetration process. So, in practical applications, ceramic is usually used as the front plate and metal or fiber-reinforced composite as the back plate, so that ceramic/metal or ceramic/composite material armor can be developed with improved ballistic performance [6]. In the process of projectile penetrating ceramic composite armor, the ceramic material first blunts and erodes the projectile to dissipate energy, and then the composite back plate absorbs the remaining kinetic energy to prevent projectile penetration [7,8]. Therefore, ceramic composite armor has a wide range of applications because of its high protection coefficient and good ballistic performance.

The conventional ceramic composite armor has good anti-penetration performance, but as projectiles have increasingly stronger penetration capabilities, to achieve the protection goal, the armor needs to have a larger mass and size. In practice, the ballistic impact on the armor is not perpendicular to its surface [9]. Therefore, researchers have carried out extensive research on the oblique penetration into the ceramic composite armor. It is found that when the projectile penetrates the ceramic composite armor with a non-ideal attitude, the projectile axis deviates from its initial impact angle under the action of the deflection moment because the axis of the projectile does not coincide with the normal line of the ceramic plate [10]. If the penetration angle keeps increasing, projectile ricochet may happen. In addition, the increase in penetration angle also leads to the increase in the effective target thickness during projectile penetration, and projectile deflection and yaw occur, making the penetration path of the projectile longer. The longer the interaction between the projectile and the ceramic armor lasts, the greater the erosion of the projectile [11] and the break-up degree of the ceramic [12], resulting in more kinetic energy dissipation [13]. In the study on the oblique penetration of the long-rod projectile into the ceramic composite armor, it has been also found that oblique penetration makes the dwell time of the long-rod projectile on the interface longer compared with vertical penetration [14]. These factors finally lead to a faster decline in projectile velocity [15], a reduction in penetration depth [16], an increase in the ballistic limit of the armor, and a significant improvement in the anti-penetration performance under the condition of oblique penetration into the ceramic composite armor.

In practical applications, such as personal protective equipment and other usage scenarios with space limitations, when the conditions do not allow the ceramic composite armor to be placed obliquely, by designing specially shaped ceramics to cover the armor surface [17] or be set inside the armor [18], the nominal normal penetration can be transformed into the actual local oblique penetration, penetration with an angle of attack or a combination of the two.

In terms of realizing the ceramic slope, S. Stanislawek [19] designed a pyramidal ceramic armor. The results showed that when the pyramids had large dimensions compared with the projectile and the impact point of the projectile was situated on the pyramid side wall, projectile turning could be especially observed, but the energy dissipation capacity of the composite armor was weaker than that of the homogeneous ceramic composite armor with equal mass. In addition, there was an obvious weak point where the pyramids adjoined, and the pyramid top was the tip of the inclined section that should be avoided, as pointed out in Ref. [15]. Based on Cohen. M's patent [20], W. L. Liu [21] carried out a study on a specially shaped ceramic composite armor with a hemispherical top and a cylindrical lower part. Compared with planar ceramics, closely arranged ceramic cylinders can transfer energy to the surrounding ceramic cylinders, and can also greatly dissipate the projectile's energy through break-up. However, it is obvious that with such a design, gaps exist between the adjacent ceramic cylinders, which may lead to anti-penetration performance reduction. J. M. Chen [22] improved the above-mentioned specially shaped ceramics by cutting the congruent circles of the cylinder into regular hexagons, so that there can be no gap between each ceramic unit. However, this also brings more seams to the whole composite armor, and the decreased ballistic performance at the seams leads to more weak areas in the improved protective structure. The positive side of the above structure is that, due to the included angle between the projectile and the ceramic surface, the projectile is subject to the longitudinal and lateral resistances that unevenly vary in size and direction. This leads to ballistic deviation, serious deformation and damage to the projectile, thus improving the ballistic performance of the ceramic plate. M. Aydin [23] and R.Z. Shao [24] proposed a protective structure using ceramic balls, which provide a large number of slopes. Unlike the fixed-shaped ceramics, the reason for the deflection, yaw and fracture of the projectile against this type of ceramic armor is the rotation, movement and brittle fracture of the ceramic balls under impact loading. In addition, under this concept, the ceramic balls that break after each impact will soon be replaced by the adjacent new balls. Therefore, the armor can continuously resist multiple rounds of impact loads, without removing the damaged ceramic balls. The authors called this armor design "self-healing armor".

Based on the advantages and disadvantages of the above-mentioned specially shaped ceramics, this paper proposes a ceramic composite armor, which is mainly divided into the following two parts: specially shaped ceramic and metal back plate. Through ballistic impact experiments combined with FEM (finite element method) and SPH (smooth particle hydrodynamics) numerical calculations, the anti-penetration performance of the composite armor structure against the 12.7 mm API projectile is mainly studied; at the same time, the penetration and deflection processes of the projectile, the deformation and failure mode of the metal back plate are analyzed.

2. Experimental and Numerical Methods

2.1. Experimental Design

2.1.1. Projectile and Target Plate

In the experiment, a 12.7 mm API projectile with a steel core coated with a copper sheath was used, with a diameter of 10.8 mm, a core length of 52 mm, and a weight of 30 g. The total weight of the core and the copper shell is about 48 g. The material of the core is T12A steel [25,26]. The target plate is divided into three parts, which are the specially shaped ceramic, the back plate, and the steel frame around the ceramic (see Figure 1). In the experiment, the diameter of the semi-cylindrical body selected in the experiment is 24 mm to ensure that the specially shaped ceramics play a deflection role, and, at the same time, to avoid obvious weak areas in the structure due to the non-uniformity of the ceramics, a 6 mm rectangular body of ceramics is set under the semi-cylindrical ceramic column to form specially shaped ceramic composed of semi-cylindrical and rectangular plates. Two materials of boron carbide and silicon carbide were selected for the ceramics. The back plate is 4 mm thick 12MnCrNi steel. The ceramic is surrounded by steel frames of equal height to constrain its displacement. Among them, the forming process of boron carbide and silicon carbide and the main mechanical properties of boron carbide, silicon carbide, and 12MnCrNi steel can be found in the literature [27].

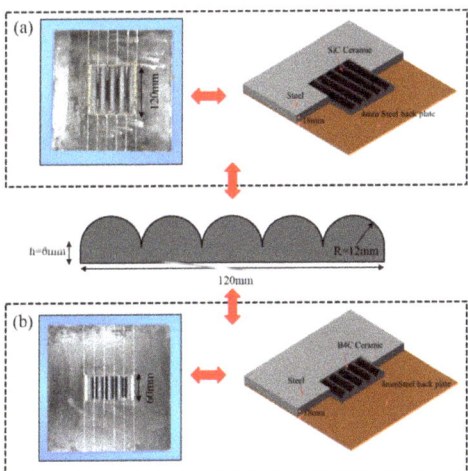

Figure 1. Schematic diagram of composite armor, (**a**) SiC Ceramic composite armor, (**b**) B$_4$C Ceramic composite armor.

2.1.2. Experimental Setup and Brief Results

Ballistic impact experiments were carried out to analyze the ballistic performance of the composite armor. The experimental arrangement is shown in Figure 2, which is mainly composed of a ballistic gun, laser velocity measurement system, support and target network velocity measurement system. In the experiment, the 12.7 mm API projectile was launched by the 12.7 mm ballistic gun, and the amount of gunpowder fired determined the projectile's velocity. We used the laser velocity measurement system to measure the initial velocity of the projectile, and used the target net velocity measurement system to measure the residual velocity of the projectile after passing through the target plate. A high-speed camera was set up at a distance of about 1 m from the right side of the target plate to observe the impact point and attitude of the projectile before hitting the target plate. The height was about the same as the trajectory, and the shooting rate was 62,000 frames /s. At the same time, the grid paper is arranged on the left side of the flight trajectory of the projectile. Perpendicular to the target plate, the grid drawn on the surface is 20 mm square. It is convenient to judge the position and flight attitude of the projectile body. The coordinate system in the experiment and numerical calculation is shown in Figure 2, where the origin of the coordinate system is the center point of the ceramic bottom surface.

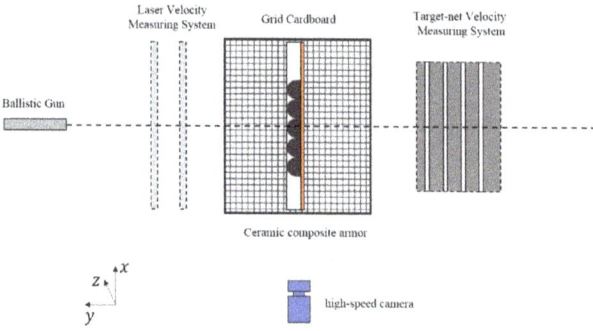

Figure 2. Experiment setup.

Four experimental tests were conducted. Figure 3 presents several high-speed photographs of the projectile before impacting the target under different working conditions. The numbering rules in the figure are as follows: the photo of the first frame when the projectile completely comes into view is photo No. 1, and then the photo of each subsequent frame is numbered in sequence. As can be observed from the figure, in Experiment 1, the projectile has an initial oblique angle of about 3.3°; in Experiment 2, the projectile has an initial angle of attack of about 4.8°; in Experiment 3 and 4, the projectile has an initial oblique angle of about 1.5°. As shown in Figure 4, the impact point of this paper is defined as follows: the intersection of the extension line of the velocity direction of the center of mass of the projectile and the surface of the composite armor. In the high-speed camera image, it can be observed that several white lines are drawn on the target plate parallel to the edge of the armor, and their positions are located at the joint of the two cylinders. According to the high-speed camera results, the impact points of the projectiles in the four experiments can be determined. The brief experimental results are shown in Table 1.

2.2. Numerical Model and Effectiveness Verification

2.2.1. Numerical Model

An adaptive coupling algorithm based on FEM and SPH in LS-DYNA [28] was adopted as the numerical model in this study. ERODING_SURFACE_TO_SURFACE contact is defined between every two components. The contact algorithm AUTOMATIC_SURFACE_TO_SURFACE_TIEBREA is employed to simulate the adhesion between the semi-cylindrical ceramic and the steel back plate, where NFLS and SFLS parameters are set to 20 MPa according to the adhesive strength provided by the manufacturer. Clamping constraints are applied to the steel back plate and the side of the rectangular plate in the specially shaped ceramic.

As shown in Figure 5, all components are modeled by a hexahedral solid element. During projectile penetration into the composite armor, the penetration effect is small due to the rapid separation of the jacket from the projectile core, so the core produces the main penetration effect [29]. Therefore, only the penetration effect of the core on the composite armor is considered in the numerical calculation [30], with a mesh size of about 0.735 mm. The mesh size of the semi-cylindrical ceramic is 0.75 mm in the radial and length directions, and the other parts continue to match it. The mesh size of the back plate and the rectangular plate of the ceramic is 0.75 mm.

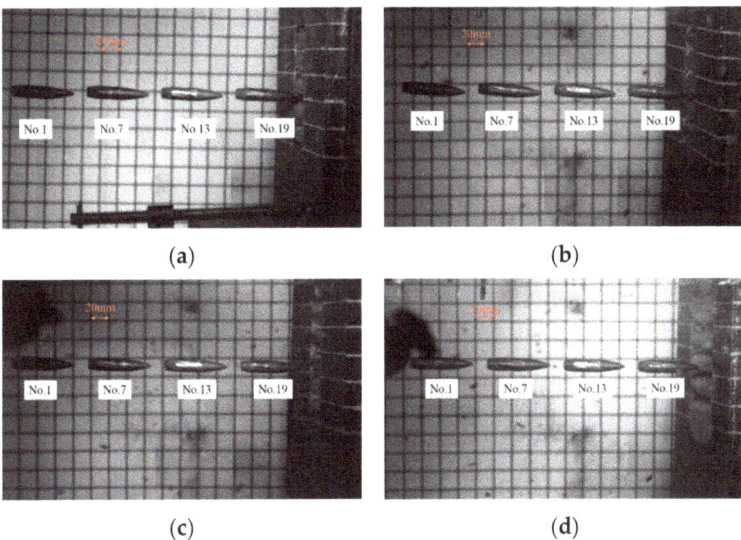

Figure 3. High-speed photographs of impact points, (**a**) Experiment 1, (**b**) Experiment 2, (**c**) Experiment 3, (**d**) Experiment 4.

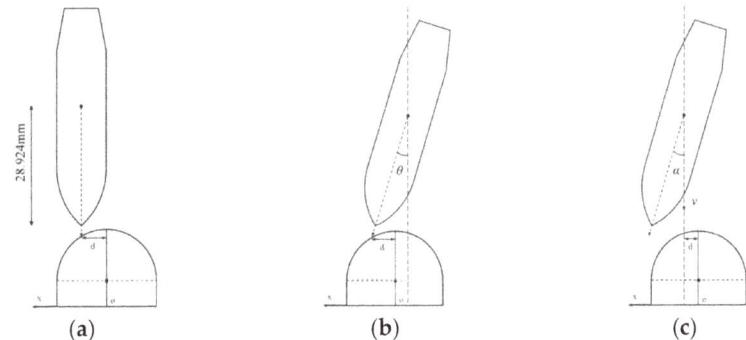

Figure 4. Schematic diagram of impact points in the case of normal penetration, oblique angle, and angle of attack. (**a**) Impact point in the case of normal penetration; (**b**) impact point in the case of oblique angle θ; (**c**) impact point in the case of angle of attack α.

Table 1. Brief experimental results.

Experiment No.	Target Plate Type	d (mm)	Oblique Angle θ (°)	Angle of Attack α (°)	Initial Velocity v_0 (m·s^{-1})	Residual Velocity v_r (m·s^{-1})
Experiment 1	B$_4$C/steel	11.0	3.3	0	802.8	/
Experiment 2	B$_4$C/steel	8.08	0	4.8	801.1	195.3
Experiment 3	B$_4$C/steel	8.30	1.5	0	812.5	342.5
Experiment 4	SiC/steel	4.00	1.5	0	810.2	0

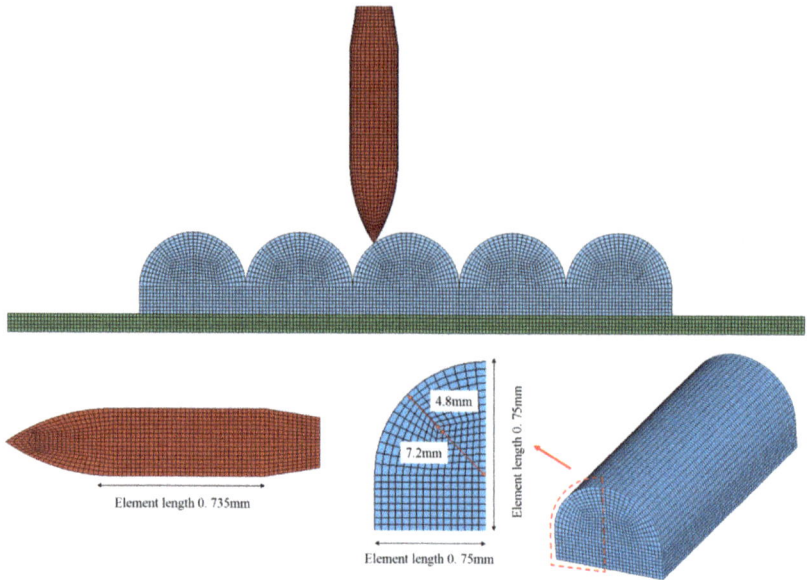

Figure 5. Finite element model.

The constitutive model of B$_4$C, SiC ceramics using the JH-2 model, metal using the Cowper–Symonds model, the introduction of two constitutive models and the detailed material parameters of B$_4$C, SiC ceramics and 12MnCrNi can be observed in the literature [27]. According to Refs. [31–33], among the material parameters of T12A steel, those related to the strain rate effect is set to 0, that is, the strain rate effect is not considered. So, the strain rate effect of T12A steel is not considered either in the numerical model in this paper, and the strain rate parameters SRC and SRP in LS-DYNA are set to 0 as well. The detailed material parameters are shown in Table 2.

Table 2. Cowper–Symonds parameters of T12A steel, adapted from Refs. [31,32].

Parameter	T12A Steel
Density, ρ (g/cm3)	7.85
Young's modulus, E (GPa)	210
Yield stress, SIGY (MPa)	1540
Tangent modulus, ETAN (MPa)	2740
Hardening parameter, BETA	1
Strain rate constant, SRC (s^{-1})	0
Strain rate exponent, SRP	0
Failure strain, F_s	0.18

2.2.2. Verification of Calculation Results

The numerical calculation is performed according to the initial projectile velocities and impact points captured by the high-speed camera in the experiments, and the residual velocities of the projectile calculated by the numerical calculation are compared with the value measured by the experiment. As shown in Table 3, in the three working conditions with residual velocity records, the maximum deviation of the residual projectile velocity is 11.11%, showing good agreement between the two methods.

Table 3. Comparison of experimental and numerically calculated residual velocities.

Experiment No.	Target Plate Type	Experimental Initial Velocity v_0(m·s^{-1})	Residual Velocity v_r(m·s^{-1})		Deviation
			Experiment	Numerical Calculation	
Experiment 1	B$_4$C/steel	802.8	/	297	/
Experiment 2	B$_4$C/steel	801.1	195.3	217	11.11%
Experiment 3	B$_4$C/steel	812.5	342.5	330	−0.03%
Experiment 4	SiC/steel	810.2	0	0	0

The damage morphology of the steel back plate after the experiments is compared with that in the numerical calculation, and the experimental and numerical results are basically consistent. As shown in Figure 6, in Experiment 1, at T = 140 μs, four approximately vertical cracks appear in the back plate in the numerical calculation. In the final numerical and experimental results, the petaling part of the back plate can be divided into four regions, and each region is separated by cracks. As shown in Figure 7, in Experiment 3, elliptical breaches occur in both the experiment and numerical calculation; two shorter cracks appear near the short axis of the elliptical breaches, and meanwhile, in the direction of the long axis of the elliptical breaches, an outwardly rolled long petal appears on one side and a shorter petal on the other side. As shown in Figure 8, in Experiment 4, the back plate is not damaged in the experiment and numerical calculation, only the bulging-dishing deformation occurs, which has the following features: a relatively prominent bulge has formed at the vertex of the deformation, and the profile of the dishing deformation regions on both sides of the bulge is approximately straight.

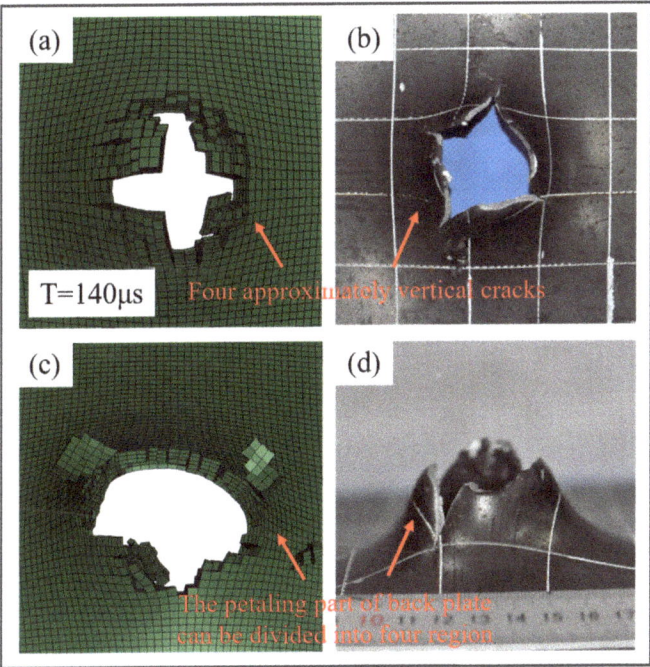

Figure 6. Comparison of back plate damage morphology of in Experiment 1.

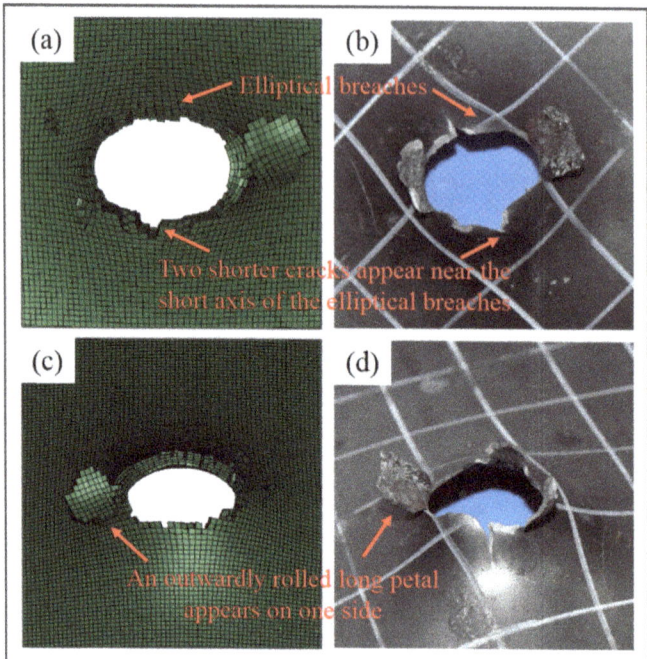

Figure 7. Comparison of back plate damage morphology in Experiment 3.

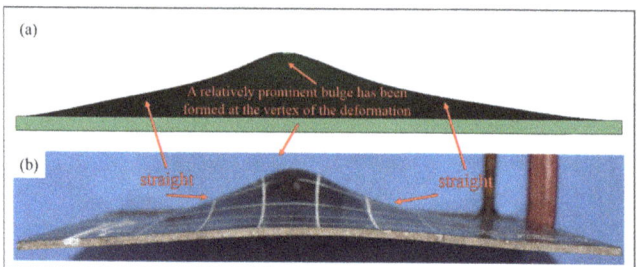

Figure 8. Comparison of back plate damage morphology in Experiment 4.

2.3. Working Conditions

To study the anti-penetration performance of the semi-cylindrical ceramic composite armor, this article adopts the method of combining experiment and numerical calculations to carry out research on 34 working conditions (Table 4). Specifically, working conditions No. 1–4 are the comparison and verification groups for the experimental and numerical methods; in working conditions, No. 5–17, only the impact point changes and the other conditions are kept unchanged. With a spacing of 1 mm, 13 impact points are selected successively from the top of the semi-cylinder to the seam (Figure 9). The impact point deviation degree η is used to represent the position of the impact point. At the top of the semi-cylinder, the impact point deviation degree is defined as $\eta = 0$, at the seam of the semi-cylinder, it is defined as $\eta = 1$, and at other positions, $\eta = d/R$, where d is the distance between the impact point and a point on the top of the nearest semi-cylinder, and R is the radius of the semi-cylindrical ceramic. In working conditions No. 18–29, only the semi-cylinder diameter and the impact point deviation degree η vary and the other conditions remain unchanged. To make the results comparable, the bottom surface area of the four semi-cylindrical ceramics with different diameters is 120 × 60 mm², and the volume is about 1.11 × 10⁵ mm³, i.e., the average areal density in this area is kept the same, and its geometric dimensions are given in Figure 10. The ratio of semi-cylindrical ceramic diameter to core diameter is represented by ζ, whose values are 0.56, 1.11, 2.22 and 2.78 for the four composite armors. In working condition No. 30, a homogeneous ceramic composite armor with equal mass is used, the areal density of the ceramic and the metal back plate remain unchanged, and a planar structure is adopted, which is for comparative analysis. In working conditions No. 31–34, only the initial velocity of the projectile varies and the other conditions remain unchanged.

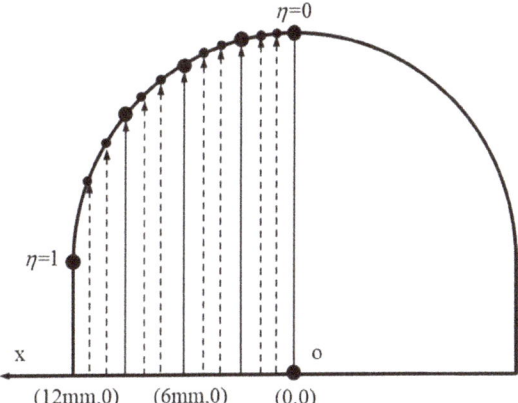

Figure 9. Impact point distribution.

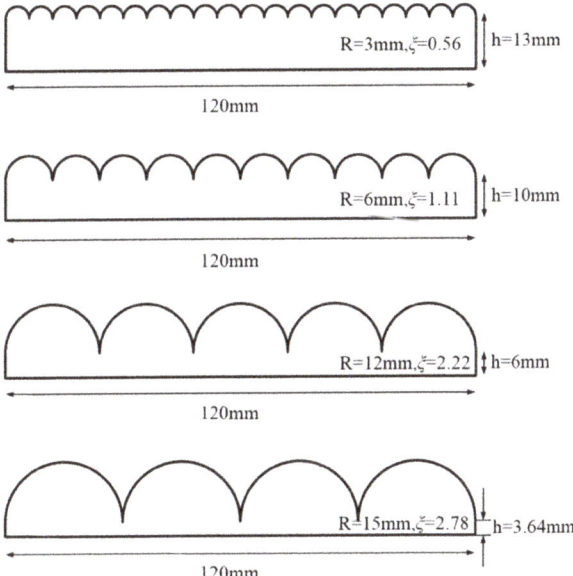

Figure 10. Schematic diagram of the cross-section of composite armors with different ratios of semi-cylindrical ceramic diameter to core diameter under equal areal density.

Table 4. Working conditions for calculation.

Condition No.	Ceramic Material	Initial Velocity of Projectile	Oblique Angle θ (°)	Angle of Attack α (°)	Ratio of Semi-Cylindrical Ceramic Diameter to Core Diameter ζ	Impact Point Deviation Degree η	Research Method
No. 1	B_4C	802.8	3.3	0	2.22	0.917	FEM-SPH/Experiment
No. 2	B_4C	801.1	0	4.8	2.22	0.673	FEM-SPH/Experiment
No. 3	B_4C	812.5	1.5	0	2.22	0.692	FEM-SPH/Experiment
No. 4	SiC	810.2	1.5	0	2.22	0.333	FEM-SPH/Experiment
No. 5	B_4C	810	0	0	2.22	0	FEM-SPH
No. 6	B_4C	810	0	0	2.22	0.083	FEM-SPH
No. 7	B_4C	810	0	0	2.22	0.167	FEM-SPH
No. 8	B_4C	810	0	0	2.22	0.25	FEM-SPH
No. 9	B_4C	810	0	0	2.22	0.333	FEM-SPH
No. 10	B_4C	810	0	0	2.22	0.417	FEM-SPH
No. 11	B_4C	810	0	0	2.22	0.5	FEM-SPH
No. 12	B_4C	810	0	0	2.22	0.583	FEM-SPH
No. 13	B_4C	810	0	0	2.22	0.667	FEM-SPH
No.14	B_4C	810	0	0	2.22	0.75	FEM-SPH
No.15	B_4C	810	0	0	2.22	0.833	FEM-SPH
No.16	B_4C	810	0	0	2.22	0.917	FEM-SPH
No.17	B_4C	810	0	0	2.22	1	FEM-SPH
No.18	B_4C	810	0	0	0.56	0	FEM-SPH
No.19	B_4C	810	0	0	0.56	0.5	FEM-SPH
No.20	B_4C	810	0	0	0.56	1	FEM-SPH
No.21	B_4C	810	0	0	1.11	0	FEM-SPH
No.22	B_4C	810	0	0	1.11	0.5	FEM-SPH
No.23	B_4C	810	0	0	1.11	1	FEM-SPH
No.24	B_4C	810	0	0	2.22	0	FEM-SPH
No.25	B_4C	810	0	0	2.22	0.5	FEM-SPH
No.26	B_4C	810	0	0	2.22	1	FEM-SPH
No.27	B_4C	810	0	0	2.78	0	FEM-SPH
No.28	B_4C	810	0	0	2.78	0.5	FEM-SPH
No.29	B_4C	810	0	0	2.78	1	FEM-SPH
No.30	B_4C	810	0	0	/	/	FEM-SPH
No.31	B_4C	400	0	0	2.22	0.5	FEM-SPH
No.32	B_4C	700	0	0	2.22	0.5	FEM-SPH
No.33	B_4C	1000	0	0	2.22	0.5	FEM-SPH
No.34	B_4C	1300	0	0	2.22	0.5	FEM-SPH

3. Results and Analysis

3.1. Analysis of Penetration Process

In order to better reveal the penetration process of the 12.7 mm API projectile into the ceramic composite armor, the penetration process of the projectile body under working condition 11 (η = 0.5) is analyzed. Figure 11 shows the velocity–time–history curve of the projectile. For the convenience of analysis, the lateral and vertical acceleration time–history curves of the projectile and three typical process pictures are also added in the velocity–time–history curve. According to Figure 11, the penetration process can be divided into the following four stages: asymmetric erosion of the projectile, ceramic cone squeezing movement, back plate failure and projectile exit.

In the asymmetric erosion stage, when the projectile hits the ceramic, the tip part of the projectile head rapidly breaks, and the ogive nose head characteristics of the projectile body gradually disappear. At the same time, due to the semi-cylindrical surface of the ceramic, the erosion failure of the projectile is asymmetric, the length of the projectile is greatly eroded, and the velocity of the projectile decreases greatly. The projectile penetrates further into the ceramic, and cracks and fragments appear in the ceramic and form ceramic cones. During the formation of ceramic cones, the vertical force of the projectile core increases sharply. At this stage, the back plate has not been squeezed by the ceramic cone, and the back plate almost has no deformation. The duration of this stage is denoted as t_1.

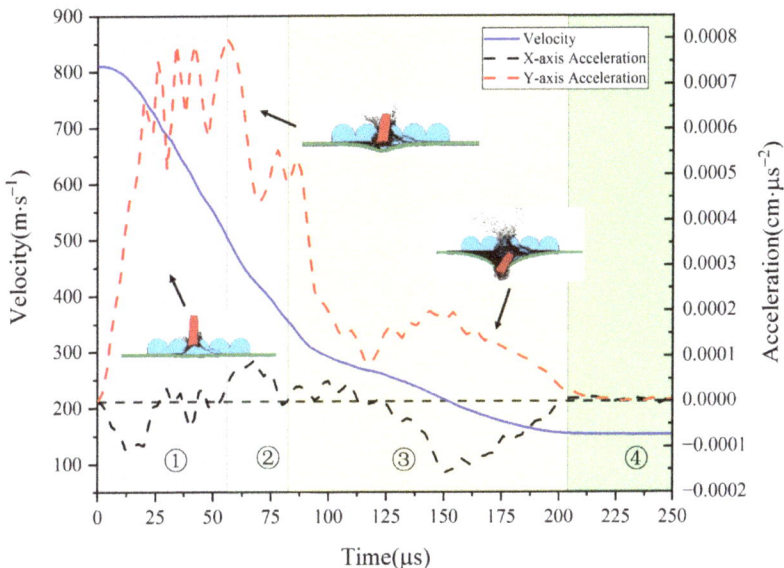

Figure 11. Projectile velocity and acceleration vs time curves ($\eta = 0.5$).

In the phase of the ceramic cone squeezing movement, the projectile penetrates further into the ceramic interior, and the asymmetric erosion failure degree of the warhead decreases. When the projectile extrudes the ceramic cone, the particles in the cone are accelerated and crushed continuously, and the diameter and thickness of the bottom of the ceramic cone become smaller and smaller. The vertical force of the projectile core starts to decrease. The projectile compresses the ceramic cone and acts on the back plate, which starts to produce local bulging deformation. When the velocity of the remaining ceramic cone reaches the velocity of the projectile, the remaining ceramic crushing cone moves with the projectile at the same speed, and the ceramic cone stops being abrasive on the projectile, and the vertical force of the projectile core continues to decrease. The plastic deformation of the back plate expands, and when the deformation of the plastic bulge area reaches a certain degree, the back plate begins to undergo tensile failure. The duration of this stage is denoted as t_2.

In the failure stage of the back plate, the projectile penetrates further. Due to the failure of the back plate, the vertical force of the projectile decreases greatly, and the velocity decrease rate decreases greatly. The left and right sides of the projectile head touch the back plate successively. At this time, the projectile has a large deflection, and the contact area between the right side of the projectile and the back plate is large, so that the velocity of the projectile decreases and the "secondary deceleration" begins. The duration of this phase is denoted as t_3.

In the phase of projectile exit, after the projectile exits out of the back plate, the attitude of the projectile is greatly deflected compared with the incident, and the velocity direction of the projectile is also changed.

When the deviation degree of the impact point changes, the penetration process will be partially different. When the deviation of the impact point is close to 0 or 1, the degree of asymmetric erosion in the first stage of the penetration process will be significantly reduced. As a result, the attitude deflection of the projectile is small during the penetration process, and the secondary deceleration process is not obvious during the penetration process of the projectile. The whole penetration process is similar to that of homogeneous ceramic composite armor.

The projectile deflection process in this study is similar to that reported in the literature [27], but in the process of projectile penetration, due to the ogive nose of the projectile head, the contact area between the projectile head and the ceramic is large at the initial moment when the projectile hits the ceramic. This allows the projectile to deflect slightly more than the flat-ended rod projectile in the first stage. In the next penetration stage, the rapid erosion failure oval head did not play a significant role. In addition, because the 12.7 mm API projectile has a long body and a relatively low velocity, the contact area between the projectile and the back plate was large for a long time in the failure stage of the back plate. As a result, the negative angular acceleration of the projectile at this stage was large, the angular velocity of the projectile decreased to a negative value, the deflection angle of the projectile decreased to 0, and then the inverse direction increased.

3.2. Backplate Failure Mode Analysis

As the 12.7 mm API projectile penetrates the ceramic part, the geometric characteristics of the ogive nose of the projectile have been completely destroyed, and the damaged nose becomes similar to that of the flat-nosed projectile. However, during the projectile penetration into the semi-cylindrical ceramic composite armor, the ceramic in front of the projectile is broken and a ceramic cone is formed; the projectile squeezes the ceramic cone to move forward, causing bulging deformation in a large area of the back plate around the projectile nose. Such an impact response is similar to that of an ogive-nosed projectile perforating a thin plate armor at a low velocity. During the penetration, at the impact center, the tensile deformation of the back plate increases sharply as it is squeezed by the projectile and the broken ceramic cone, so necking occurs, and cracks are generated and gradually expand to form petals. The projectile and the broken ceramic cone continue to press forward, and the petals bend and turn outward, forming the dishing deformation-petaling failure (Figure 12). If the projectile does not perforate the back plate, the back plate displays asymmetric bulging-dishing deformation (Figure 13).

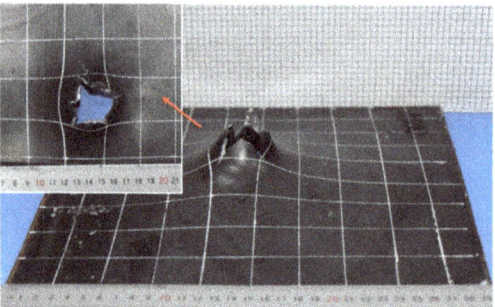

Figure 12. Dishing deformation-petaling failure.

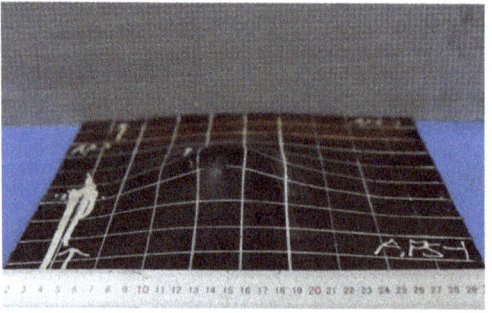

Figure 13. Bulging-dishing deformation.

In the working conditions where the projectile attitude and trajectory apparently change during penetration, the forces on the broken ceramic cone and the deformation area of the back plate bulge are obviously asymmetric. Meanwhile, due to the large projectile deflection, the breach shape is elliptical. The long axis direction of the ellipse is consistent with the axis direction of the projectile, and petals appear on both sides of the elliptical breaches (Figure 14). Taking working condition 11 as an example, the failure mode is analyzed as follows. When the projectile perforates and exits the back plate, the projectile first hits one side of the back plate obliquely, and further moves in this direction, resulting in a strip crack in the back plate in the direction parallel to the Z-axis (Figure 15 (T = 75 μs)), and the crack keeps expanding at both ends. When the crack expands to a certain degree, the projectile squeezes the back plate on the side of the crack (Figure 15 (T = 88 μs)), making the petals on one side of the back plate higher, whose width is roughly equal to that of the projectile, but less than the length of the original strip crack. Then, two smaller cracks appear near the short axis of the ellipse (Figure 15 (T = 114 μs)). When one side of the projectile tail hits the other side of the back plate, the petals on the other side also become higher. However, on this side, the projectile moves in the direction of gradually moving away from the petals; both the height and size of the petals on the projectile tail side are smaller than those on the nose side (Figure 15 (T = 194 μs)).

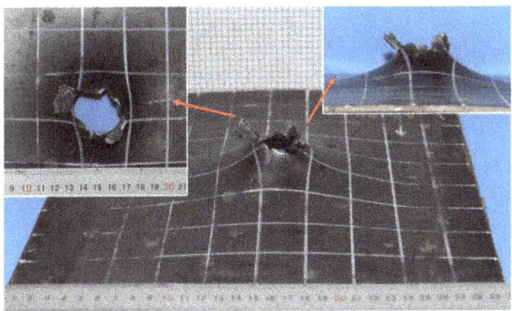

Figure 14. Fracture morphology of back plate in Experiment 3.

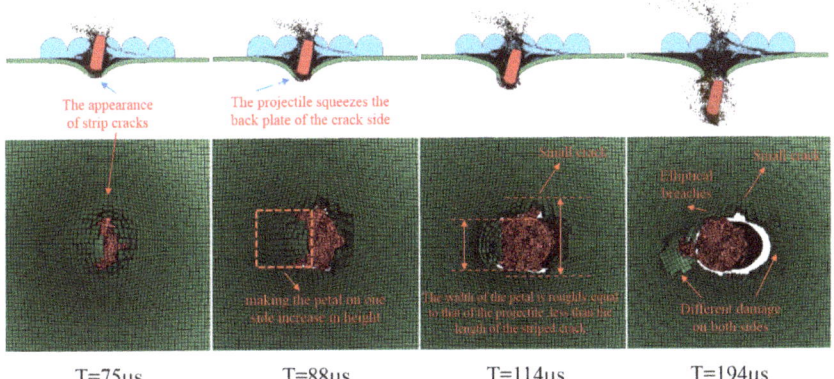

Figure 15. Deformation and failure process of the back plate under working condition 11.

In the working conditions where the projectile attitude and trajectory change little during penetration, the petals on the back plate appear more evenly, the breach shape is relatively regular, and the petal height is roughly the same (Figure 16). Taking working condition 17 as an example, when petaling failure occurs to the back plate and since there are many ceramic fragments in the front of the projectile and the deflection is small, the back plate is subjected to a relatively uniform force, resulting in a relatively uniform distribution of cracks. As the back plate damage continues along the crack, petals with large areas and good integrity are formed (Figure 17).

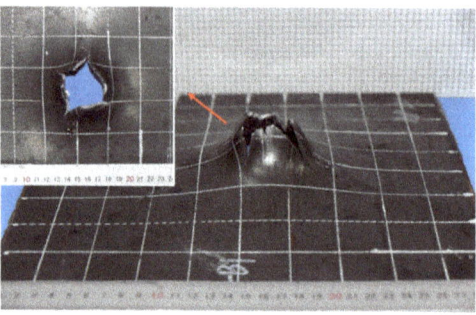

Figure 16. Fracture morphology of back plate in Experiment 1.

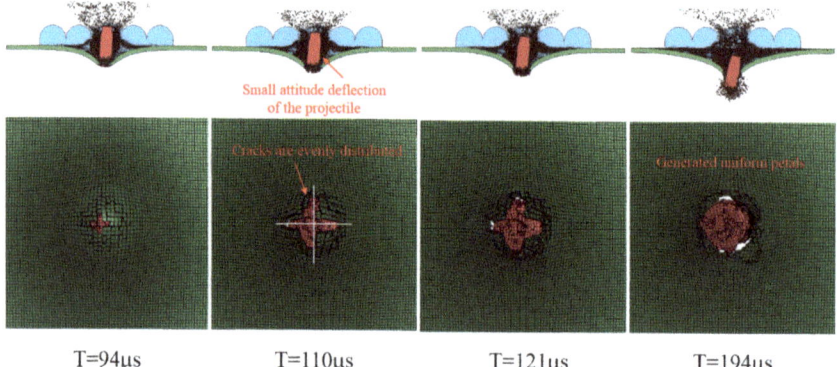

Figure 17. Deformation and failure process of the back plate under working condition 17.

3.3. Influence of Impact Point on Anti-Penetration Performance

The ceramic composite armor designed in this paper has periodic inhomogeneity in the direction perpendicular to the length of the ceramic column, so that the anti-penetration performance of the composite armor will be affected by the change in the impact point. In this section, under the condition that the initial velocity of the projectile is 810 m/s and the ratio of the diameter of the semi-cylindrical ceramic to the core is unchanged, the influence of the impact point on the anti-penetration performance of the semi-cylindrical ceramic composite armor is explored by changing the deviation degree η of the impact point.

Figure 18 shows the relationship between the residual projectile velocity v_r and the impact point deviation degree η. As shown in the figure, with the gradual increase η, the residual velocity v_r displays an overall decreasing trend first and then increases. For the residual velocity v_r, the change in impact point leads to the variation in two influencing factors, with the first being (1) ceramic thickness. The influence of the ceramic thickness below the projectile on the residual projectile velocity is mainly reflected in the change in the time $t_1 + t_2$ from the beginning of projectile penetration to back plate failure, when projectile velocity experiences a significant reduction. The thicker the ceramic, the longer $t_1 + t_2$, and the greater the velocity reduction. After the back plate is damaged, the velocity reduction rate of the projectile slows down rapidly. In the working conditions with large ceramic thicknesses, when the back plate starts to be damaged, the projectile velocity is smaller than that in the conditions with small ceramic thicknesses, so the projectile and the composite armor have an increased action time, causing greater velocity reductions in the third and fourth stages of penetration than that in the conditions with small thicknesses. Therefore, without considering projectile deflection, the thicker the ceramic below the projectile, the greater the deceleration capability of the composite armor to the projectile. (2) The second influencing factor is the maximum attitude deflection angle β_{max} of the projectile. As shown in Figure 19, with the increase η, the β_{max} increases first and then decreases. The change in β_{max} mainly affects the projectile's deceleration process in the third stage. The greater the β_{max}, the greater the effect of the back plate on the projectile and the greater the deceleration degree in the third stage. Therefore, with the increase η, although the ceramic material below the projectile decreases gradually, the β_{max} in the penetration process gradually increases first, and projectile deceleration in the third stage of penetration is obvious, resulting in the decrease in the residual projectile velocity. When η is further increased and the impact point is gradually close to the seam, the β_{max} in during penetration is gradually decreased, and meanwhile, the ceramic material below the projectile is further reduced, so the residual projectile velocity is significantly increased.

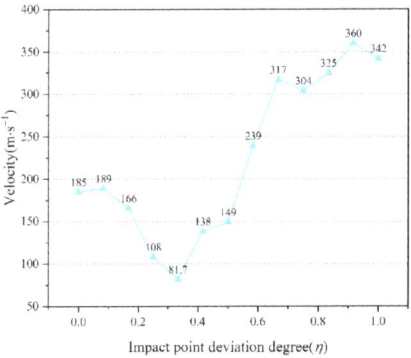

Figure 18. Relationship between residual projectile velocity and impact point deviation degree η.

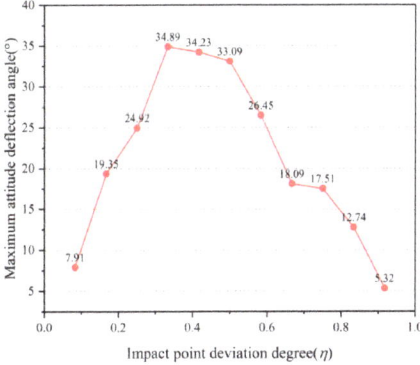

Figure 19. Relationship between the maximum projectile attitude deflection angle β_{max} and impact point deviation degree η.

In order to reflect the anti-penetration performance of the semi-cylindrical ceramic composite armor more directly, the area of the composite armor is divided according to the different anti-penetration performance. The division criteria are the same as in the literature [27], and the remaining characteristics of the projectile body are shown in Table 5, where E_{plate} is the residual kinetic energy of the projectile with the same initial velocity when penetrating the equivalent homogeneous ceramic composite armor. According to the division criteria and the above residual characteristics of the projectile body, the composite armor can be divided into three areas, whose penetration resistance decreases successively, which are named as the strong protection zone, secondary protection zone, and weak protection zone. The distribution is shown in Figure 20, accounting for 66.67%, 16.67% and 16.67%, respectively.

Table 5. Residual characteristics of projectile at different impact points.

Condition No.	Impact Point Deviation Degree η	Residual Velocity v_r(m·s^{-1})	Exit Attitude Deflection Angle β_{ri}(°)	Residual Kinetic Energy E_i(J)	E_i/E_{plate}
No. 5	0	185	/	334	0.30
No. 6	0.083	189	3.13	328	0.29
No. 7	0.167	166	8.5	280	0.25
No. 8	0.25	108	9.74	123	0.11
No. 9	0.333	81.7	14.5	76.2	0.07
No. 10	0.417	138	11	215	0.19
No. 11	0.5	149	10.3	234	0.21
No. 12	0.583	239	15.7	598	0.53
No. 13	0.667	317	13.9	1070	0.95
No. 14	0.75	304	13.4	975	0.86
No. 15	0.833	325	9.7	1110	0.98
No. 16	0.917	360	4.75	1400	1.24
No. 17	1	342	/	1190	1.05
No. 30	/	330	/	1130	1

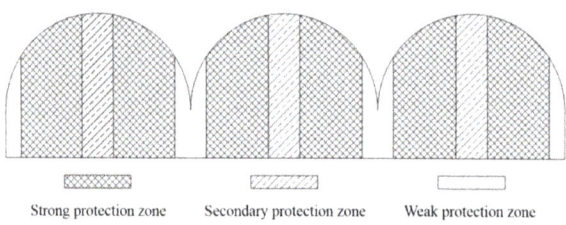

Figure 20. Area division of specially shaped ceramic composite armor.

3.4. Influence of the Ratio of Semi-Cylindrical Ceramic Diameter to Core Diameter on Anti-Penetration Performance

In order to investigate the effect of the ratio of the semi-cylindrical ceramic diameter to core diameter ξ on the anti-penetration performance, the areal density of the semi-cylindrical ceramic and the initial projectile velocity are left unchanged, and $\xi = 0.56, 1.11, 2.22$ and 2.78 are used. For the three typical impact points when $\eta = 0, 0.5$ and 1, a total of twelve working conditions are calculated respectively to study the anti-penetration performance of the semi-cylindrical ceramic composite armor under different semi-cylindrical ceramic diameters.

Figure 21 gives a comparison of the residual velocities of the projectile when penetrating ceramic composite armors with different ξ values, under the above three impact point conditions. By comparing the residual velocities of the projectile with the same diameter but different impact point deviation degrees η, it can be concluded that when $\xi = 0.56, 1.11$, the residual velocity difference under the three typical impact point deviation degrees is relatively small. This is mainly because the semi-cylindrical ceramic diameter is too small; in the first stage of penetration, it does not cause large asymmetric erosion to the projectile, resulting in small projectile deflection during penetration (Figure 22). In the third stage of penetration, the secondary deceleration of the projectile in the case of $\eta = 0.5$ is not apparent.

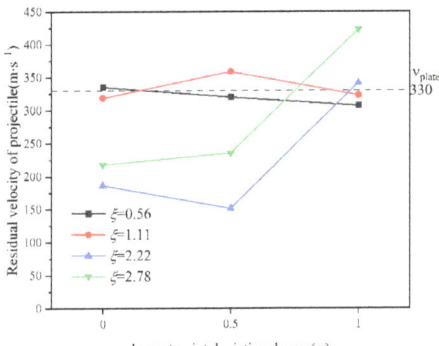

Figure 21. Comparison of residual velocities of the projectile when penetrating the ceramic composite armor with different ξ values when $\eta = 0, 0.5, 1$.

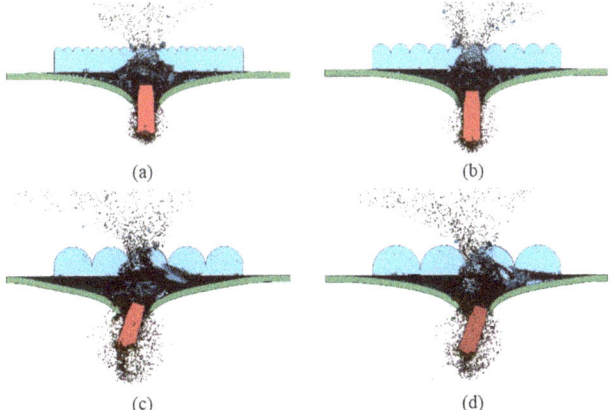

Figure 22. Damage produced when 12.7 mm API projectile penetrates three semi-cylindrical ceramic composite armors with different ξ values, when $\eta = 0.5$, (**a**) $\xi = 0.56$, (**b**) $\xi = 1.11$, (**c**) $\xi = 2.22$, (**d**) $\xi = 2.78$.

Regarding the anti-penetration performance of composite armors with different ξ values, as can be observed from Figure 21, when $\xi = 0.56, 1.11$, the residual velocities under the three typical impact point deviation degrees are slightly less than or slightly greater than the residual velocity v_{plate} in the case of homogeneous ceramic composite armor with equal areal density, and the projectile attitude deflection angle does not change significantly. So, the semi-cylinder design does not play a positive role in improving the anti-penetration performance of the armor. When $\xi = 2.78$, under the same areal density, the distribution of the ceramic material in the X direction is extremely uneven, resulting in the fact that the residual velocities in the case of $\eta = 0, 0.5$ are smaller than v_{plate}, but the residual velocity with $\eta = 1$ is much greater than v_{plate}; meanwhile, the theoretical value of the projectile attitude deflection angle is 0 in this case, which is unfavorable for armor protection. The anti-penetration performance of the ceramic composite armor with $\xi = 2.22$ has been described in Section 3.3. The comprehensive comparison shows that the composite armor with $\xi = 2.22$ performs the best in resisting penetration among the designed four types of composite armor with different semi-cylindrical ceramic diameters, under the areal density conditions adopted in this paper.

3.5. Influence of Projectile Initial Velocity

In order to explore the anti-penetration performance of the semi-cylindrical ceramic composite armor against the projectile with different initial velocities, the initial velocities of v_0 = 400, 700, 1000 and 1300 m/s are used, while keeping the deviation of the impact point of the projectile at η = 0.5. Under different initial velocity conditions, the maximum attitude deflection angle (before it completely hits the target) and residual mass of the projectile are shown in Figure 23. As can be observed from the figure, when the initial velocity of the projectile increases, the erosion degree of the projectile increases continuously, and the maximum attitude deflection angle decreases.

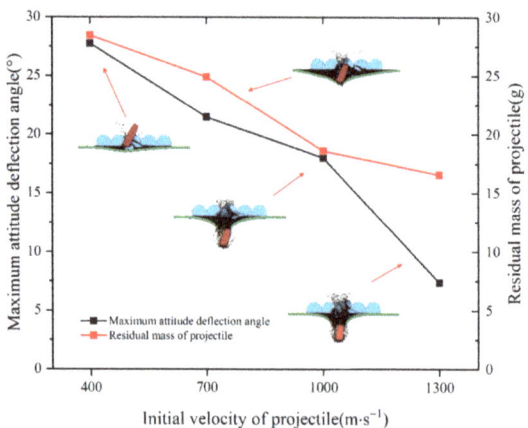

Figure 23. Maximum attitude deflection angle and residual mass of projectile under different initial velocities.

The increase in the initial projectile velocity directly leads to the decrease in its target penetration time. Under the working condition of v_0 = 1000 m/s, at t = 34 μs, the displacement of the center of mass of the projectile is $\Delta y = -3.22$ cm. Figure 24 presents the Y-direction displacement–time curve of the node at the center of mass of the projectile under different initial velocities. As illustrated, $\Delta y = -3.22$ cm under the working conditions of v_0 = 700, 1000 and 1300 m/s corresponds to 52 μs, 34 μs and 25 μs, respectively. In the working condition of v_0 = 400 m/s, the deflection of the projectile is too large, which cannot be displayed in the figure, and the corresponding time is t = 112 μs. Figure 25 shows the penetration under different working conditions with the same displacement of $\Delta y = -3.22$ cm. It can be observed from the figure that the projectile nose is near the bottom of the ceramic half cylinder in the three working conditions, i.e., the asymmetric erosion stage of the projectile ends. In other words, due to the increase in the initial projectile velocity, the time required for the projectile to go through this stage is shortened, and the angular acceleration differences between the three working conditions at this stage are small, all within the range of 0~0.008 rad·μs^{-1}. Therefore, the time of the projectile in this phase is proportional to its attitude deflection angle, which is 13.74°, 4.42°, 1.37°, and 0.37° in the three conditions. The magnitude of the attitude deflection angle of the projectile has a positive feedback effect on its further deflection. The larger the attitude deflection angle, the stronger the asymmetric action on the projectile, resulting in a further increase in the deflection of the projectile, which further promotes its self-deflection; otherwise, the further deflection will be smaller. Therefore, as the velocity increases, the deflection effect of the projectile decreases continuously. At the same time, the increase in the initial projectile velocity leads to the increase in the projectile-target acting force during the projectile's penetration into the composite armor, which then results in an increased degree of projectile erosion.

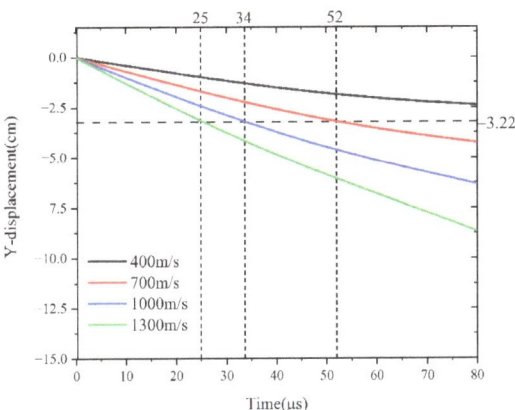

Figure 24. The time-history curve of Y-direction displacement of the projectile center of mass under different initial velocities.

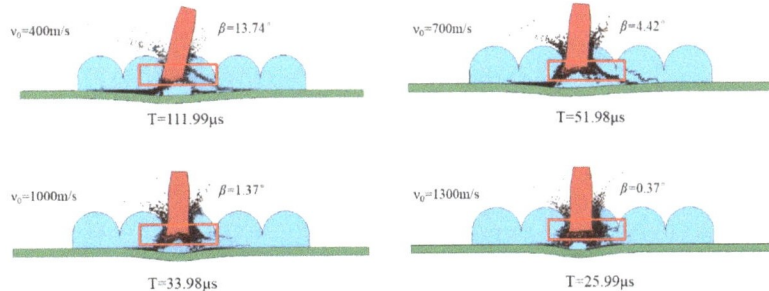

Figure 25. Penetration under different working conditions.

To sum up, the anti-penetration performance of the semi-cylindrical ceramic composite armor against projectiles with different initial velocities shows the following trend: under low initial projectile velocities, the composite armor mainly exhibits large deflection and a relatively small erosion effect on the projectile; as the initial velocity increases, the erosion effect of the composite armor on the projectile gradually increases, while the deflection effect decreases.

4. Conclusions

A semi-cylindrical ceramic composite armor structure is designed in this paper. The processes of projectile penetration deflection and the deformation and failure mode of the back plate are analyzed by the method of combining experiment and numerical calculations. The protective effect of the composite armor with different ratios of semi-cylindrical ceramic to core diameter on 12.7 mm armor-piercing firebombs in different areas and under the condition of equal areal density is mainly explored, and the trend of anti-penetration performance of the composite armor to projectiles with different initial velocities is reported.

(1) The process of the 12.7 mm API projectile penetrating ceramic composite armor can be divided into four stages. Projectile attitude deflection during penetration is mainly attributed to the non-axial force on the projectile, due to the asymmetric erosion of the projectile nose by the specially shaped ceramic and the non-ideal attitude of the projectile when penetrating the back plate. Due to projectile deflection, the projectile velocity experiences a significant secondary decline at the stage of back plate failure. After exiting the target plate, the projectile has an attitude deflection angle and angular velocity.

(2) After the projectile perforated and exited the target plate, dishing deformation-petaling failure occurred at the back plate, which is similar to the deformation/failure mode of an ogive-nosed projectile perforating a thin plate at a low velocity. When the projectile deflected to a certain extent during penetration, the breaches in the back plate showed elliptical shapes of varying degrees, and the size and height of the petals were also significantly different. When the attitude deflection was small, the breaches were regular, and the difference in the size and height of petals was relatively small.

(3) Due to the periodic inhomogeneity of the structure in the direction perpendicular to the length of the semi-cylinder, the impact point of the projectile has a great influence on the anti-penetration performance of the composite armor to 12.7 mm API projectiles. According to the remaining characteristics of the projectile, the composite armor can be divided into strong protection area, secondary protection area and weak protection area, and the proportion of each area is 67.67%, 16.67% and 16.67%, respectively.

(4) For the semi-cylindrical ceramic composite armor designed in this paper, when the ratio of semi-cylindrical ceramic diameter to core diameter ξ is less than 2, it leads to a small deflection during projectile penetration, and the design of the semi-cylinder contributes very little to improving the anti-penetration performance of the composite armor. Among the tested composite armor with ξ = 0.56, 1.11, 2.22 and 2.78, the armor with ξ = 2.22 has the best anti-penetration performance.

(5) The anti-penetration performance of the semi-cylindrical ceramic composite armor against projectiles with different initial velocities shows the following trend: under low initial projectile velocities, the composite armor mainly exhibits large deflection and a relatively small erosion effect on the projectile; as the initial velocity increases, the erosion effect of the composite armor on the projectile gradually increases, while the deflection effect decreases.

Author Contributions: Literature review, A.J. and Y.L.; writing—original draft preparation, A.J. and Y.L.; writing—review and editing, A.J., D.L. and H.H.; experiment performance, Y.L. and A.J.; experiment analysis, A.J. All authors have read and agreed to the published version of the manuscript.

Funding: This research was funded by the National Natural Science Foundation of China, grant numbers 51679246 and 52101378.

Institutional Review Board Statement: Not applicable.

Informed Consent Statement: Not applicable.

Data Availability Statement: Not applicable.

Acknowledgments: The authors would like to thank the editor, associate editor, and the anonymous reviewers for their helpful comments and suggestions that have improved this paper.

Conflicts of Interest: The authors declare no conflict of interest.

References

1. Montgomery, J.S.; Wells, M.G. Titanium armor applications in combat vehicles. *JOM* **2001**, *53*, 29–32. [CrossRef]
2. Gooch, W.A. An overview of ceramic armor applications. In Proceedings of the 6th Technical Conference IDEE 2004, Trenčín, Slovakia, 6–7 May 2004.
3. Forquin, P.; Tran, L.; Louvigne, P.F.; Rota, L.; Hild, F. Effect of aluminum reinforcement on the dynamic fragmentation of SiC ceramics. *Int. J. Impact Eng.* **2003**, *28*, 1061–1076. [CrossRef]
4. Serjouei, A.; Chi, R.; Zhang, Z.; Sridhar, I. Experimental validation of BLV model on bi-layer ceramic-metal armor. *Int. J. Impact Eng.* **2015**, *77*, 30–41. [CrossRef]
5. Roeder, B.A.; Sun, C.T. Dynamic penetration of alumina/aluminum laminates: Experiments and modeling. *Int. J. Impact Eng.* **2001**, *25*, 169–185. [CrossRef]
6. Hou, H.; Zhu, X.; Kan, Y. The advance of ballistic performance of light ceramic composite armor under the impact of projectile. *Acta Armamentarii* **2008**, *29*, 208–216. (In Chinese)
7. Zhang, Z.; Wang, M.; Song, S.; Min, L.; Sun, Z. Influence of panel/back thickness on impact damage behavior of alumina/aluminum armors. *J. Eur. Ceram. Soc.* **2010**, *30*, 875–887. [CrossRef]

8. Madhu, V.; Ramanjaneyulu, K.; Bhat, T.B.; Gupta, N.K. An experimental study of penetration resistance of ceramic armour subjected to projectile impact. *Int. J. Impact Eng.* **2005**, *32*, 337–350. [CrossRef]
9. Gunes, R.; Hakan, M.; Apalak, M.K.; Reddy, J.N. Numerical investigation on normal and oblique ballistic impact behavior of functionally graded plates. *Mech. Adv. Mater. Struct.* **2020**, *28*, 2114–2130. [CrossRef]
10. Petrudi, A.M.; Vahedi, K.; Rahmani, M.; Petrudi, M.M. Numerical and analytical simulation of ballistic projectile penetration due to high velocity impact on ceramic target. *Frat. Integrità Strutt.* **2020**, *14*, 226–248. [CrossRef]
11. Fawaz, Z.; Zheng, W.; Behdinan, K. Numerical simulation of normal and oblique ballistic impact on ceramic composite armours. *Compos. Struct.* **2004**, *63*, 387–395. [CrossRef]
12. Wang, H.; Wang, J.; Tang, K.; Chen, X.; Li, Y. In Investigation on the Damage Mode and Anti-penetration Performance of B4C/UHMWPE Composite Targets for Different Incident Velocities and Angles. In *Journal of Physics: Conference Series*; IOP Publishing: Bristol, UK, 2021; p. 012010.
13. Khan, M.; Iqbal, M.A.; Bratov, V.; Morozov, N.; Gupta, N. An investigation of the ballistic performance of independent ceramic target. *Thin-Walled Struct.* **2020**, *154*, 106784. [CrossRef]
14. Luo, D.; Wang, Y.; Wang, F.; Cheng, H.; Zhu, Y. Ballistic behavior of oblique ceramic composite structure against long-rod tungsten projectiles. *Materials* **2019**, *12*, 2946. [CrossRef]
15. Li, J.C.; Chen, X.W. Theoretical analysis of projectile-target interface defeat and transition to penetration by long rods due to oblique impacts of ceramic targets. *Int. J. Impact Eng.* **2017**, *106*, 53–63. [CrossRef]
16. Salimi, B.; Vahedi, K.; Petrudi, A.M.; Rahmani, M. Optimization and Numerical Analysis of the Ricochet of Conical Nose Projectile in the Collision with Ceramic- Aluminum Armor. *J. Adv. Mech. Eng. Appl.* **2020**, *2*, 53–63. [CrossRef]
17. Wang, Q.; Chen, Z.; Chen, Z. Design and characteristics of hybrid composite armor subjected to projectile impact. *Mater. Des.* **2013**, *46*, 634–639. [CrossRef]
18. Zhou, B.K. An experimental study on anti-penetration characteristics of concretes shielded with single layer of tightly arrayed corundum spheres. *Explos. Shock. Waves* **2003**, *23*, 173–177. (In Chinese)
19. Stanisławek, S.; Morka, A.; Niezgoda, T. Pyramidal ceramic armor ability to defeat projectile threat by changing its trajectory. *Bull. Pol. Acad. Sci. Tech. Sci.* **2015**, *63*, 843–849. [CrossRef]
20. Cohen, M.; Etzion, K.K. Composite Armor. U.S. Patent US6203908B1, 20 March 2001.
21. Liu, W.; Chen, Z.; Cheng, X.; Wang, Y.; Amankwa, A.R.; Xu, J. Design and ballistic penetration of the ceramic composite armor. *Compos. Part B Eng.* **2016**, *84*, 33–40. [CrossRef]
22. Chen, J.; Zeng, Y.; Liang, X.; Hou, Y.; Wang, Y.; Sun, Z.; Cui, S. Lightweight Design and Experimental Study of Ceramic Composite Armor. *Processes* **2022**, *10*, 1056. [CrossRef]
23. Aydin, M.; Soydemir, M. Ballistic protection performance of a free ceramic particle armor system: An experimental investigation. *Ceram. Int.* **2021**, *47*, 11628–11636. [CrossRef]
24. Shao, R.; Wu, C.; Su, Y.; Liu, Z.; Liu, J.; Chen, G.; Xu, S. Experimental and numerical investigations of penetration resistance of ultra-high strength concrete protected with ceramic balls subjected to projectile impact. *Ceram. Int.* **2019**, *45*, 7961–7975. [CrossRef]
25. Wang, X.D.; Yu, Y.L.; Jiang, Z.X.; Ma, M.H.; Gao, G.F. Dynamic fragmentation and failure of the hard core of a 12.7 mm API projectile against SiC/6061T6Al composite armor with various impact velocities. *Explos. Shock. Waves* **2022**, *42*, 023303. (In Chinese)
26. Wang, P. Research of Capacity for Penetrating Resistance of SiC Ceramic. Ph.D. Thesis, Nanjing University of Science and Technology, Nanjing, China, 2012.
27. Jiang, A.; Li, Y.; Li, D.; Hou, H. Deflection effect and mechanism of semi-cylindrical ceramic composite armor for flat-ended rod projectile. *Ceram. Int.* **2022**, *48*, 31023–31040. [CrossRef]
28. Jamroziak, K. Comparison of Numerical Simulation Techniques of Ballistic Ceramics under Projectile Impact Conditions. *Materials* **2021**, *15*, 18. [CrossRef]
29. Zhang, Y.; Dong, H.; Liang, K.; Huang, Y. Impact simulation and ballistic analysis of B4C composite armour based on target plate tests. *Ceram. Int.* **2021**, *47*, 10035–10049. [CrossRef]
30. Guo, L. The Thickness Effect of Metal Armor with Poerstructure Penetrating by Small Caliber Projectile. Ph.D. Thesis, Nanjing University of Science and Technology, Nanjing, China, 2012.
31. Hou, E.Y. Investigation of Mechanism and Performance of Spaced Ceramic Target under Impact of 12.7 mm Armor Piercing Projectile. Ph.D. Thesis, National University of Defense Technology, Changsha, China, 2008.
32. Wang, C.; Chen, A.; Li, Z.; Gong, C.; Wang, S.; Yan, W.-M. Experimental and numerical investigation on penetration of clay masonry by small high-speed projectile. *Def. Technol.* **2021**, *17*, 1514–1530. [CrossRef]
33. Tian, C.; An, X.; Sun, Q.; Dong, Y. Experimental and numerical analyses of the penetration resistance of ceramic-metal hybrid structures. *Compos. Struct.* **2019**, *211*, 264–272. [CrossRef]

Article

Influential Factors of a Reactive Materials Projectile's Damage Evolution Behavior

Xiangrong Li [1,2], Cong Hou [1], Huan Tong [1], Lei Yang [1] and Yongkang Chen [1,*]

1. Department of Arms and Control, Academy of Army Armored Forces, Beijing 100072, China
2. College of Mechatronic Engineering, North University of China, Taiyuan 030051, China
* Correspondence: 3120215138@bit.edu.cn; Tel.: +86-155-1019-2099

Abstract: To determine the mechanism of penetration of multi-layer aluminum targets (MLAT) by a reactive materials projectile (RMP), AUTODYN-3D numerical simulations and experimental tests were carried out. The Powder Burn equation of the state ignition model was introduced for the reactive core activation under different projectile–target interaction conditions, which effectively simulated the deflagration reaction damage effects behavior of the RMP and the damage evolution behavior of the MLAT. The activation rate of the reactive core increased significantly when the thickness of the steel target was 8–15 mm; a significant combined destructive effect of kinetic and chemical energy was produced on the MLAT. The initial velocity was proportional to the penetration and destruction effect of the front-layer aluminum target. For the rear-layer aluminum target, the detonation damage showed a tendency to increase and then decrease. If the head metal block was too thick, the penetration ability would be improved at the same time, and the deflagration reaction damage effects ability of the steel target would be significantly reduced. In order to achieve good battlefield damage efficacy, all of the influencing factors should be comprehensively considered.

Keywords: reactive materials projectile; multi-layer aluminum target; damage evolutionary behavior; deflagration reaction

1. Introduction

Reactive materials (RM) are ignited or detonated in the process of penetrating a target, and their joint destruction via the penetration of kinetic energy and chemical energy release is a cutting-edge development in the field of advanced and efficient destruction technology [1–3]. Compared with traditional standard projectiles, the significant technological and damage efficiency advantages of reactive materials projectile (RMP) are mainly reflected in their high-efficiency destruction, versatility, high safety, long storage lives, simple structures, and low costs [4–7].

At present, many achievements have been made in the field of RM warheads around the world [8–15]. Mock et al. [16] studied the impact detonation threshold of polytetrafluoroethylene/aluminum (PTFE/Al) RMs through crash tests; based on the test results, an empirical formula relating the impact activation response time and impact pressure was proposed. The Beijing Institute of Technology [17] conducted a study on the detonation and energy release caused by the impact of PTFE RMs, and they analyzed the influence of the post-target implosion overpressure effect and impact velocity of the RMs on the implosion overpressure and compared the energy release behaviors of the RMs of different formulations.

Due to requiring a dynamic impact process with a high strain rate of the plastic deformation to provide the initiation energy, in general, their impact-induced initiation mechanism is not well understood, and the initiation properties have not been well characterized. One means by which to investigate their impact initiation behavior is through conducting a Taylor impact test. Studies with the PTFE/Al projectile have shown that it first passes through a brittle fracture stage, and then, this progresses to ignition. This initiation phenomenon indicates that while the fracture process is likely important to the ignition, it is not sufficient to produce the initiation alone. Thus, an additional amount of energy that is possibly related to the crack propagation properties and fracture surface energies must be deposited into the fractured materials before the initiation takes place. More detailed studies show that the first signs of ignition are visible in the regions of intensive shear during the Taylor impact test, and both the finite element analysis and high-speed videos provide further evidence for the intensive shear-induced ignition mechanism [18].

Moreover, the traditional energy release characterization techniques such as bomb calorimetry only measure the energy release behavior of energetic materials that are ignited in a static configuration. However, the initiation efficiencies of these RMs strongly depend upon the impact conditions, which makes the measurement of their energy release characteristics difficult, and the research methods on the mechanism of this explosion damage behavior are still relatively limited.

In view of this situation, based on the analysis of the mechanical behavior of the projectile target, the damage energy release behavior is transformed into the damage evolution behavior of the MLAT. As part of this effort, our research group carried out work on the target by matching the characteristics and the impact velocity [19,20]. This study began with the construction of the numerical simulation models, combined with a theoretical analysis and an experimental verification. The mechanical behavior of the impact action and the evolution behavior of the MLAT under different target action conditions were studied, and the flight dispersion behavior and the change of the perforation caliber of the after-effect target fragments were studied accordingly.

2. Numerical Models

On the AUTODYN-3D platform, the projectile penetration of the reactive substance was numerically simulated by the Lagrange algorithm. All of the materials were modeled by an erosion model, in which the reactive substance which was activated by the impact pressure obeyed a two-phase powder combustion equation of state (EOS). Gases and solids in the model existed at the same time, simulating the materials' deflagration reaction, and the relevant parameters were taken from the literature [21]. The unreactive materials were modeled with the state impact equation to simulate the mechanical behavior of the RMs under pressure and expansion [22,23].

2.1. Theoretical Model Building

The action process of the projectile–target interaction was mainly divided into four stages: the penetration of the target and the RM compression deformation fragmentation stage, the RM pressure relief dispersion and local ignition stage after penetrating the target, the RM debris cloud deflagration propagation and chemical energy release stage, and the stage where the shell fragments form a fragment power field. These are shown in Figure 1. When the RMP collides with the steel target, the shock wave generated by the impact compresses the projectile to produce different degrees of radial expansion of the core and the shells, and the Poisson's ratio of the reactive core is greater than that of the shells materials, so the radial expansion of the core is more significant, resulting in a mechanical radial expansion effect. Once the radial stress in the shells exceeds its fragmentation limit, a large amount of shell fragments are generated, and these fragments have a certain radial velocity. At the same time, high strain rates can cause the core of the RM to partially fragment and form a cloud of reactive debris after the steel target arrived. In a debris cloud, small reactive fragments are activated first due to their high specific surface area and surface ignition energy, and they further trigger intensity deflagration. With the increase in the number of aluminum targets, the length of the reactive core will gradually decrease, the radial expansion of the shells will be weakened, and the damage ability of the after-effect aluminum target will be greatly reduced.

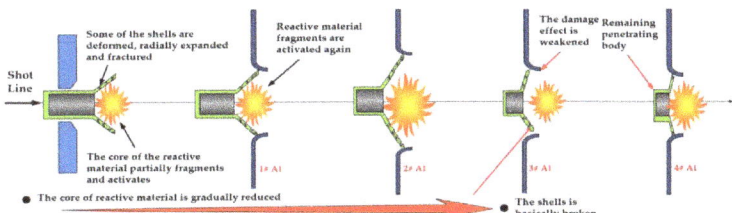

Figure 1. The process of action of RMP impacting a MLAT.

To obtain the activation state of the reactive substance during the target penetration process, it was necessary to analyze the first impact, that is, the state of the reactive substance during the impact of the steel target. When the RMP hits the steel target, based on the Ranking–Hugoniot relationship, the conservation of the mass, momentum, and energy can be expressed, respectively as follows:

$$\rho_1/\rho_0 = (U - u_0)/(U - u_1), \tag{1}$$

$$P_1 - P_0 = \rho_0(u_1 - u_0)/(U - u_0) \tag{2}$$

$$e_1 - e_0 = (P_1 u_1 - P_0 u_0)/\rho_0(U - u_0) + \left(u_1^2 - u_0^2\right)/2 \tag{3}$$

where ρ is the density, U is the shockwave velocity, u is the particle velocity, P is the impact pressure, and e is the specific internal energy. Subscripts 0 and 1 represent the undisturbed and affected states, respectively. In addition, the linear relationships between the shockwave velocity of the RM and the steel target particle velocity can be expressed as:

$$U_p = c_p + s_p u_p \tag{4}$$

$$U_t = c_t + s_t u_t \tag{5}$$

where c and s are the speed of sound and the materials coefficient, respectively, and the subscripts p and t represent the RM and the target, respectively. The values of c and s of the steel target were 4610 m·s^{-1} and 1.73 [24], respectively, and the corresponding values for the RM were 1350 m·s^{-1} and 2.26 [25,26].

At the impact interface, the shockwave is transmitted to the RM and the steel target, respectively, so the particle velocity and pressure compatibility relationship of the entire target surface can be expressed, respectively, as:

$$v_0 = u_p + u_t \tag{6}$$

$$\rho_{P_0}(c_p + s_p u_p)u_p = \rho_{t_0}(c_t + s_t u_t)u_t \tag{7}$$

We define Pc as the critical detonation pressure. By default, at this pressure, the RM can meet the size of the ignition and detonation fragments, while the fragments below this size could undergo a chemical energy deflagration reaction. Based on the shockwave propagation attenuation characteristics, we can effectively describe the activation behavior of the core. This activation length can be described as:

$$x_1 = -(1/\alpha)\ln(P_c/P_0) \tag{8}$$

where P_0 is the maximum stress and α is the correlation coefficient for the properties of the RM. According to a previous study [27,28], its value is 0.036 mm^{-1}.

2.2. Numerical Simulation Model Building

The failure of the materials is closely related to the state under the load. In order to better describe the whole process of the projectile target action of the active materials, the principal stress failure standard was adopted for all of the materials in the numerical simulation, and when the maximum tensile principal stress and shear stress exceeds the ultimate stress value of the materials, the materials will fail. Tables 1 and 2 list the main model parameters used for numerical simulations [29–31]; the erosion algorithm was used for the relevant materials.

Table 1. Main materials model.

Parts	Materials	Equation of State	Intensity Model	Invalidation Model
Shells	30CrMnSiA	Shock	Johnson Cook	Principal Stress
Core (reactive)	PTFE/AL	Powder Burn	Johnson Cook	Principal Stress
Core (unreactive)	PTFE/AL	Shock	Johnson Cook	Principal Stress
Steel target	RHA	Shock	von Mises	Principal Stress
Post-effect target	AL 2024	Shock	Johnson Cook	Principal Stress

Table 2. Main materials parameter table.

Materials	Density (g/cm^3)	Shear Modulus (GPa)	Yield Strength (MPa)	Specific Heat (J/kg·K)	Tensile Strength (MPa)
W.ALLOY	17	160	1506	172	2210
Core	2.76	7	120	/	240
AL 2024	2.78	28.6	260	1220	720
RHA	7.86	64.1	1500	/	2000
30Cr	7.86	80.8	1800	460	1200

To facilitate the analysis of the problem, the RMP was simplified into three parts: the high-strength shells (30CrMnSiNi2A), a reactive core (PTFE/Al), and a head metal block (a wolfram alloy). Since the symmetry condition was satisfied under positive penetration conditions, the calculations were performed using a 1/4 model, and the three-dimensional simplified model of the RMP is shown in Figure 2. The caliber of the RMP is 30 mm, the ratio of the inner to the outer diameter is 0.6, and the length is 100 mm. The high-strength shells mainly played the role of restraining the reactive core and penetrating the target. The reactive core mainly underwent impact compression and expansion as well as pressure explosion, deflagration, and destruction. The head metal block mainly played a role in enhancing the penetration ability of the warhead. The target was a rolled homogeneous armor (RHA), and the after-effect target was a spaced aluminum target with a thickness of 3 mm. The penetration analysis model is shown in Figure 3.

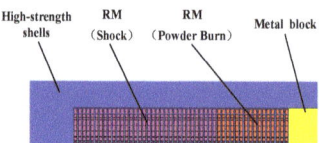

Figure 2. Schematic diagram of the RMP.

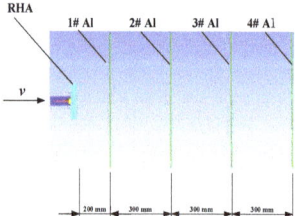

Figure 3. Three-dimensional penetration analysis model.

3. MLAT Destruction Law Study

Based on the construction of the above numerical model, this section illustrated how we carried out the study of the MLAT destruction mechanisms of the RMP under different projectile–target interaction conditions, especially the activation behavior caused by the penetration and the evolution behavior of MLAT destruction, in order to obtain a clearer mechanism of RMP penetration-initiation combined damage effects.

3.1. Steel Target Thickness Impact Analysis

3.1.1. Numerical Models

The internal stress peak analysis was performed on the projectile penetration of the RHA steel target, as shown in Figure 4. When the thickness of the RHA steel target was small, the thickness of the steel target had a significant impact on the internal stress value of the projectile. When the target thickness was greater than 15 mm, the internal stress value of the RM was not much different from the case with a target thickness of 15 mm, but when the target thickness was 10 mm or lower, the internal stress value of the RM decreased significantly.

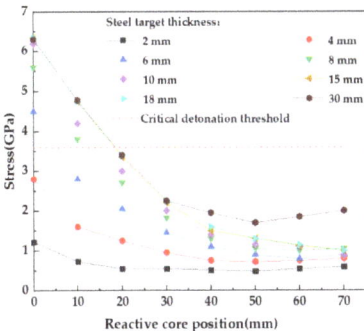

Figure 4. Peak stresses of the projectiles penetrating steel target.

In light of the above, the speed was set to 1000 m/s to impact the 2, 8, 15, and 30 mm RHA steel targets. The RMP activation models for different target thicknesses are shown in Figure 5. The red RM near the head of the metal block in Figure 5 was modeled using the Powder Burn EOS, and the pink region away from this area was modeled using the Shock EOS. Figure 5a indicates that the RM was not effectively activated with the 2 mm RHA steel target.

Figure 5. RMP model of impacting RHA steel target with different thicknesses: (**a**) 2 mm, (**b**) 8 mm, (**c**) 15 mm, and (**d**) 30 mm.

3.1.2. Results and Discussion

The multi-layer target damage pairs with different target thicknesses are shown in Figure 6. For the 2 and 30 mm RHAs, especially the 2 mm RHA, the multi-layer-targeted damage effect was significantly reduced, and the fragment dispersion area and the maximum perforation size were slightly larger than the caliber size, without causing any surface damage effects. For the 8 and 15 mm RHAs, the MLAT damage effect was significant. For the 15 mm RHA, the fragment dispersion area on target No. 2 reached about 750 cm^2, the maximum perforation area was about 64 cm^2, and the damage and maximum perforation areas reached 27 and 2.5 times the projectile caliber, respectively. The analysis showed that this was due to the thickening of the RHA steel target, which prolonged the target action time. The reactive core was completely compressed and broken, and due to the instantaneous unloading of the steel target pressure, many reactive debris clouds were generated behind the steel target. This produced a violent chemical energy explosion and deflagration, causing large-scale fragmentation damage and a significant reaming effect on the MLAT. However, if the RHA steel targets had been too thick, this would also cause a large amount of RM energy release reactions to occur inside the steel target so that penetration-initiation combined damage effects would not be able to effectively act on the subsequent multi-layer aluminum target.

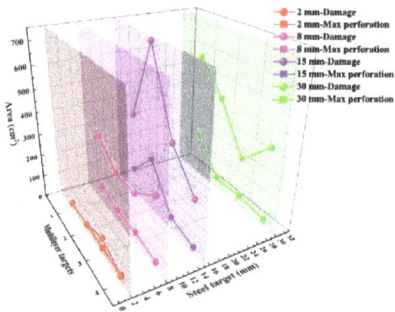

Figure 6. Comparison of MLAT target damage effects for different thicknesses of the RHA steel target.

3.2. Analysis of Impact Velocity

3.2.1. Numerical Models

The activation behavior of the 20 mm RHA target plates damaged by the projectile at different impact velocities as shown in Figure 7. The peak internal stress increases with the increase in the impact velocity. With the increase in the impact velocity, the curvature of the area near the impact point increased significantly, and the internal stress of the reactive core increased significantly. When the impact velocity was at 1000 m/s, the activation length of the reactive core was only about 11.62 mm, and the activation rate was about 15.5%. With the increase in the impact velocity, the activation rate of the reactive substance gradually increased, and when the impact velocity reached 1800 m/s, the activation length of the reactive core reached about 51.94 mm, and the activation rate increased to about 69.3%.

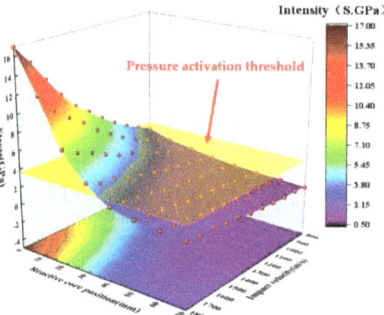

Figure 7. Changes in internal stress at different impact velocities(red ball: peak internal stress of the reactive core).

Based on the influence of the impact velocity from 800 to 1800 m/s on the activation behavior of the RMP, Figure 8 shows the numerical simulation results of the projectile at different impact velocities and the fracture state diagrams of the shells. When the velocity was 800–1200 m/s, only the head of the projectile showed a slight radial expansion and valgus deformation. No reactive core fracture was evident, and the projectile body remained long. When the impact velocity was greater than 1400 m/s, the entire shells were destroyed after penetrating the RHA steel target, the radial expansion effect was significant, and the head of the unreactive core was also greatly fragmented, ejecting a cone-shaped cloud of reactive debris at the opening.

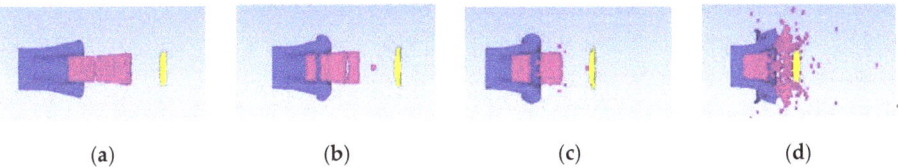

(a)   (b)   (c)   (d)

Figure 8. Warhead deformation and fragment distributions after penetrating RHA steel target at different impact velocities: (**a**) v_0 = 800 m/s, (**b**) v_0 = 1200 m/s, (**c**) v_0 = 1400 m/s, and (**d**) v_0 = 1600 m/s.

3.2.2. Results and Discussion

The damage status of the double-layer aluminum target at two typical velocities is shown in Figure 9. The perforation produced by the 1# aluminum target in Figure 8a was relatively flat, and the reaming effect was not evident. Only the 2# aluminum target had evident irregular petal-like perforations, and some pits, small holes, and other debris appeared near the perforation. As shown in Figure 8b, after the impact velocity was greater than 1200 m/s, the 1# aluminum target showed a significant reaming effect. The petal-like perforation diameter and damage area were significantly increased when they compared to how they were before. More target plate fragments formed, and the perforation diameter of the 2# aluminum target also increased. However, the damaged area showed a trend of increasing first, and then decreasing.

The damage effects of the RMP on the target plate at different impact velocities are shown in Figure 10. When the impact velocity was 800 m/s, the damage diameters of the 1# and 2# targets were minimized, and that of the 1# target was slightly less than that of the 2# target. This was because when the impact velocity was small, with the extension of the impact action time, the energy released by the projectile kinetic energy was gradually unloaded; the instantaneous stress value inside the projectile did not reach the activation threshold of the reactive substance, and it was not effectively activated. The slight radial expansion effect of the projectile caused the perforation diameter of the 2# target plate to be slightly greater than that of the 1# target plate, mainly based on kinetic energy destruction. When the impact velocity increased to 1400 m/s, the diameter of the 2# target damage reached the maximum value, and the diameter of the 1# target damage gradually decreased. This occurred because when the impact velocity reached a certain value, the projectile and RHA steel target action time was shorter, the initial stress increased, the degree of squeezing of the RM increased, the shells had a significant radial expansion effect, and the deflagration propagation rate of the RM was slower than the detonation wave of the explosive. When the reactive core hit it to form a debris cloud, a violent deflagration reaction occurred in front of the 1# target, causing a large damage area on the target, while only a small amount of RM acted on the 2# target through the 1# target. When the impact velocity reached 1800 m/s, the growth trend of the damage diameter of the 1# target was greatly slowed. If the velocities were too high, the initial stress would increase significantly, the RM would be activated prematurely, and more RMs would react inside the RHA steel target. This would result in less reactive debris overflowing from the shells port, a smaller dispersion radius of the fragments, and a weakening of the MLAT destruction ability.

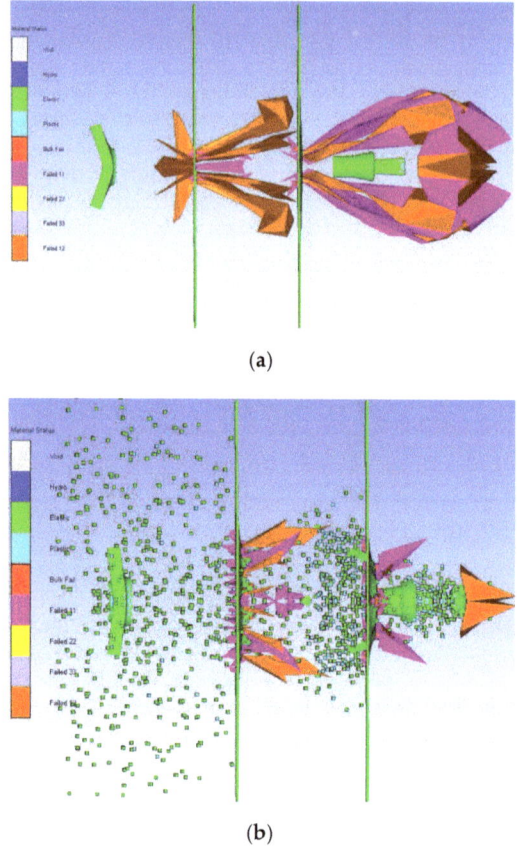

(a)

(b)

Figure 9. Result of damage caused by warheads to double-layer target plates at different impact velocities: (**a**) v_0 = 800 m/s and (**b**) v_0 = 1400 m/s.

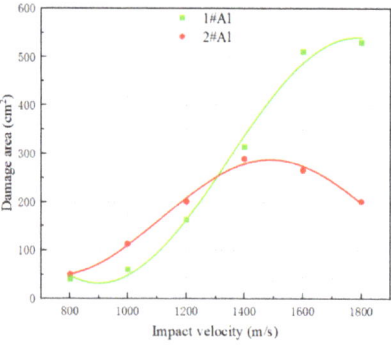

Figure 10. Damage area of the double-layer target at different impact velocities.

In summary, under the conditions described in this section, when the impact velocity was at 1400–1600 m/s, it could produce a good joint damage effect on the MLAT and result in efficient damage to the battlefield target.

3.3. Metal Block Thickness Effect Analysis
3.3.1. Numerical Models

The stress variation distribution along the thickness of a typical head metal block is shown in Figure 11. The presence or absence of the head metal block had a significant impact on the penetration performance of the RMP. Furthermore, the change in the thickness of the head metal block also had a greater impact on the penetration performance of the RMP. When the head metal block was not present, the deflagration reaction length of the warhead RM could reach about 45 mm. When the warhead metal block thickness was increased to 20 mm, the deflagration reaction length of the warhead RM was reduced to less than 20 mm.

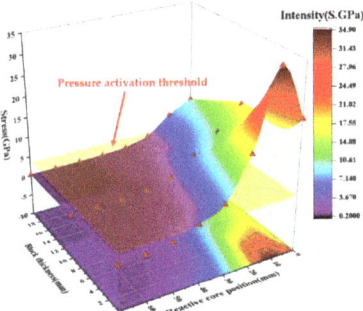

Figure 11. Changes in internal stress under different metal block thicknesses(red triangles: peak internal stress of the reactive core).

Adding a metal block to the head of the RMP could effectively increase the penetration capacity of the projectile, but it would correspondingly weaken its internal stress value, resulting in a significant reduction in the activation rate. To study the effect of the head metal block thickness on the MLAT damage effect, a corresponding activation model was constructed for the numerical simulation calculations, as shown in Figure 12. The head metal block materials was a tungsten alloy, and for comparative analysis, the thicknesses were set to 0, 10, and 20 mm.

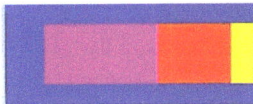

Figure 12. Metal block thickness activation model.

3.3.2. Results and Discussion

The pressure cloud of an RHA steel target colliding with head metal blocks of different thicknesses is shown in Figure 13. when the head was not equipped with a metal block, the stress on the steel target was the greatest, the reaming effect was the most significant, and the target produced significant ductile reaming holes. With the increase in the thickness of the metal block, the reaming effect was significantly weakened. The analysis suggests that this was due to the increase in the thickness of the head metal block, resulting in a significant decrease in the activation rate of the reactive substance. The energy released by the chemical energy explosion was also sharply reduced. At the same time, due to the increase in the thickness of the metal block, the projectile penetration ability of the RM was enhanced, the target action time was shortened, and the ductility reaming phenomenon was gradually weakened.

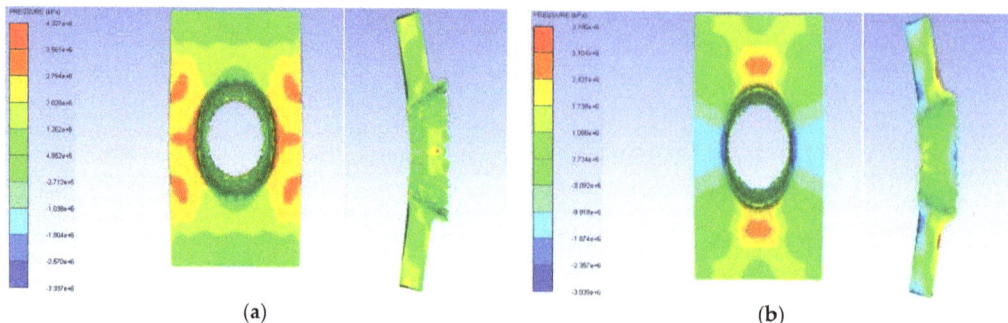

Figure 13. Pressure cloud diagrams of different head metal blocks colliding with RHA steel target at 0.1 ms, (**a**) 0 mm, and (**b**) 10 mm.

The MLAT damage state pairs for different head metal block thicknesses are shown in Figure 14. When the head metal block was not installed, the damage effect area of the 1# Al target fragment reached 620 cm^2, and the perforation area was about 145 cm^2. The damage effect area on the 2# Al target fragment reached 706 cm2, and the perforation area reached about 72 cm^2. The damage effect area on the surface of the 2# Al target was significantly greater than that of the head metal block, and the average perforation diameter was significantly smaller than the diameter of the head metal block. The improvement of the damage effect when the head metal block was installed was mainly reflected in the 2# and 3# Al targets, and the improvement of the damage effect when the head metal block was not installed was mainly reflected in the 1# and 2# Al targets. The analysis showed that this was due to the installation of thicker head metal blocks, which could effectively enhance the penetration ability of the RMP, reduce the pressure inside the projectile, and weaken the degree of fragmentation of the reactive nucleus. The size of the fragmentation increased, and the activation process occurred slowly, which could act more effectively on the multi-layer target plate.

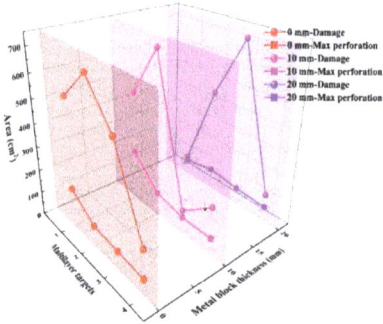

Figure 14. Damage results of MLAT with different metal block thicknesses.

4. Impact Experiments

Aiming to solve the problem of the mechanism of intrusion and destruction of the RMP, MLAT destruction effect experiments were carried out, and the evolutionary law of the MLAT damage under different projectile–target interaction conditions and the mechanism of the activation initiation of the RMP were verified. This provides reference data for the improvement of the damage efficiency of subsequent RMPs.

4.1. Experimental Setup

The schematic diagram of the experimental site layout and the actual site layout are shown in Figure 15. The reactive projectile was prepared by using a PTFE/Al powder with a zero-oxygen ratio by mixing cold-pressed sinter hardening materials. The density after sintering was 2.4 g/cm^3, and the multi-layer spacer board was composed of an RHA homogeneous armored steel target and an MLAT. The dimensions of the RHA steel target were 500 mm (length) × 500 mm (width), and the thickness dimensions were 2, 8, 15, 20, and 30 mm, respectively. The dimensions of the AL2024 aluminum target were 1000 mm (length) × 1000 mm (width), and the thickness was 3 mm. The spacing between the steel target and the MLAT was 200 mm, and the spacing between the MLAT was 300 mm. The impact speed was about 1000 m/s, and a Tianmu tachymeter and a velocity radar system which was installed in front of the steel target were used to measure the projectile velocities. Two target holders were used to secure the steel target and the MLAT.

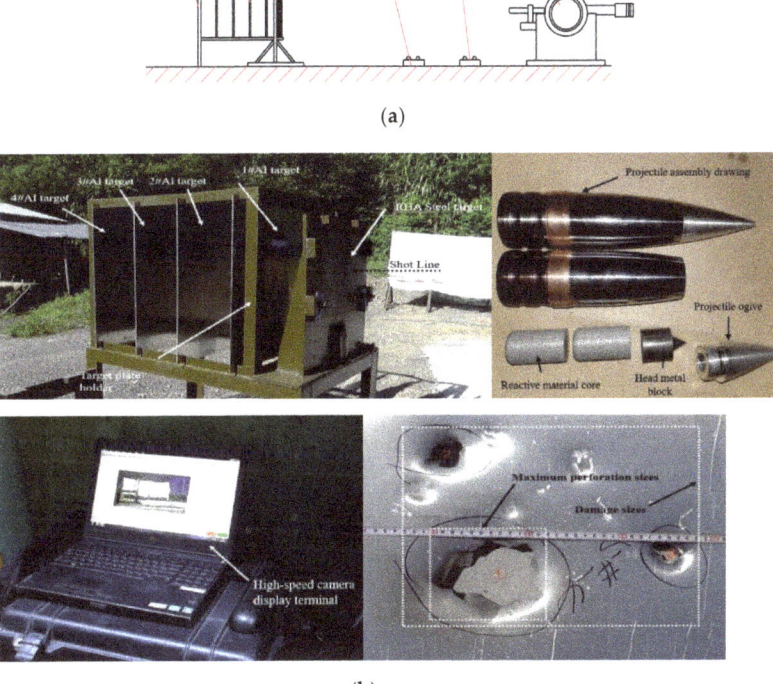

Figure 15. Experimental setup (**a**) Schematic diagram of the experimental layout: 1: 30 caliber artillery; 2, 3: Tianmu tachymeter; 4: RHA steel target; 5: MLAT; 6: target frame. (**b**) On-site layout.

4.2. Experimental Results and Discussion

4.2.1. Analysis of Influencing Factors of Metal Block Thickness

The effects of three typical metal block (w alloy) thicknesses on the multi-layer metal target are shown in Figure 16. The experimental results are shown in Table 3. The damage comparison experiment was carried out with a 20 mm RHA steel target without the installation of the head metal block, a 10 mm head metal block, and a 20 mm head metal block. Through verification experiments, it was shown that the thickness of the metal block of the head had a significant effect on the activation effect of the RM, and the reactive core also had a strong penetration ability. When the head metal block was not installed, its damage enhancement effect performed well in the aluminum target, especially in the steel target and the first two layers of the MLAT. Because there was no head metal block, the internal pressure of the projectile was greatly increased, resulting in a sharp increase in the pressure of the RM, and thus, more fragmentation occurred. Most of the fragments broke in the steel target and were activated by the ignition, and the reactive fragment cloud gathered behind the steel target, causing a significant reaming effect and a fragmentation effect on the first and second layers of the aluminum target. After the installation of the head metal block, due to the increase in the thickness of the head metal block, the reactive core was impact compression expansion effect reduced when the projectile collided with the steel target. Thus, fewer reactive fragments reached the ignition size, and most of the activation ignitions occurred in front of the 3# aluminum target, resulting in better fragmentation surface destruction and point penetration. Because the influence of the deflagration reaction of the RM on the activation effect could not be considered in the numerical simulation, the activation ignition behavior was analyzed only by the preliminary impact kinetics behavior. Thus, the damage effect was not significant, but the damage characteristics had a certain reference value. To ensure that the RM could reach the activation threshold, the thickness of the metal block of the head was appropriately reduced, and the length of the reactive core was increased to improve the comprehensive penetration and destruction ability of the RMP intrusion to the multi-layer spacer board.

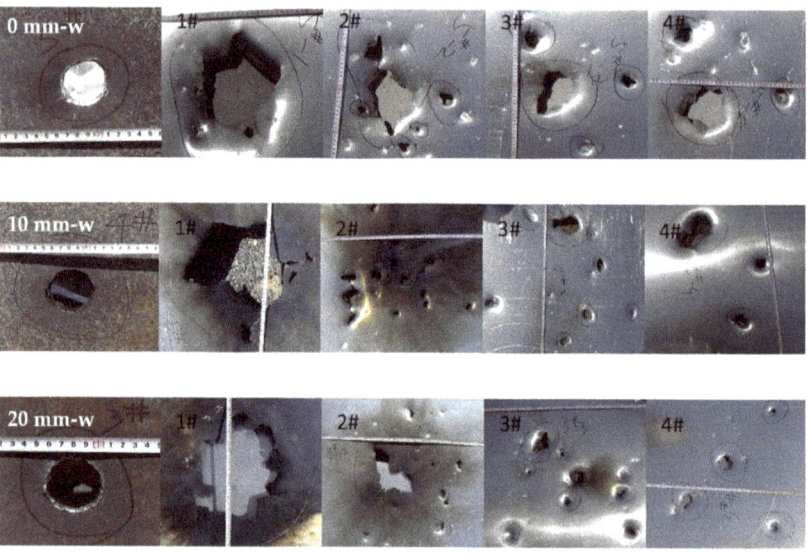

Figure 16. Results of multi-layer spacer board damage with different metal block thicknesses.

4.2.2. Analysis of Influencing Factors of Steel Target Thickness

To study the influence of the steel target thickness on the damage behavior of the RMP endpoint, comparative experiments of RHA steel target of different thicknesses were conducted. The RMP caliber was 30 mm, the ratio of the internal to the external diameters was 0.6, and the RHA steel thicknesses were 2, 8, 15, 20, or 30 mm. Only some of the results were taken for analysis. The high-speed images of the RMP impact multi-layer spacer board at the same time are shown in Figure 17, and the damage is shown in Figure 18.

Table 3. Experimental results.

Shot #	Thickness of RHA	Thickness of Metal Block	Impact Velocity (m/s)	Maximum Perforation Sizes (mm) Damage Sizes (mm) Perforation Number				
				RHA	1#Al	2#Al	3#Al	4#Al
1-1	2 mm	10 mm	936	40 × 45	35 × 35	40 × 45	50 × 55 75 × 60 3	70 × 70 140 × 80 4
1-2	8 mm	10 mm	942	40 × 45	200 × 190 230 × 220 3	180 × 120 340 × 220 4	280 × 210	350 × 410
1-3 3-3	15 mm	10 mm	940	40 × 40	230 × 210 340 × 210 2	270 × 300	190 × 220 280 × 310 7	130 × 190 290 × 230 4
1-4 2-3	20 mm	10 mm	952	45 × 45	180 × 140	210 × 210 320 × 470 9	90 × 60 490 × 460 9	80 × 60 600 × 390 6
1-5	30 mm	10 mm	946	55 × 50	100 × 130	110 × 80 80 × 90	80 × 80 115 × 65	100 × 80 150 × 80
2-1	20 mm	0 mm	970	50 × 52	360 × 290	540 × 625 13	600 × 440 9	580 × 280 4
2-2	20 mm	20 mm	932	45 × 45	180 × 140 220 × 180 2	210 × 214 400 × 260 6	170 × 160 560 × 350 9	190 × 200 700 × 190 5
3-1	15 mm	10 mm	862	40 × 40	240 × 200	330 × 300	240 × 160	210 × 120 260 × 150 2
3-2	15 mm	10 mm	907	40 × 42	240 × 220 270 × 220 2	280 × 200	240 × 200 260 × 320 5	120 × 80 280 × 150 2

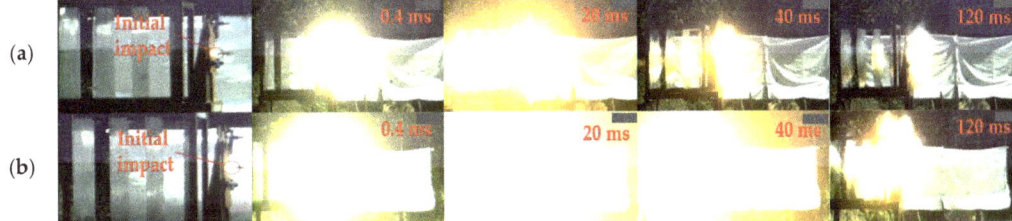

Figure 17. Comparison of high-speed images of RMP hitting MLAT: (**a**) hitting a 2 mm RHA steel target at a speed of 936 m/s and (**b**) hitting a 15 mm RHA steel target at a speed of 940 m/s.

The experimental results of the multi-layer aluminum target damage effect are shown in Table 3. As the thickness of the target plate increased to 30 mm, higher requirements were placed on the armor-piercing ability of the RMP. It can be seen from the damage of the MLAT that the RMP could penetrate the 30 mm steel target, but when the 15 mm steel target was reached, the damage effect was the best, especially when the third layer of the aluminum target achieved seven piercings. The dimensions of the fragment distribution area were 280 mm × 310 mm, and the maximum perforation sizes were 190 mm × 220 mm, the number of perforations reached seven, and the damage area and maximum perforation size reached 13.5 times and 6.5 times the caliber of the projectile, respectively. The analysis showed that this was due to the increase in the action time of the projectile–target interaction when the steel target was too thick, resulting in more reactive substances being prematurely ignited and detonated during the penetration of the target. The premature release of energy caused a certain reaming effect on the steel target. At the same time, for a thin target plate with a thickness of 2 mm, it can be seen that the damage effect of the RMP was poor, the number of perforations was small, the fragmentation distribution area was small, and the maximum perforation sizes was only 1–2 times the caliber of the projectile. This was consistent with the numerical simulation results. Combined with high-speed photographs, when the steel target was too thin, the resulting projectile–target interaction time was too short; moreover, the impact pressure did not reach the threshold and the reactive core was not effectively broken. Only kinetic energy penetration occurred, and the chemical action of ignition did not occur in the RM.

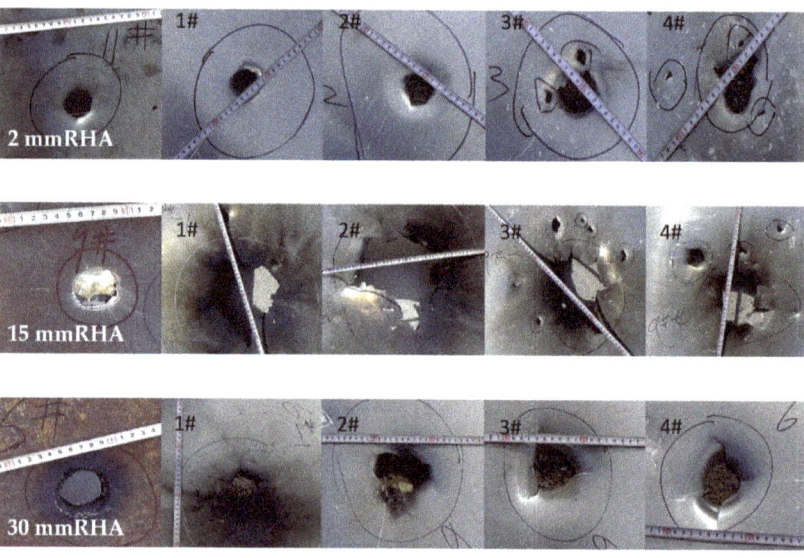

Figure 18. Damage results of MLAT with different RHA steel target thicknesses.

4.2.3. Comparative Discussion

Based on the above findings, some numerical simulations were compared with the experimental results, as shown in Figure 19. The maximum perforation sizes of the four verified target plates were used to validate the numerical model, and the evolution and mechanism of the multi-layer target plate failure were analyzed.

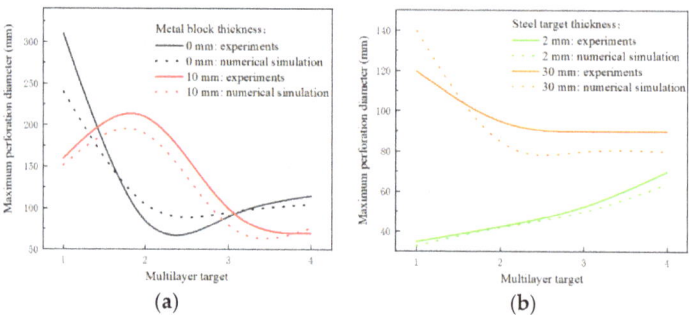

Figure 19. Numerical simulation validation: (**a**) Metal block thickness and (**b**) Steel target thickness.

Without installing the head metal block (that is, the 0 mm head metal block), the numerical simulation result of the No. 1 verification target plate was significantly lower than the experimental result, and the numerical simulation results of other verified target plates had certain deviations from the experimental values. The comprehensive destruction effect of the numerical simulations was slightly lower than that of the experiments. For the thickness of the steel target plate, when the 2 mm RHA steel target plate was impacted, the numerical simulation results of the damage effect of the MLAT were slightly lower than the experimental results, but when the 30 mm RHA steel target plate was impacted, the numerical simulation results and experimental results had significant differences. The analysis showed that, because a series of chemical energy-releasing reactions—such as explosion and deflagration—were involved in the destruction process, the damage process was more complicated; that is, the activation model built in the numerical simulation of this paper was only based on the one-dimensional shock wave theory. The first step of the projectile–target interaction triggered the deflagration effect of a small amount of active materials, and the subsequent partial activation reaction that occurred due to the local stress wave strength caused by the chemical energy release effect was promoted again. Because the reactive core in the experiment still broke in the process of impacting the MLAT, resulting in a smaller fragmentation size of the reactive core again, when the fragmentation activation size was reached, the activation reaction occurred again, so the experimental effect of the post-layer spacer aluminum target was better than the numerical simulation results; however, the overall error was controlled to be within 20%. Thus, we believe the simulation results were reasonable.

5. Conclusions

In this paper, an AUOTDYN-3D numerical simulation and experimental verification study of the penetration mechanism of the RMP multi-layer metal target was carried out. In addition, the deflagration reaction behavior of the RMP after the penetration of the steel target plate was effectively simulated using the SPH-Lagrange algorithm under different projectile–target interaction conditions. A gunpowder combustion EOS was introduced in the activation part of the reactive core. The scattering behavior of the shell target plate fragments was analyzed along with the subsequent penetration and destruction behaviors. The main conclusions are as follows:

1. Aiming to solve the problem of the large deformation of the projectile when the RMP collides with the MLAT, the SPH-Lagrange algorithm has been proposed. The SPH algorithm was used to calculate the pressure expansion and fragmentation behavior of the reactive core, and the activation behavior of the reactive core after the RMP penetrates the steel target was effectively simulated.
2. Aiming to solve the problem of deflagration reaction behavior caused by the RMP, the Powder Burn model was introduced to effectively simulate the deflagration reaction process of the reactive core when it was colliding with the MLAT. The damage evolution law of collision speed, the steel target thickness and the head metal block thickness of the MLAT were obtained.
3. Aiming to solve the problem of battlefield target damage assessments of the RMP, the deflagration reaction behavior of the reactive core and the radial expansion behavior of the shell after the RMP penetrates the steel target were reasonably characterized, and the damage mechanism of the RMP has been revealed. A new method is proposed for the rapid construction of damage prediction engineering models under different projectile–target interaction conditions.
4. The experimental results show that the combined damage and damage evolution behavior of the RMP on the MLAT was basically consistent with the results of the numerical simulation, and the error was within a reasonable range. This indicated that the SPH-Lagrange algorithm used in this numerical simulation has a high degree of accuracy for the study of the activation of a reactive core, the radial expansion of the shells, the scattering of the fragments, and other behaviors caused by the penetration. This provides a certain reference value for the study of the penetration-initiation combined damage effects mechanism of the RMP.

Author Contributions: Conceptualization, X.L. and L.Y.; software, C.H.; validation, Y.C.; formal analysis, C.H.; investigation, X.L.; resources, H.T.; writing—original draft preparation, C.H.; writing—review and editing, X.L. and Y.C.; supervision, H.T.; project administration, X.L.; funding acquisition, X.L. All authors have read and agreed to the published version of the manuscript.

Funding: This research was funded by the CST Foundation Strengthening Project, and the project No. is 2019JQJCZD01.

Conflicts of Interest: The authors declare no conflict of interest.

References

1. Jazon, B.; Backofen, J., Jr.; Brown, R.E.; Cayzac, R.; Giraud, M.; Held, M.; Diederen, A. The future of warheads, armour and ballistics. In Proceedings of the 23rd International Symposium on Ballistics, Tarragona, Spain, 16–20 April 2007; Volume 1, pp. 3–27.
2. Xu, F.Y.; Yu, Q.B.; Zheng, Y.F.; Lei, M.A.; Wang, H.F. Damage effect of doublespaced aluminum plates by reactive material projectile impact. *Int. J. Impact Eng.* **2017**, *4*, 1320.
3. Liu, S.; Zheng, Y.; Yu, Q.; Ge, C.; Wang, H. Interval rupturing damage to multi-spaced aluminum plates impacted by reactive materials filled projectile. *Int. J. Impact Eng.* **2019**, *130*, 153–162. [CrossRef]
4. Ames, R.G. Energy Release Characteristics of Impact-Initiated Energetic Materials. In Proceedings of the Materials Research Society Symposium, Boston, MA, USA, 27 November–1 December 2006.
5. Ames, R.G. Vented Chamber Calorimetry for Impact-Initiated Energetic Materials. In Proceedings of the 43rd AIAA Aerospace Sciences Meeting and Exhibit, Reno, NV, USA, 10–13 January 2005.
6. Valluri, S.-K.; Schoenitz, M.; Dreizin, E. Fluorine-containing oxidizers for metal fuels in energetic formulations. *Def. Technol.* **2019**, *15*, 1–22. [CrossRef]
7. Wang, H.-f.; Zheng, Y.-f.; Yu, Q.-b.; Liu, Z.-w.; Yu, W.-m. Initiation behavior of covered explosive subjected to reactive fragment. *J. Beijing Inst.Technol.* **2012**, *21*, 143–149.
8. Steven, N. Energetic Materials to Meet Warfighter Requirements: An Overview of Selected US Army RDECOM-ARDEC. *Energetic Mater. Programs* **2007**, *724*, 3016.
9. Ge, C.; Yu, Q.-b.; Lu, G.-c.; Zheng, Y.-f.; Wang, H.-f. Igniting effects and mechanism of diesel oil tank by projectile with reactive core. *Trans. Beijing Inst. Technol.* **2020**, *40*, 1072–1080, 1087. (In Chinese)
10. Lutz, B.; Wolfgang, S. 105/120/125 mm PELE Firing Results. In Proceedings of the NDIA–40th Annual Armament Systems: Guns-Ammunition-Rocket-Missiles Conference & Exhibition, New Orleans, LA, USA, 25–28 April 2005.
11. Xiao, J.; Nie, Z.; Wang, Z.; Du, Y.; Tang, E. Energy release behavior of Al/PTFE reactive materials powder in a closed chamber. *J. Appl. Phys.* **2020**, *127*, 165106. [CrossRef]
12. Zhang, X.; Xiao, J.; Yu, Q.; Zheng, Y.-f.; Wang, H.-f. Damage effect of concrete target under the action of shaped charge with reactive material charge. *Trans. Beijing Inst.Technol.* **2016**, *36*, 1211–1215. (In Chinese)
13. Jun, H.E.; Wei-xia, W.; Zhong-hua, D.U. Study on the implementation of reinforced concrete target in PELE with hood. *IOP Conf.Ser. Earth Environ.Sci.* **2021**, *791*, 012025. [CrossRef]
14. Cheng, C.; Du, Z.H.; Chen, X.; Xu, L.Z.; Du, C.X.; Han, J.L.; Wang, X.D. Damage of multi-layer spaced metallic target plates impacted by radial layered PELE. *Def. Technol.* **2020**, *16*, 201–207. [CrossRef]
15. Schirm, V.; Paulus, G. Penetrator Concept with Enhanced Lateral Efficiency (PELE). In Proceedings of the DEA 1060 A/AA—Workshop 2004, ARL, Aberdeen, MD, USA, 14–17 September 2004.
16. Mock, W., Jr.; Holt, W.H. Impact Initiation of Rods of Pressed Polytetrafluoroethylene (PTFE) and Aluminum Powders. In Proceedings of the American Institute of Physics Conference, Baltimore, MD, USA, 31 July–5 August 2005; pp. 1097–1100.
17. Yu, Q.; Guo, Z.; Zhong, S.; Zhao, H.; Ge, C. Analysis of penetration and blast combined damage effects of reactive material jet. *Trans. Beijing Inst.Technol.* **2021**, *41*, 465–473. (In Chinese)
18. Wang, H.; Zheng, Y.; Yu, Q.; Liu, Z.; Yu, W. Impact-induced initiation and energy release behavior of reactive materials. *J. Appl. Phys.* **2011**, *110*, 074904:1–074904:6.
19. Zhang, J.; Hou, C.; Li, X. Numerical simulation of the target matching characteristics of active material projectiles. *J. Proj. Arrows Guid.* **2021**, *41*, 112–117.
20. Li, X.; Hou, C.; Zhang, J. Study on Impact Speed of Active Projectile on Multilayer Target. *J. Ordnance Equip. Eng.* **2021**, *42*, 66–72.
21. Jimmy, V. Analytical and numerical description of the PELE fragmentation upon impact with thin target plates. *Int. J. Impact Eng.* **2015**, *76*, 196–206.
22. Century Dynamics Inc. *AUTODYN Theory Manual (Revision 4.3)*; Century Dynamics Inc: Concord, CA, USA, 2005.
23. Jiansheng, Z. Functional mechanism of penetrator with enhanced lateral effect. Ph.D. Thesis, Nanjing University of Science and Technology, Nanjing, China, 2008.
24. ANSYS Inc. *AUTODYN. User Manual Version 12*; ANSYS Inc.: Canonsburg, PA, USA, 2009.
25. Colby, C.; Swan, I.K. Voigt-Reuss topology optimization for structures with nonlinear material behaviors. *Int. J. Numer. Methods Eng.* **1997**, *40*, 3785–3814.
26. Jiang, J.W.; Wang, S.Y.; Zhang, M.; Wei, Q. Modeling and simulation of JWL equation of state for reactive Al/PTFE mixture. *J. Beijing Inst. Technol.* **2012**, *21*, 150–156.
27. Ge, C.; Yu, Q.; Zhang, H.; Qu, Z.; Wang, H.; Zheng, Y. On dynamic response and fracture-induced initiation characteristics of aluminum particle filled PTFE reactive material using hat-shaped specimens. *Mater. Des.* **2020**, *188*, 108472. [CrossRef]

28. Charlet, K.; Saulnier, F.; Dubois, M.; Beakou, A. Improvement of wood polymer composite mechanical properties by direct fluorination. *Mater. Des.* **2015**, *74*, 61–66. [CrossRef]
29. Paulus, G.; Schirm, V. Impact behaviour of PELE projectiles perforating thin target plates. *Int. J. Impact Eng.* **2006**, *33*, 566–579. [CrossRef]
30. Liangliang, D.; Jingyuan, Z.; Wenhui, T.; Xianwen, R.; Ye, C. Damage characteristics of PELE projectile with gradient density inner core material. *Materials* **2018**, *11*, 2389.
31. Yu, Q.B.; Zhang, J.H.; Zhao, H.W.; Xiao, Y.W.; Wang, H.F. Behind-plate overpressure effect of steel-encased reactive material projectile impacting thin aluminum plate. *Def. Technol.* **2022**, *18*, 723–734. [CrossRef]

Article

Study on the Formation of Reactive Material Shaped Charge Jet by Trans-Scale Discretization Method

Guancheng Lu [1], Chao Ge [1], Zhenyang Liu [1], Le Tang [2] and Haifu Wang [1,*]

[1] State Key Laboratory of Explosion Science and Technology, Beijing Institute of Technology, Beijing 100811, China; 3120195177@bit.edu.cn (G.L.); gechao@bit.edu.cn (C.G.); 3120215137@bit.edu.cn (Z.L.)
[2] Beijing Institute of Space Launch Technology, Beijing 100076, China; letang.cn@gmail.com
* Correspondence: wanghf@bit.edu.cn

Abstract: The formation process of reactive materials shaped charge is investigated by X-ray photographs and numerical simulation. In order to study the formation process, a trans-scale discretization method is proposed. A two-dimensional finite element model of shaped charge and reactive material liner is established and the jet formation process, granule size difference induced particle dispersion and granule distribution induced jet particle distribution are analyzed based on Autodyn-2D platform and Euler solver. The result shows that, under shock loading of shaped charge, the Al particle content decreases from the end to the tip of the jet, and increases as the particle size decreases. Besides, the quantity of Al particles at the bottom part of the liner has more prominent influence on the jet head density than that in the other parts, and the Al particle content in the high-speed section of jet shows inversely proportional relationship to the ratio of the particle quantity in the top area to that in the bottom area of liner.

Keywords: reactive material; shaped charge jet; trans-scale discretization; formation

1. Introduction

Reactive material, fabricated by pressing/sintering fluoropolymer and active metal powders, characterized by its metal-like strength and impact-initiated energy release, has been widely researched since 2000s [1–3]. The reactive jet, generated by shaped charge with reactive material liner, causing catastrophic damage to the armor due to its penetration and internal explosion properties [4], provides a novel application of reactive material and has received dramatically raised concern in recent years [4].

The present researches on reactive material jets involve jet formation [5], penetration capability [6], energy release characteristics [7], as well as the enhanced terminal effect [8,9]. Experimental and numerical methods, such as shock loading experiments by shaped charge [10], X-ray photographs investigation [5], and macroscopic modeling [11] are general methods applied to study the formation process and terminal effects of reactive material jets.

The reactive liner forms a jet in an extremely short duration. Apart from this, the reactive material jet would undergo an initiation and explosion process during its formation. Although numerical simulation has been widely used to investigate this issue, the reactive material was generally set to be homogeneous without regarding the granular metal particles. Recently, mesoscopic simulation on composite materials impacted by dynamic loading has gained much progress [12,13]. Researchers propose a two-dimensional real microstructure-based modeling technology to describe reactive material in much smaller scale [14], and appropriate equation of state parameters are given [15]. Mesoscopic numerical simulation is also introduced into the study of metal composites jets formation [16,17], while previous work [18,19] has also demonstrated the feasibility of the mesoscale simulation method for the PTFE-Al granular composites. Besides, the previous study also presented the damage enhancement behavior for typical PTFE-Al reactive material liner [11] and double-layered liner shaped charges [10], and reveals that the jet density and material

ratio have great influence on damage ability using macroscopic modeling. However, the jet formation process of reactive material liner under shock loading, regarding the mesoscale material characteristics, evolution of structures, and effect on terminal damage, are of great complexity, and are not reported so far.

In this research, a trans-scale discretization method is proposed. Two-dimensional finite element model of shaped charge and reactive material liner is established. Additionally, the validity of the trans-scale numerical simulations method is proved by the photographs of the X-ray experiment. Based on Autodyn-2D platform and Euler solver, the jet formation process, granule size difference induced particle dispersion, and granule distribution induced jet particle distribution are studied. The results would provide a valuable guide for the design and application of reactive material liner and shape charge.

2. Experiment Setup and Simulation Method

2.1. Experiment Setup

In this research, a typical shaped charge structure with a mono-cone reactive material liner is presented to study the influence of mesoscopic reactive material characteristics on the jet formation process. The warhead, depicted in Figure 1, mainly consists of an explosive and a mono-cone reactive material liner. The high explosive 8701 is poured into the press mold and a pressure load of 200 MPa is applied at room temperature, thus the diameter and length of the charge are both 40 mm, while the density of the charge is 1.71 g/cm^3. The cone angle and thickness of the liner are 60° and 4 mm, respectively. Detonation point is located at the center of the bottom of the charge.

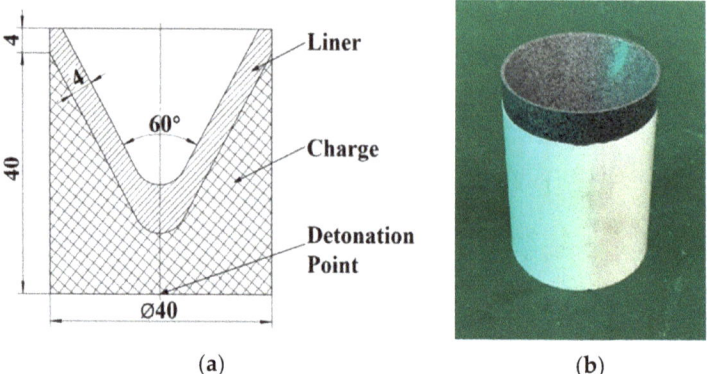

Figure 1. Structure and photograph of shaped charge for X-ray experiment: (**a**) structure and (**b**) photograph.

The PTFE/Al (70 wt.%/30 wt.%) reactive material liners are prepared by typical mixing/pressing/sintering method. Firstly, Al particles with an average diameter of 70 μm are dried and uniformly mixed with the PTFE matrix. Component mixtures are then pressed via a mold to shape a liner structure with particularly designed geometrical characteristics. Finally, the pressed liner would undergo a sintering cycle to further improve the mechanical and chemical properties.

The schematic diagram of the experimental setup is shown in Figure 2a, and the corresponding photograph is shown in Figure 2b. As can be seen, the warhead is positioned on a standoff (an experiment device), and an X-ray system is used to capture the formation of the reactive material jet. In the testing process, the two X-ray tubes were placed in different positions at the same distance from the negative, and the two rays intersected with the axis of the shaped charge at a certain angle. According to the prior simulation estimation, the times of jet head reaching 1 CD (charge diameter) standoff (the distance between the jet head and the initial position of the liner bottom) and 2 CD standoff were obtained. It should be noted that, since the damage effect of the reactive material jet is greatly influenced by the standoff, and the optimum standoff of reactive material jet is about 1 CD–2 CD [20], thus the times corresponding to these two standoffs are selected. The two different times were then set for the two ray tubes as ray emission delay times, where the initial time corresponds to the moment of the detonator being initiated. Thus, two appearances of the jet at different times were obtained on the negative films. The protective plate is also used to prevent the jet from impacting the cement floor.

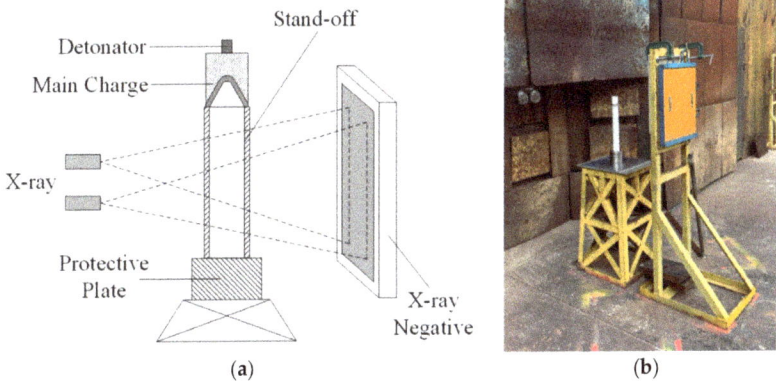

Figure 2. Experimental principle and setup of X-ray: (**a**) experiment principle and (**b**) setup of X-ray.

2.2. Trans-Scale Discretization Model

On this basis of the above mixing/pressing/sintering process of the reactive material liner, the Al particles could be assumed to have ideal roundness with diameters conforming to the lognormal hypothesis, and their positions are random without overlap, and the particle diameter and initial distribution are used respectively as control groups.

Therefore, the major issue of developing the discretization model is to fill the Al particles randomly into polymer matrix liner. The generation method of random circular particles is developed using Python programming and includes the following steps:

(1) Calculate the liner area A_{liner}, then the total Al particle area A_{Al} can be obtained from:

$$A_{Al} = \frac{W_{Al}}{\rho_{Al} \cdot \left(\frac{W_{Al}}{\rho_{Al}} + \frac{1-W_{Al}}{\rho_{PTFE}}\right)} A_{liner} \quad (1)$$

where ρ_{Al} and ρ_{PTFE} is the density of Al and PTFE respectively, and W_{Al} is the mass fraction of aluminum particles.

(2) Obtain a random particle diameter according to the lognormal distribution of Al particle diameter with consideration of the mean and standard deviation [14]. In this paper, the standard deviation of random particle diameters in all cases are set as 10% average particle diameter, and typical distribution of aluminum particle diameters is shown in Figure 3.

(3) Generate a random particle coordinate (x, y) in the liner region, and regenerate another one if the edge of the particle exceeds the liner boundary or the particle overlaps with the existing particles.

(4) Repeat steps 2 and 3 until the Al particle area meets the following condition:

$$\sum_{i=1}^{n} \pi R_i^2 \geq A_{Al} \quad (2)$$

where R_i represents the radius of each Al particle.

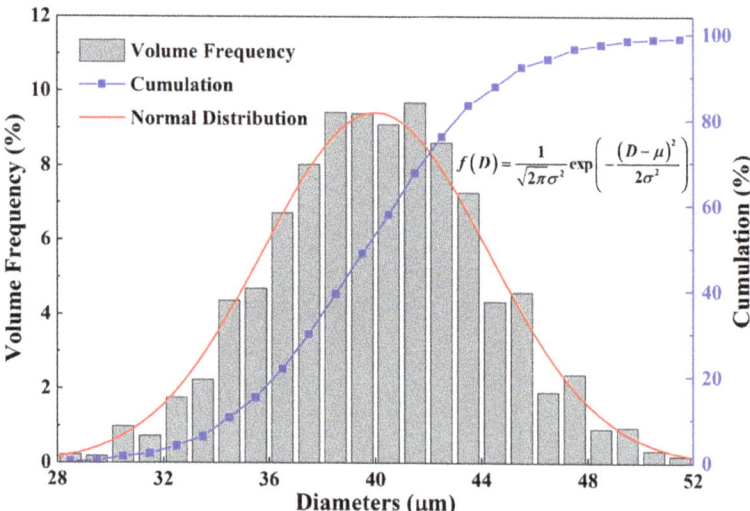

Figure 3. Distribution of aluminum particle diameters.

Finally, the discretization model of typical PTFE-Al liner could be obtained, as described in Figure 4. The diameters and coordinates of all the particles are finally converted into node data information for further characterization.

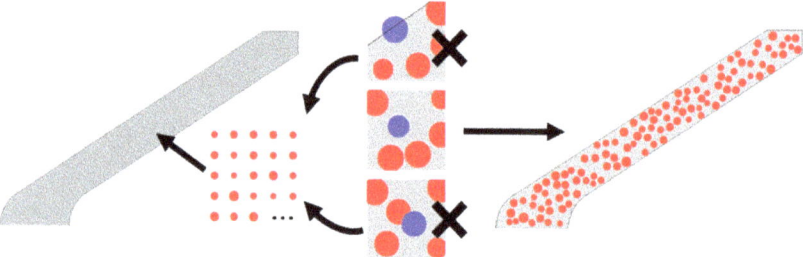

Figure 4. Generation process of typical PTFE-Al liner discretization model.

2.3. Finite Element Model

The finite element models are developed using Euler solver based on the Autodyn-2D platform. As shown in Figure 5, the model consists of a bigger air part (9 mm × 3.5 mm) joined with a smaller one (4 mm × 1.5 mm), which aims to shorten the computation duration. However, meshing is quite difficult because on one hand, fewer nodes lead to lower resolution and loss of accuracy, thereby losing information at the interface between matrix and particles; on the other hand, the number of nodes cannot be excessive due to computing power limitations. Therefore, considering that the average diameter of the Al particles is 40–100 μm, the two-dimensional axisymmetric numerical models are developed with a mesh size of 5 μm × 5 μm. Lastly, the flow-out (ALL EQUAL) boundary is set to eliminate the boundary effect.

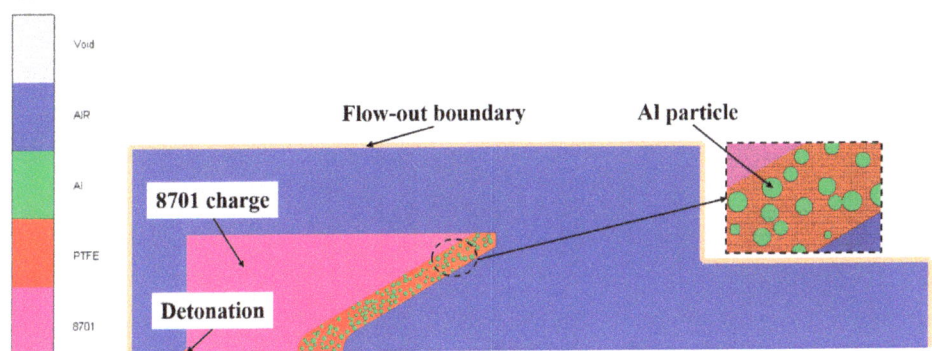

Figure 5. 2D finite element model of the PTFE/Al liner shaped charge.

In this paper, the expansion of the detonation product for 8701 explosive is described by the JWL equation of state (EOS), according to the following form:

$$P = A\left(1 - \frac{\omega}{R_1 V}\right)e^{-R_1 V} + B\left(1 - \frac{\omega}{R_2 V}\right)e^{-R_2 V} + \frac{\omega E_0}{V} \tag{3}$$

where A, B, R_1, R_2, and ω are material constants, E_0 represents the detonation energy per unit volume, and V is the relative volume. The corresponding parameters of 8701 explosive are from reference [6], in which $\rho_0 = 1.71$ g/cm^3, $A = 524.23$ GPa, $B = 7.678$ GPa, $R_1 = 4.2$, $R_2 = 1.1$, $\omega = 0.34$, $E_0 = 8.499$ GPa, CJ detonation pressure $P_{CJ} = 28.6$ GPa, and detonation velocity $D = 8315$ m/s.

The shock EOS is used to describe the behavior of Al and PTFE matrix. In the Autodyn-2D program, the shock EOS is established from the Mie–Gruneisen form of EOS based on shock Hugoniot,

$$P = P_H + \Gamma \rho (E - E_H) \tag{4}$$

where it is assumed that $\Gamma \rho = \Gamma_0 \rho_0 = $ constant and

$$P_H = \frac{\rho_0 c_0 u(1+u)}{(1-(s-1)u^2)} \tag{5}$$

$$E_H = \frac{1}{2}\frac{P_H}{\rho_0}\left(\frac{u}{1+u}\right) \tag{6}$$

where Γ_0 is the Gruneisen coefficient, $u = (\rho/\rho_0) - 1$, ρ is the current density, ρ_0 is the initial density, s is a linear Hugoniot slope coefficient, and c_0 is the bulk sound speed. The Johnson–Cook strength model, which combines the strain hardening, strain rate strengthening, and temperature softening effect, is used to represent the strength behavior of Al particles and PTFE matrix. The model defines the yield stress σ as:

$$\sigma = \left(A + B(\bar{\varepsilon}^p)^n\right)\left(1 + C \ln \dot{\varepsilon}^*\right)(1 - T^{*m}) \tag{7}$$

where $\bar{\varepsilon}^p$ is the effective plastic strain, $\dot{\varepsilon}^* = \dot{\varepsilon}/\dot{\varepsilon}_0$ is the normalized effective plastic strain rate for $\dot{\varepsilon}^* = 1.0$ s^{-1}, T^{*m} is the homologous temperature, where $T^{*m} = (T - T_{room})/(T_{melt} - T_{room})$, and A, B, C, n, and m are material constants. Detailed material parameters of the Al and PTFE are listed in Tables 1 and 2 [15].

Table 1. EOS parameters of the materials.

Material	ρ_0 (g/cm^3)	c_0 (m/s)	s (−)	γ_0
Al	2.71	5250	1.370	2.00
PTFE	2.15	1680	1.123	0.59

Table 2. Strength parameters of the materials.

Material	A (MPa)	B (MPa)	C	m	n	T_m (K)
Al	265	426	0.015	1.00	0.34	775
PTFE	11	44	0.120	1.00	1.00	350

2.4. Treatment of Mixed Material Grid

After calculation based on the finite element model, the PRT history file containing all grid data for each selected time is obtained through the print function of Autodyn software, then the data is analyzed using Python programming. However, in the Euler solver, multiple materials are mapped onto the Euler grid through a volume fraction technique [21], while all variables are grid centered. This characteristic leads to the appearance of the mixed material grids, which affects the tracking and calculation of the material distribution. Thus, the mixed grid is transformed into a specific material element according to the following method:

(1) According to the data of 8 grids around the hybrid grid, the average density of the same material grid except the mixed ones is calculated respectively:

$$\rho_{material,i} = \frac{\sum_1^n \rho_{unit,n}}{n} \quad (8)$$

where $\rho_{material,i}$ represents the average density of each substance in the surrounding grids, $\rho_{unit,n}$ is the density of a unit corresponding to the substance, and n is the total number of the units of this material.

(2) The material of the target grid is replaced by the surrounding material whose average density is closest to the target one.

For example, as shown in Figure 6, the typical partial material grid graphic (Figure 6a) consists of the PTFE/Al/mixed grids. Following the order from top to bottom and from left to right, the mixed grid (i = 2, j = 2) is processed first. According to the above method, the mixed grid (i = 3, j = 3) is ignored, only the density of the other five PTFE grids and two Al grids is considered (Figure 6b). Due to the density of the mixed grid (i = 2, j = 2) is closer to the average density of PTFE, the material of the mixed grid (i = 2, j = 2) is transferred into PTFE. Repeat these steps (Figure 6c) to continue processing the mixed grid (i = 3, j = 3) until all the grids are searched. Finally, all the mixed material grids are converted into specific material grids, and then all the grid data can be analyzed by Python programming.

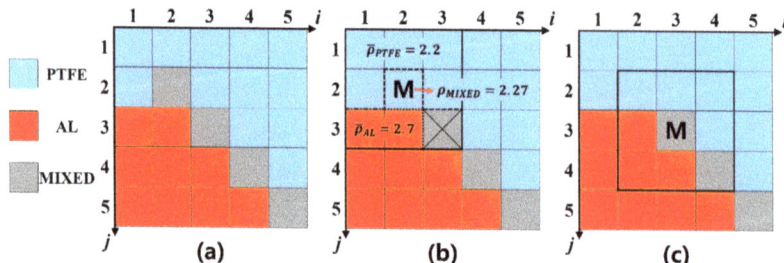

Figure 6. The processing method of the mixed material grids: (**a**) initial grids, (**b**) treatment of mixed material grid (i = 2, j = 2) and (**c**) repeat the treatment for next mixed material grid (i = 3, j = 3).

3. Result and Discussion

3.1. Comparison between Experimental and Simulated Results

The comparison between the experimental and simulated results is presented in Figure 7. The exposure times of the two X-ray pictures are 16 μs and 21 μs, corresponding to the situation when the jet tip reaches 1 CD and 2 CD standoff, respectively. It can be seen from the X-ray that the jet is constantly elongated in the forming process, and the morphology shape of the jet is clear when it reaches 1 CD standoff. At the 2 CD standoff situation, the tip and neck of the jet are not very clear, though the shape and contour of the jet still exist. That may indicate that some materials react at the tip of the jet though the overall shape of the jet is relatively stable at a 2 CD standoff.

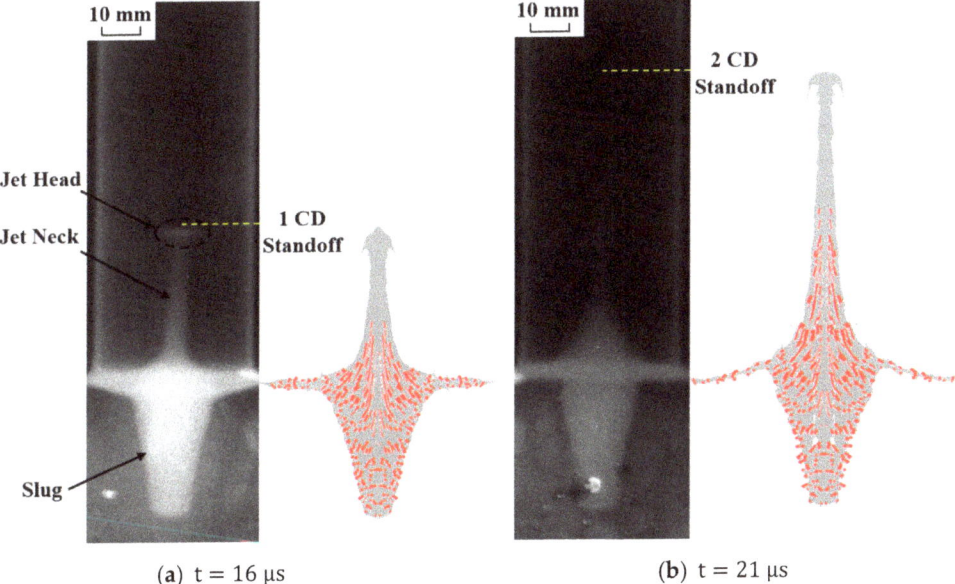

Figure 7. Comparison between experimental and simulated results.

At the same time, the morphologies of the reactive material jet formation process shown in the X-ray photographs agree well with the numerical simulation results. In the simulation image, the red part represents Al and the gray part represents the PTFE matrix, respectively. By comparison, it can be found that when the standoff is 1 CD, the head and neck of the jet in the X-ray photo are clearer than the slug, while the Al particles in the jet head and neck are less than the part of the slug in the simulation. It indicates that low content of Al particles in the jet (low material density) would reduce the clarity of the X-ray image. When the standoff is 2 CD, the part with more Al particles in the simulation also has higher definition in the X-ray film. At the same time, compare the two X-ray images, the 1 CD standoff film is brighter than the 2 CD one. That is most likely because the materials in the jet are more diffuse in the 2 CD situation, which leads to lower density than the 1 CD jet, thus the brightness on the 2 CD picture is darker. On the other hand, in the 2 CD standoff case, some materials may react, thus reducing the brightness of the picture. However, according to reference [6], the average initiation delay time is about 120 μs, thus the reaction of the reactive materials at 21 μs is not the main reason for the darker X-ray image. Therefore, it is necessary to analyze the dispersion characteristics of Al particles in the jet forming process for further analysis of the penetration and reaction performance of the reactive material jet.

3.2. Formation of Reactive Material Jet

Figure 8 shows the formation process of a typical reactive material jet. After explosion of the main charge, the reactive material liner is integrally accelerated by the shock wave and detonation products. At 5 µs and 10 µs, the inner wall of the liner collapses and gathers into the symmetry axis to form the jet due to its relatively higher speed. Because of the lower density, the PTFE in the liner accelerates faster and becomes the main component of the jet. With the increase of time, the jet containing both PTFE and Al gradually elongates, but the content of PTFE in the jet is significantly higher than that of Al.

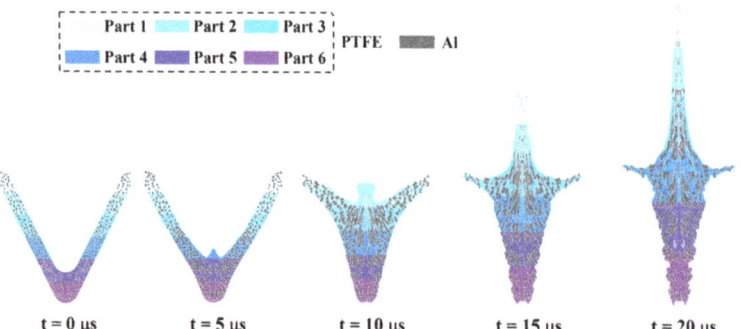

Figure 8. Formation of typical reactive material jet.

To investigate the velocity difference between PTFE and Al during the forming process, the liner and jet in Figure 8 are divided into six equal-length parts. The curves of the velocity difference between the two components from part 2 to part 6 over time are shown in Figure 9a. Since the jet head is composed of PTFE with a higher relative velocity, from 5 µs to 10 µs, there is a period of time during the jet forming process when the jet head (part 1 when t = 10 µs in Figure 8) has no Al particles. Thus, the velocity difference in part 1 corresponding to the tip area of the jet is not steady and the data is not included in Figure 9a. The average velocities of Al and PTFE in the reactive material jet at different times are shown in Figure 9b.

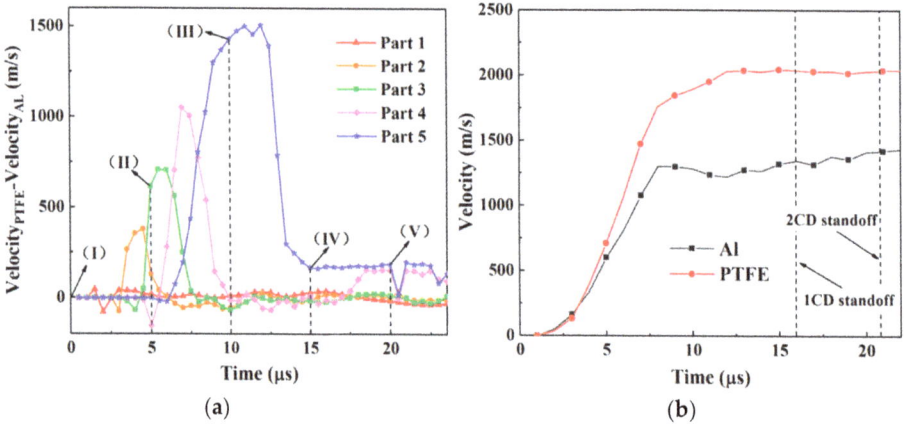

Figure 9. Velocity–time curve of the Al and PTFE in the reactive material jet: (**a**) the velocity difference between PTFE and Al of each part versus time; (**b**) the velocity of total Al and PTFE versus time.

Overall, the velocity of Al and PTFE increases simultaneously at the initial acceleration process. However, after 5 µs, the PTFE accelerates faster, up to about 2000 m/s, while the velocity of Al increases more slowly to 1500 m/s. This may be the difference between the density of the two components that contributes to the different particle velocities under the same impact load. However, the two components cannot transfer their stress stably due to the extremely short period of the jet formation process. Therefore, the relative displacement of the two components appears and finally forms the density gradient along the axis of the jet.

For the different parts, the velocity difference curves reach the peak in sequence from part 6 to part 2. It suggests that as the liner collapses, the two components are accelerated and form relative displacement. For each part, the PTFE component moves faster than Al particles, and thus flows forward relative to the Al particles. At the same time, the velocity of the part closer to the jet head is much higher than the backward parts, leading to a larger velocity difference and relative displacement. Finally, fewer aluminum particles remain in the part with higher velocity.

In Figure 10, the reactive material jet at a standoff of 2 CD is equidistantly divided into 10 parts and the Al particle mass fraction of each part is shown in the calculated Al content curve. The result shows that the Al particle content gradually decreases along the symmetry axis from the slug to the jet tip, which corresponds with the previous statement about the gradient density.

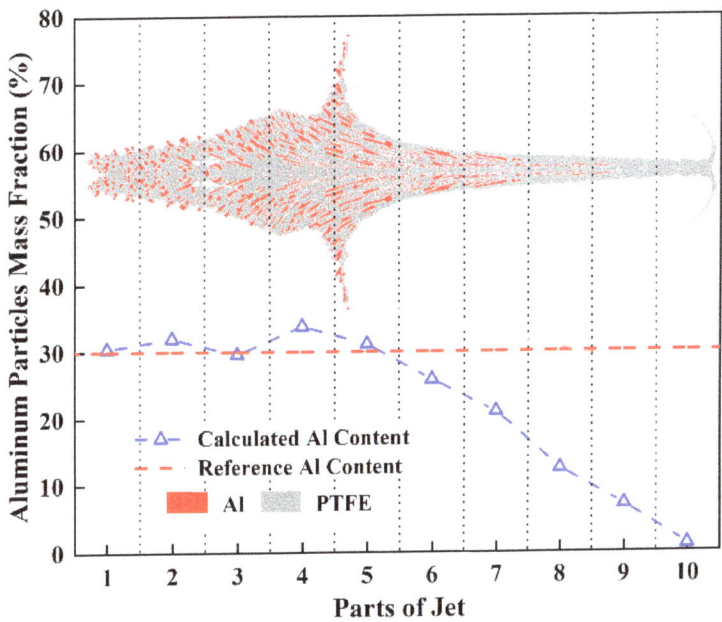

Figure 10. Aluminum particles mass distribution along the jet axis at 2 CD standoff.

3.3. Granule Size Difference Induced Particle Dispersion

In order to study the influence of the granule size on the particle distribution characteristics of the reactive material jet, four granule sizes (40 µm, 60 µm, 80 µm, and 100 µm) are chosen to build the discrete models. At the same time, to distinguish the different parts of the jet, the threshold velocity is defined as 2000 m/s in this article to divide the jet into a high-speed section (HSSJ) and a low-speed section (slug). As shown in Figure 11a, the velocities of all the materials (include Al and PTFE) in HSSJ are over 2000 m/s. Besides, Figure 11b shows the parts whose velocities are over 2000 m/s and below 2000 m/s in the jets of different standoffs, and the region shape of HSSJ in all standoff cases are basically the same except for stretching over time.

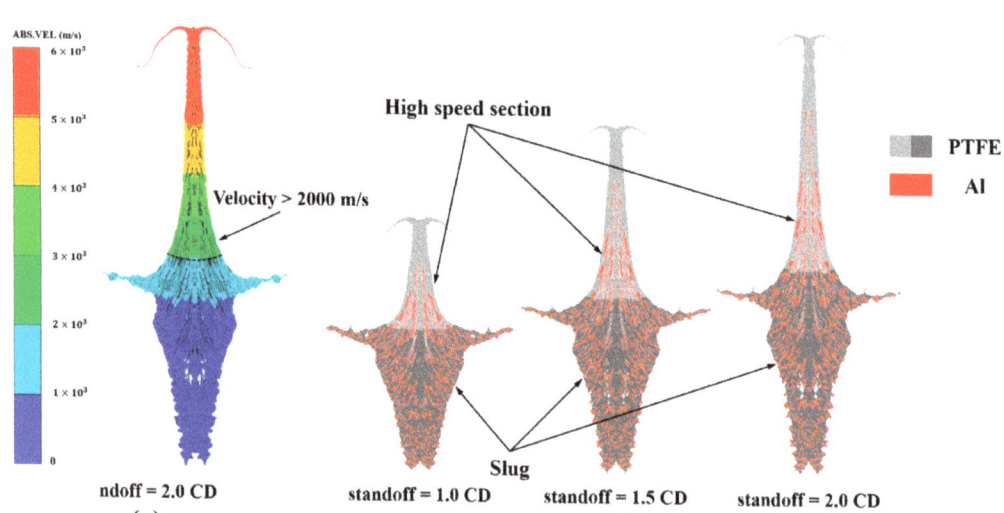

Figure 11. The velocity profile and region division of jets: (**a**) velocity profile of 2 CD-standoff jet; (**b**) the HSSJ and slug of different standoffs.

The aluminum content in the HSSJ is analyzed for different granule sizes, at different standoff values. The results are shown in Table 3. The data in Table 3 illustrates that the particle content of the HSSJ is rarely affected by the standoffs when the value is more than 1.25 CD. Therefore, the focus of the following analysis will be mainly on the jet under the 2.0 CD standoff condition.

Table 3. List of computation conditions and results.

Sample	Al (wt.%)	Granule Diameter (µm)	Al Content in HSSJ (wt.%)				
			1.0 CD	1.25 CD	1.5 CD	1.75 CD	2 CD
1	30	40	20.36	20.33	20.08	20.27	20.38
2	30	60	18.80	18.48	17.95	17.53	17.58
3	30	80	13.57	13.23	15.15	16.21	16.23
4	30	100	8.28	7.99	10.84	10.40	11.64

Geometrical morphology of jets formed by granules of different size at 2 CD standoff are shown in Figure 12. By comparison, smaller granules are more easily dispersed in the jet formed by the reactive material liner. When the granule size is 40 µm, the aluminum particle can enter the tip area of the jet, while the 100 µm situation has no particles entering the jet tip. Thus, the dispersion of the particles in the jets increases as the granule size decreases. This effect is related to the different velocities of the granules, as explained in the next sections.

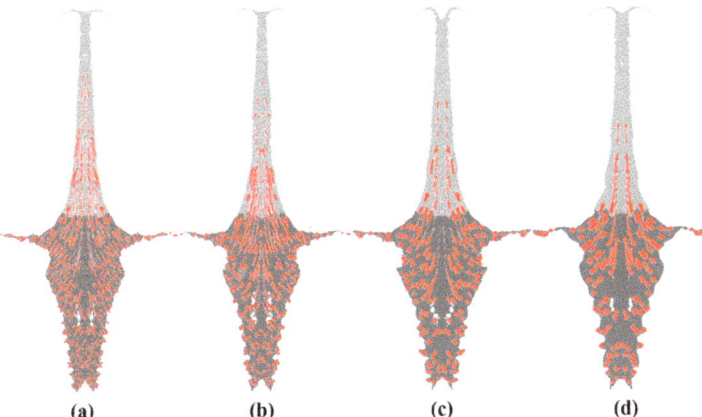

Figure 12. Geometrical morphology of jets formed by Al granules of different average size at 2 CD standoff: (**a**) $\Phi_{Al} = 40$ μm, (**b**) $\Phi_{Al} = 60$ μm, (**c**) $\Phi_{Al} = 80$ μm and (**d**) $\Phi_{Al} = 100$ μm.

In order to describe the change of aluminum content in each part of the jet, the following Equation (9) is proposed to calculate the relative aluminum content Δ_{AL} [13]:

$$\Delta_{AL} = \frac{m_{AL} - m_{0,AL}}{m_{0,AL}} \times 100\% \tag{9}$$

where m_{AL} is the aluminum mass fraction in the corresponding part of jet, $m_{0,AL}$ is the initial aluminum mass fraction in the reactive material liner.

In Figure 13, the reactive material jets of different granule diameters are equidistantly divided into 10 parts (the same division method as Figure 10) and the relative aluminum content Δ_{AL} distribution in each part of the cases is shown for each granule size. The plot shows that, as a whole, the aluminum content in the reactive material jet increases at first and then decreases from the head to the tail independently of the granule size. For the four defined granule sizes, there is a maximum Al content in the middle region (parts 4–6) of the jet, increasing with the granule size. On the contrary, for the upper region of the jet (parts 7–10), the Al content decreases and it is lower for the larger granule size condition.

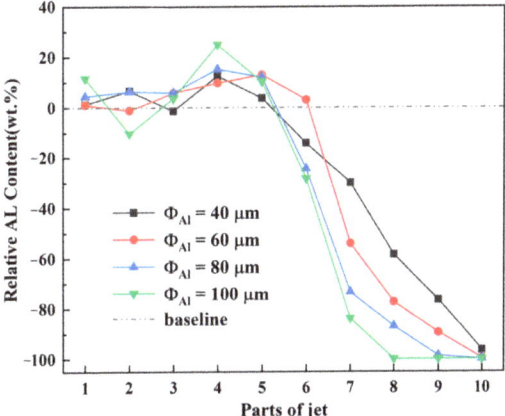

Figure 13. Relative Al content in each part of jet at 2 CD standoff. Note: The parts are defined using the same division method as Figure 10; the baseline corresponds to the original aluminum content in the liner.

The above results indicate that, in the jet formation process, smaller particles are more likely to enter the front part of the jet (the part with higher velocity), leading to a higher aluminum content in the head of the jet, which is consistent with the particle distribution characteristic in HSSJ in the table above. This is due to the fact that under the same loading condition, larger particles gain lower velocity compared with smaller particles, making it more easily to generate relative displacement. Assuming that the load on aluminum particles in the jet formation process is in a single direction and is proportional to the projection area of the particles, the acceleration of aluminum particles can be calculated by:

$$a_{granule} = \frac{3q}{4\rho r} \quad (10)$$

where q is the stress per unit area, ρ is the granule density, and r is the granule diameter. The product (ρr) is related to mass, therefore the size and mass of granule have the same dependence. Meanwhile, the mass and acceleration are related inversely. Accordingly, the acceleration of the particle is inversely proportional to the granule diameter, and finally results in a higher velocity of the smaller granule. Figure 14 shows the aluminum particle velocity and the velocity difference ($V_{PTFE} - V_{AL}$) profiles over time. The Al particle velocity curves show that the aluminum particles in all cases first accelerate integrally, then decelerate for a period of time, and finally accelerate again. At the same time, larger particles decelerate more than the smaller ones, and finally lead to the increase of the average velocity difference among the four kinds of particles. It should be noted that, in order to show the real speed difference, all the average velocities at each moment in Figure 14 only calculate material with velocities higher than 0 m/s. Additionally, after 8 μs, the material in the middle and bottom of the liner is accelerated and calculated into the average velocity of aluminum particles, resulting in a decline section in the velocity curve. Therefore, the deceleration corresponding to the granule size actually reflects its acceleration ability. The velocity difference curves in the Figure 14 show that, velocity difference increases with the particle diameter, leading to larger relative displacement between the two components. Finally, the content of aluminum granules in the HSSJ decreases while the content of aluminum granules in the slug increases.

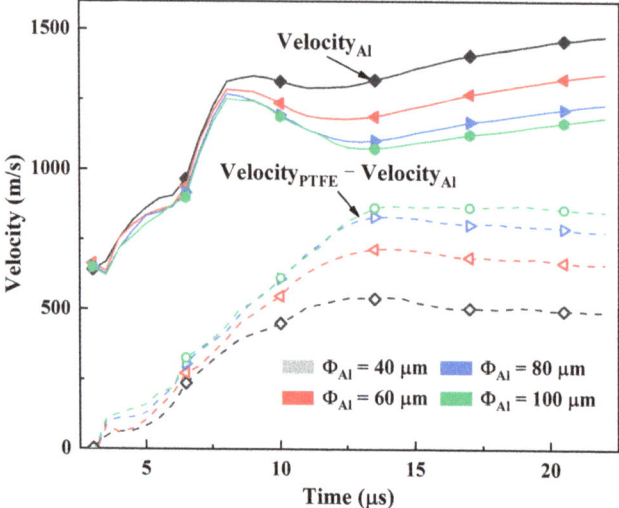

Figure 14. Velocity–time profiles of Al and PTFE-Al. Note: All the average velocities at each moment only calculate material with velocities higher than 0 m/s.

The result suggests that because of its weaker speed-up-ability, larger granules would have a higher proportion of particles which concentrate in the middle and bottom area of the jet. On the contrary, smaller particles accelerate faster and can disperse more into the HSSJ. Therefore, the smaller Al particles could be used in the engineering design to increase the density of the jet head which leads to a higher penetration performance. However, a previous study [10] showed that smaller Al particles speed up the reaction of the reactive material resulting in a weaker penetration performance. Thus, the diameter of the Al particles should be controlled within a relatively appropriate range to obtain a more balanced penetration performance and reaction capacity.

3.4. Evolution of Granules from Liner to Jet

The influence of the initial particle distribution in the liner on the particle distribution characteristics of the jet is studied in this section. A uniform distribution of discrete model is conducted and the reactive material liner is divided into three parts from top to bottom. Figure 15 shows the movement of the particles from different parts of the liner. Firstly, the particles from the top parts of the liner move to the symmetry axis, and the PTFE matrix accelerates faster than the Al particles due to its lower density. The PTFE then gathers into the jet ahead of the Al particles at 6 μs. With the collapse of the liner and the elongation of the jet, the particles from the top part of the liner are distributed to the head and tail of the jet, and the particles from the middle and bottom section move to the head and neck of the jet.

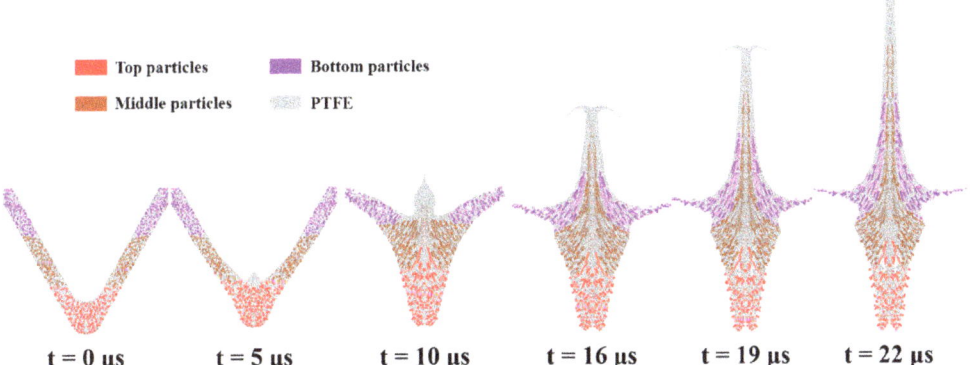

Figure 15. Typical uniform-distributed particles movement in the jet formation process.

The mass fraction and velocity charts of the aluminum particles from different parts of the liner are shown in Figure 16. In Figure 16a, the jet formed by the liner with uniform distributed particles is divided into 10 parts, and the mass fractions of the particles from each part are shown as three curves, respectively. As a whole, the particles from three parts of the liner are distributed in sequence from the head to the slug. The bottom particles are mainly distributed in the top and middle area of the jet, and the top particles are mainly distributed in the slug. In HSSJ, most of the aluminum particles are from the bottom and the middle area of the liner, while the top particles can hardly enter the HSSJ. Figure 16b shows the velocity–time profile of the Al particles from three parts of the liner. The figure shows that, in the formation process, the top particles are first accelerated, and then quickly drop to a relatively low speed. The middle and bottom particles are subsequently accelerated to a higher speed. At the same time, both the curves of the middle and the top particles contain a deceleration period. That is because when the wall of the liner collapses and closes, pressure is transferred from the back forward to the particles closer to the axis, and forcing them into the area with higher velocity. As a result, the particles from other areas slow down. As the jet continues elongating, this energy transfer gradually decreases and the velocity of the middle and top particles tends to be stable.

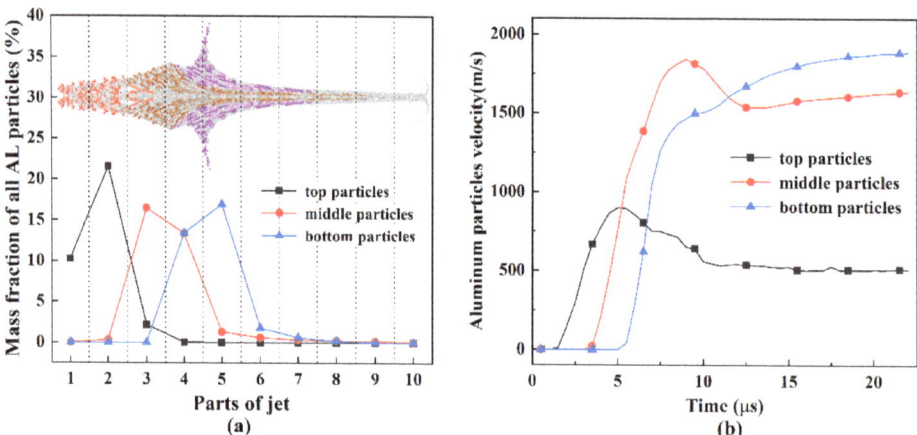

Figure 16. Velocity and mass fraction of Al particles from different parts: (**a**) mass distribution of Al particles from different parts along the jet axis at 2 CD standoff; (**b**) average velocity of Al particles from different parts.

3.5. Granule Distribution Induced Jet Particle Distribution

According to the above result, the particles from the middle and the bottom part of the liner have a significant impact on the density of the HSSJ. Besides, the powder-pressed manufacturing process also causes uneven density of each part of the reactive material liner, which is due to the concentration effect of the particles [22,23]. Figure 17 shows the liner models of different particle distributions. Four liner models with uneven particle distribution (Figure 17a,b,d,e) are studied and compared with the uniform density one (Figure 17c). All the liner models are equally divided by area into three parts, and each part is filled with different quantities of Al particles until the overall aluminum particle mass in the liner is m, so that all the cases have the same Al particle mass fraction ($wt_{Al} = 30\%$) in the liner. Besides, the Al particle diameter in all samples is set as 40 μm. At the same time, the particle distribution characteristic of each sample is represented by $\rho_{liner} = m_{top}/m_{bottom}$, where the middle region of the liners maintains the same concentration in all cases. The five models correspond to configurations with maximum granulate concentration in the bottom region ($\rho_{liner} = 0.5$), up to the top region ($\rho_{liner} = 2$).

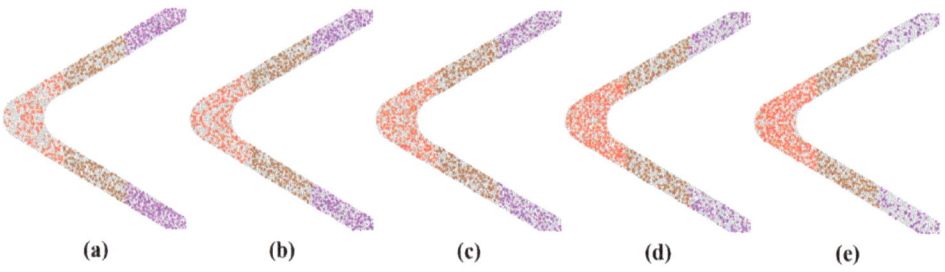

Figure 17. Liner models of different particle distributions: (**a**) $\rho_{liner} = 0.5$, (**b**) $\rho_{liner} = 0.75$, (**c**) $\rho_{liner} = 1$, (**d**) $\rho_{liner} = 1.5$ and (**e**) $\rho_{liner} = 2$. Note: The material colors are the same as in Figure 15.

The aluminum mass ratio in HSSJ for different liner granule distributions are shown in Table 4. Figure 18 shows the typical images of 2 CD-standoff jet with different initial particle distribution. In all the cases in Figure 18, the top particles are mainly distributed in the slug, the middle particles are mainly distributed in the back and middle of the jet, and the bottom particles are distributed in the front of the jet. The simulations show clearly that the middle and bottom particles are the main components of aluminum particles in the HSSJ.

Table 4. Al content in HSSJ for different liner granule distributions according to ρ_{liner}.

Sample	ρ_{liner}	Initial Al Content (wt.%)			Al Content in HSSJ (wt.%)		
		Top	Middle	Bottom	Top	Middle	Bottom
1	0.5	22.22	33.33	44.44	0	25.17	74.83
2	0.75	28.57	33.33	38.10	0.19	25.99	73.82
3	1	33.33	33.33	33.33	0.33	29.44	70.23
4	1.5	40	33.33	26.67	0.53	35.96	63.51
5	2	44.44	33.33	22.22	3.57	51.70	44.73

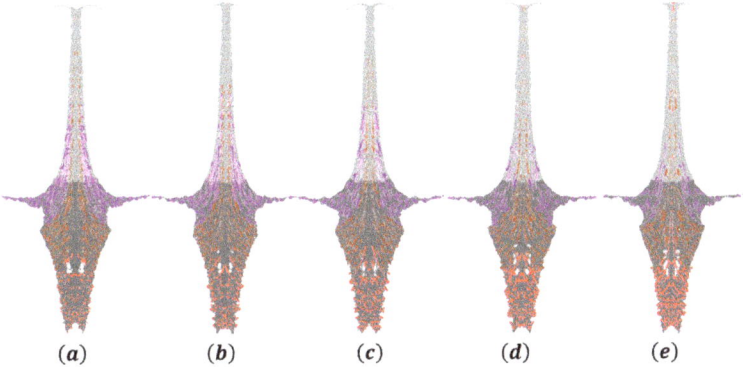

Figure 18. Typical images of 2 CD-standoff jet with different initial particle distribution: (**a**) $\rho_{liner} = 0.5$, (**b**) $\rho_{liner} = 0.75$, (**c**) $\rho_{liner} = 1$, (**d**) $\rho_{liner} = 1.5$ and (**e**) $\rho_{liner} = 2$. Note: The particle colors are the same as in Figure 15. The light gray corresponds to HSSJ while the dark gray corresponds to slug.

Table 4 shows the mass ratio of the particles from different parts of liner to all aluminum material in HSSJ. The results show that with the increase of ρ_{liner} (top particles content in liner increases, bottom particles content in liner decreases), the content of bottom particles in HSSJ decreases while the content of the middle particles increases. However, there is a small increase of the content of top particles in HSSJ. As shown in Figure 18e, the top particles are only distributed in the tip of the jet (the red material) in HSSJ. That is because only a small portion of the top particles which is close to the jet axis can flow into the tip in the formation process. The increase of the portion of the middle particles is due to the decrease of the bottom particle content since the top particles can rarely enter the HSSJ. Besides, the mass ratio of the middle and bottom particles in HSSJ is still above 95%, which means the middle and bottom particles are the key components of the Al in HSSJ. However, in general, the bottom particle density has the greatest influence on the Al particle content in HSSJ.

Figure 19 shows the relative Al content in each part of the jet of different samples. The illustration shows that, with the increase of the ρ_{liner}, the position where the maximum relative Al content appears in each sample moves from the middle area to the tail. In the tip of the jet (parts 9 and 10), the sample with more initial particles in the top area of the liner gains more relative aluminum content. That is because the content of the top particles in HSSJ rises with the quantity of the initial top particles in the liner. In the middle parts (parts 4–8), the relative aluminum content in the jet of each sample decreases successively with the increase of ρ_{liner}. This is because the bottom particles are mainly distributed in the middle part of the jet, therefore more particles are distributed in the middle part of the jet with the increase of the quantity of the initial bottom particles. On the contrary, the relative aluminum content in the jet of the samples decreases successively with the increase of ρ_{liner} in the slug (parts 9 and 10). This is because the top particles are mainly distributed in the slug. As the quantity of particles in the top area of the liner increases, more particles are distributed in the slug.

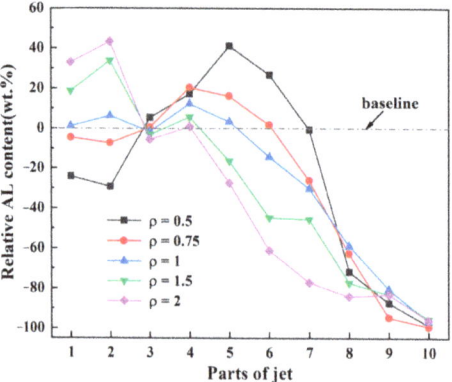

Figure 19. Relative Al content in each part of jet.

The change of the Al particle distribution characteristics in the jet with the initial particle distribution in the liner is qualitatively discussed above. It should be noted that, on the premise that the overall Al quantity in the liner remains unchanged, the aluminum content in the HSSJ should be related to the ρ_{liner}, which is the ratio of the particle quantity in the top area to that in the bottom area of the liner. Assuming that the relationship between the two can be expressed by $wt_{AL} = k \times \rho_{liner} + b$. The fitted curve according to the simulation data is shown in Figure 20, and the constant k and b are fitted as -10.01 and 28.57, respectively. For the fitted curve, it shows a downward trend with the increasing of ρ_{liner}.

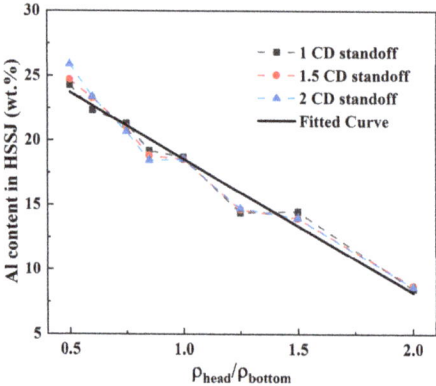

Figure 20. Fitted curve of the Al content in HSSJ of different initial particle distribution in liner.

4. Conclusions

In this paper, a trans-scale discretization method for analyzing the formation of reactive material shaped charge jet is proposed. Additionally, an X-ray experiment is conducted to confirm the validity of the trans-scale numerical simulations. Formation process of PTFE/Al reactive material jet, Al granule size difference induced particle dispersion, and granule distribution induced jet particle distribution are obtained and discussed. The main conclusions are as follows:

(a) Due to the difference of densities, the PTFE matrix accelerates faster than the Al particles under shock loading. The relative displacement results in a density gradient along the axis of the jet and PTFE becomes the main component of the jet.

(b) Because of the weaker speed-up-ability, larger Al particles would mainly concentrate in the middle and bottom area of the jet. On the contrary, smaller particles accelerate faster and mainly disperse in the high-speed section of the jet.

(c) The initial granule distribution in the liner has great influence on the particle distribution in the jet. The particle quantity in the top area of the liner has little impact on the Al content in the HSSJ, while the particles from the middle and bottom area of the liner influence that significantly. Furthermore, the aluminum content in the HSSJ is inversely proportional to the ratio of the particle quantity in the top area to that in the bottom area of the liner.

Author Contributions: G.L., conceptualization, investigation, writing—original draft, writing—review and editing; C.G., conceptualization, writing—original draft; Z.L., resources, conceptualization; L.T., investigation; H.W., conceptualization, writing—review and editing. All authors have read and agreed to the published version of the manuscript.

Funding: This research is supported by the Natural Science Foundation of Beijing, China (No. 1214022).

Institutional Review Board Statement: Not applicable.

Informed Consent Statement: Not applicable.

Data Availability Statement: Not applicable.

Conflicts of Interest: The authors declare no conflict of interest.

References

1. Ames, R. Vented Chamber Calorimetry for Impact-Initiated Energetic Materials. In Proceedings of the 43rd AIAA Aerospace Sciences Meeting and Exhibit, Reno, NV, USA, 10–13 January 2005.
2. Wang, L.; Liu, J.; Li, S.; Zhang, X. Investigation on reaction energy, mechanical behavior and impact insensitivity of W-PTFE-Al composites with different W percentage. *Mater. Des.* **2016**, *92*, 397–404. [CrossRef]
3. Xu, F.Y.; Yu, Q.B.; Zheng, Y.F.; Lei, M.A.; Wang, H.F. Damage effects of double-spaced aluminum plates by reactive material projectile impact. *Int. J. Impact Eng.* **2017**, *104*, 13–20. [CrossRef]
4. Xiao, J.; Zhang, X.; Wang, Y.; Xu, F.; Wang, H. Demolition Mechanism and Behavior of Shaped Charge with Reactive Liner. *Propellants Explos. Pyrotech.* **2016**, *41*, 612–617. [CrossRef]
5. Wang, Y.; Yu, Q.; Zheng, Y.; Wang, H. Formation and Penetration of Jets by Shaped Charges with Reactive Material Liners. *Propellants Explos. Pyrotech.* **2016**, *41*, 618–622. [CrossRef]
6. Guo, H.; Zheng, Y.; Yu, Q.; Ge, C.; Wang, H. Penetration behavior of reactive liner shaped charge jet impacting thick steel plates. *Int. J. Impact Eng.* **2019**, *126*, 76–84. [CrossRef]
7. Zheng, Y.; Su, C.; Guo, H.; Yu, Q.; Wang, H. Behind-Target Rupturing Effects of Sandwich-like Plates by Reactive Liner Shaped Charge Jet. *Propellants Explos. Pyrotech.* **2019**, *44*, 1400–1409. [CrossRef]
8. Xiao, J.; Zhang, X.; Guo, Z.; Wang, H. Enhanced Damage Effects of Multi-Layered Concrete Target Produced by Reactive Materials Liner. *Propellants Explos. Pyrotech.* **2018**, *43*, 955–961. [CrossRef]
9. Zheng, Y.; Su, C.; Guo, H.; Yu, Q.B.; Wang, H.F. Chain damage effects of multi-spaced plates by reactive jet impact. *Def. Technol.* **2021**, *17*, 393–404. [CrossRef]
10. Wang, H.F.; Guo, H.G.; Geng, B.Q.; Yu, Q.; Zheng, Y. Application of reactive materials to double-layered liner shaped charge for enhanced damage to thick steel target. *Materials* **2019**, *12*, 2768. [CrossRef] [PubMed]
11. Su, C.H.; Wang, H.F.; Xie, J.W.; Ge, C.; Zheng, Y.F. Penetration and damage effects of reactive material jet against concrete target. *Acta Armament.* **2019**, *40*, 1829–1835.

12. Austin, R.A.; Mcdowell, D.L.; Benson, D.J. Mesoscale simulation of shock wave propagation in discrete Ni/Al powder mixtures. *J. Appl. Phys.* **2012**, *111*, 123511. [CrossRef]
13. Wagner, F.; Ouarem, A.; Gu, C.F.; Allain-Bonasso, N.; Toth, L.S. A new method to determine plastic deformation at the grain scale. *Mater. Charact.* **2014**, *92*, 106–117. [CrossRef]
14. Ge, C.; Dong, Y.X.; Maimaitituersun, W. Microscale Simulation on Mechanical Properties of Al/PTFE Composite Based on Real Microstructures. *Materials* **2016**, *9*, 590. [CrossRef] [PubMed]
15. Qiao, L.; Zhang, X.F.; He, Y.; Shi, A.S.; Guan, Z.W. Mesoscale simulation on the shock compression behaviour of Al-W-Binder granular metal mixtures. *Mater. Des.* **2013**, *47*, 341–349. [CrossRef]
16. Wang, F.; Ma, D.; Wang, P.; Liu, J.; Jiang, J. Experimental and numerical study on the meso-scopic characteristics of metal composites jets by a shaped charge. *J. Appl. Phys.* **2019**, *126*, 095901. [CrossRef]
17. Wang, F.; Jiang, J.W.; Men, J.B. Mesoscopic Numerical Simulation on the Formation of Tunsgsten-Copper Shaped Charge Jet. *Acta Armament.* **2018**, *39*, 245–253.
18. Tang, L.; Ge, C.; Guo, H.G.; Yu, Q.B.; Wang, H.F. Force chains based mesoscale simulation on the dynamic response of Al-PTFE granular composites. *Def. Technol.* **2021**, *17*, 56–63. [CrossRef]
19. Tang, L.; Wang, H.; Lu, G.; Zhang, H.; Ge, C. Mesoscale study on the shock response and initiation behavior of Al-PTFE granular composites. *Mater. Des.* **2021**, *200*, 109446. [CrossRef]
20. Guo, H.; Xie, J.; Wang, H.; Yu, Q.; Zheng, Y. Penetration Behavior of High-Density Reactive Material Liner Shaped Charge. *Materials* **2019**, *12*, 3486. [CrossRef] [PubMed]
21. Century Dynamics, Inc. *AUTODYN Theory Manual (Revision 4.3)*; ANSYS Inc.: Canonsburg, PA, USA, 2005.
22. Zhao, Z.Y. *Research of Penetration Properties and Penetration Mechanism of W-Cu Alloy Shaped Charge Liner*; Beijing Institute of Technology: Beijing, China, 2016.
23. Sippel, T.R.; Son, S.F.; Groven, L.J. Aluminum agglomeration reduction in a composite propellant using tailored Al/PTFE particles. *Combust. Flame* **2014**, *161*, 311–321. [CrossRef]

Article

The Effect of Aluminum Particle Size on the Formation of Reactive Jet

Mengmeng Guo [1], Yanxin Wang [2], Yongkang Chen [3], Jianguang Xiao [2,4] and Haifu Wang [1,*]

[1] State Key Laboratory of Explosion Science and Technology, Beijing Institute of Technology, Beijing 100081, China
[2] College of Mechatronic Engineering, North University of China, Taiyuan 030051, China
[3] Department of Weapons and Control, Academy of Armored Force Engineering, Beijing 100072, China
[4] Science and Technology on Transient Impact Laboratory, No.208 Research Institute of China Ordnance Industries, Beijing 102202, China
* Correspondence: wanghf@bit.edu.cn

Abstract: In order to study the morphology characteristics of the PTFE/Al reactive shaped charge jet and the chemical reaction during the jet formation, PTFE/Al reactive liners with aluminum particle sizes of 5 μm and 100 μm were prepared. The parameters of the Johnson–Cook constitutive model of PTFE/Al reactive materials (RMs) were obtained through quasi-static compression experiments and SHPB (Split Hopkinson Pressure Bar) experiments. X-ray imaging technology was used to photograph the shape of reactive shaped charges jet at two different time points. The AUTODYN secondary development technology was used to simulate the jet formation, and the simulation results are compared with the experimental results. The results show that the simulation results are close to the experimental results, and the error is in the range of 4–8%. Through analysis, it is observed that the RMs reacted during the PTFE/Al reactive shaped charge jet formation, and due to the convergence of the inner layer of the liner during the jet formation, the chemical reaction of the jet is from inside to outside. Secondly, the particle size of aluminum powder has an influence on the chemical reaction and morphology of the jet. During the jet formation, there were fewer RMs reacted when the PTFE/Al reactive liners were prepared with 100 μm aluminum powder. Compared with 5 μm aluminum powder, when the aluminum powder is 100 μm, the morphology of the jet is more condensed, which is conducive to generating greater penetration depth.

Keywords: PTFE/Al; Johnson-Cook constitutive model; reactive shaped charge jet; numerical simulation

1. Introduction

With ongoing progress in the development of novel materials, protective technologies, structural designs, and protective devices, target defensive performance and survivability have been continuously improved [1]. It has become difficult for traditional metal jets to effectively damage targets. For example, for airstrips, metal jets only cause perforation damage to the runways and have little impact on fighter takeoffs, so it is necessary to seek new technologies. Unlike the traditional metal-liner shaped charge that produces penetration only through kinetic energy (KE) [2,3], when the RMs-liner shaped charge penetrates the target, the RMs undergo a chemical reaction, thereby increasing the damaging effect of the combination of KE and chemical energy (CE) of the reaction bounce [4,5], thus making up for the lack of after-effect damage.

Polytetrafluoroethylene (PTFE) is a common halogen polymer with good thermal stability and high fluorine content, and the RMs prepared by mixing PTFE with metals have excellent properties. The popular PTFE-based RMs are the mixture of 73.5 wt% PTFE and 26.5 wt% Al powders by mass matched ratios. Baker [6], Daniels [7], and Xiao [8] studied the enhanced damage effect on concrete, and found that PTFE-based reactive liner shaped charged jets produce dramatically catastrophic structural damage to concrete.

Zhang [9,10] studied the overpressure behind armor penetration produced by PTFE based reactive jet, and found that the overpressure decays parabolically with the thickness of plate, and a model was developed to describe the overpressure behind armor penetration. Wang [11] studied the PTFE/Al reactive jet formation by Euler algorithm, and found that compared with metal shaped charge jets, reactive shaped charge jets have a larger diameter and lower ductility. Guo [12,13] studied the penetration behavior of reactive shaped charge jets, and found that compared with metal shaped charge jets, the reactive jets produced larger holes in steel targets, but the penetration depth was lower. The relationship between the initiation delay time of reactive jets and the penetration depth was analyzed. Zheng [2] studied the behind-target rupturing effects of sandwich-like plates by reactive liner shaped charge jets. Then, the interaction mechanism between the reactive jet and target is discussed in three phases. Li [14] studied the forming cohesion of reactive shaped charge jets of PTFE-based liners, and found that reactive shaped charge jets would constantly undergo reaction expansion in the jet formation. With the passage of time, the contour of the jet would gradually blur and its density would decrease. Liu [15] studied the effect of aluminum particle size and molding pressure on the impact reaction of Al/PTFE, and found that the impact ignition of the reactive material is related to the microscopic defects, the propagation of the stress wave in the SHPB device, the amplitude of the stress pulse, and the destruction process of the material. Mao [16] studied the effect of aluminum particle size on the impact behavior of PTFE/Al RMs with a mass ratio of 50:50. The results show that aluminum particle size has significant effects on the shock-reduced reaction diffusion, reaction speed, and degree of reaction of the PTFE/Al reactive material.

Judging from the published papers both domestic and international, there are many works on the enhancement damage effect and mechanical properties of RMs, but few on the forming characteristics of reactive shaped charge jets, especially the effect of Al particle size on the forming and the reaction during the jet formation. In this paper, the jet morphology is obtained by X-ray photography experiment. The jet formation of reactive liners prepared with different Al particle size was simulated by numerical simulation method. The simulation results are compared with the experimental results. The research results have important reference value for the design of reactive liners.

2. X-ray Experiment

2.1. Experimental Composition

In this paper, X-ray imaging method was adopted to photograph the morphology of reactive shaped charge jets. The layout of the experiment equipment is shown in Figure 1. In this experiment, the stand-off was calibrated with a paper cylinder. There was an X-ray opening on each side of the left and right, which can take pictures at two moments of the same jet. The jet's morphology of the two moments photographed by X-ray were formed on two negatives at the rear of the shaped charge. The X-ray inspection system in the experiment is Sweden Scandiflash A B company 1200 KV.

In this paper, the ratio of PTFE/Al reactive liners used in the experiment was the mixture of 73.5 wt% PTFE and 26.5 wt% Al powders by mass matched ratios. The PTFE/Al reactive liners used in the experiment were divided into two kinds, namely the mixture of 34 μm PTFE and 5 μm aluminum powder and the mixture of 34 μm PTFE and 100 μm aluminum powder. The 8701 explosives, consisting of RDX, Polyvinyl acetate, DNT and Calcium stearate, which were used as the main charge in the experiment. The detonator is LD8 (No. 8 electric detonator). Figure 2 shows a structure sketch of the shaped charge.

2.2. Experimental Results

2.2.1. The Morphology of Reactive Jets

Through X-ray photography technology, two groups of reactive shaped charge jets were obtained. The measured time of the two groups were (a) 19.1 μs and 29.8 μs, (b) 18.9 μs and 29.9 μs, respectively. The aluminum particle size used in group (a) is 5 μm, and that used in group (b) is 100 μm.

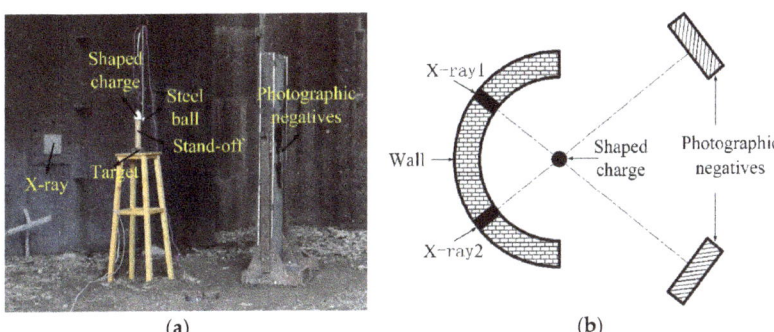

Figure 1. (a) Composition of experiment (b) schematic diagram of experiment.

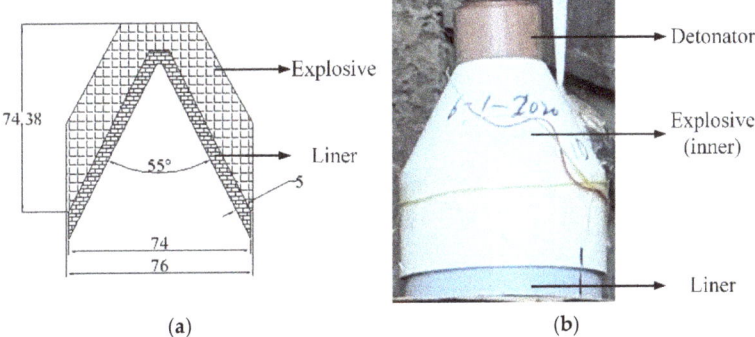

Figure 2. (a) Structure sketch of the shaped charge; (b) Picture of real products.

As shown in Figure 3, the reactive jet morphologies of the two groups were obtained by X-ray photography technology. As can be seen from the figure, neither group (a) nor group (b) can capture the morphology of the jet head. The reason is that during the jet formation, the RMs of the jet head first initiated the chemical reaction, which leads to the low density of the jet head and fails to show the morphology of the jet head in the negatives. Secondly, it can be seen from the figure that, at the same time, the jet morphology of group (b) is clearer than that of group (a). At the second moment, the morphology of the slug in group (a) expanded, while the morphology of the slug in group (b) was more condensed. These conditions are caused by the different sizes of aluminum used in the preparation of the liners. Under the same conditions, the liner prepared with 5 μm aluminum powder more quickly and easily undergoes chemical reaction than the liner prepared with 100 μm aluminum powder, so that the density of the jet in group (a) is lower than that of group (b), and the morphologies taken are not clear.

In addition to the above results, the contour of the shaped charge appeared in the second picture of group (a) in Figure 3, because before the explosives had detonated, one of the X-ray tubes was triggered prematurely so that a contour of the shaped charge appeared on the negative.

2.2.2. Energy Release Behavior of Reactive Jet

High-speed photographic equipment was used to record the firelight generated during the experiment. The firelight can reflect the energy release of the PTFE/Al reactive jet during the jet formation. Figure 4 is the firelight situation obtained by high-speed photography. They correspond to two groups: (a) and (b) in Figure 3. The high-speed photographic system used in the experiment is America Phantom V7.1, the experimental sampling frequency is 15,037 fps, and the exposure time is 60 μs.

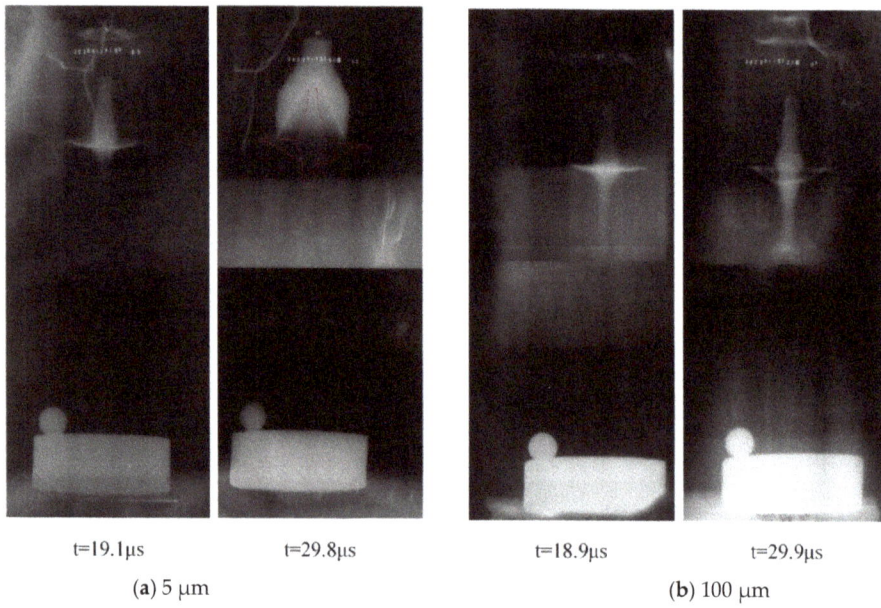

(a) 5 μm   (b) 100 μm

Figure 3. Graph of X-ray experimental results.

As can be seen from Figure 4, black smoke was found at 627 μs and 703 μs in group (a) and (b), respectively. The main reaction equation of PTFE/Al RMs is: $4Al + 3C_2F_4 = 4AlF_3 + 6C$. Throughout the chemical reaction, the polymer first breaks down into a monomer, and then the monomer decomposes into active small molecules (for example CF_3, CF_2, CF, COF_2, COF). The metal then progressively strips the F ions in the C-F bond through a redox reaction to form metal fluoride, and Al deprives F ions to produce AlF_3 mainly through the following five reaction channels [17].

Under oxygen deficient conditions

$$Al + CF_3 \rightarrow CF_2 + AlF$$

$$Al + CF_2 \rightarrow CF + AlF$$

$$Al + CF_3 \rightarrow C + AlF$$

Under oxygen rich conditions

$$Al + COF_2 \rightarrow COF + AlF$$

$$Al + COF \rightarrow CO + AlF$$

Then AlF will continue to deprive F ions to form AlF_3.

The production of black smoke can be seen in Figure 4, which may be the deflagration products. The PTFE/Al liner prepared with 5 μm aluminum powder produced black smoke earlier than the PTFE/Al liner prepared with 100 μm aluminum powder. This means that when the aluminum powder particle size is 5 μm, the RMs react earlier during the jet formation.

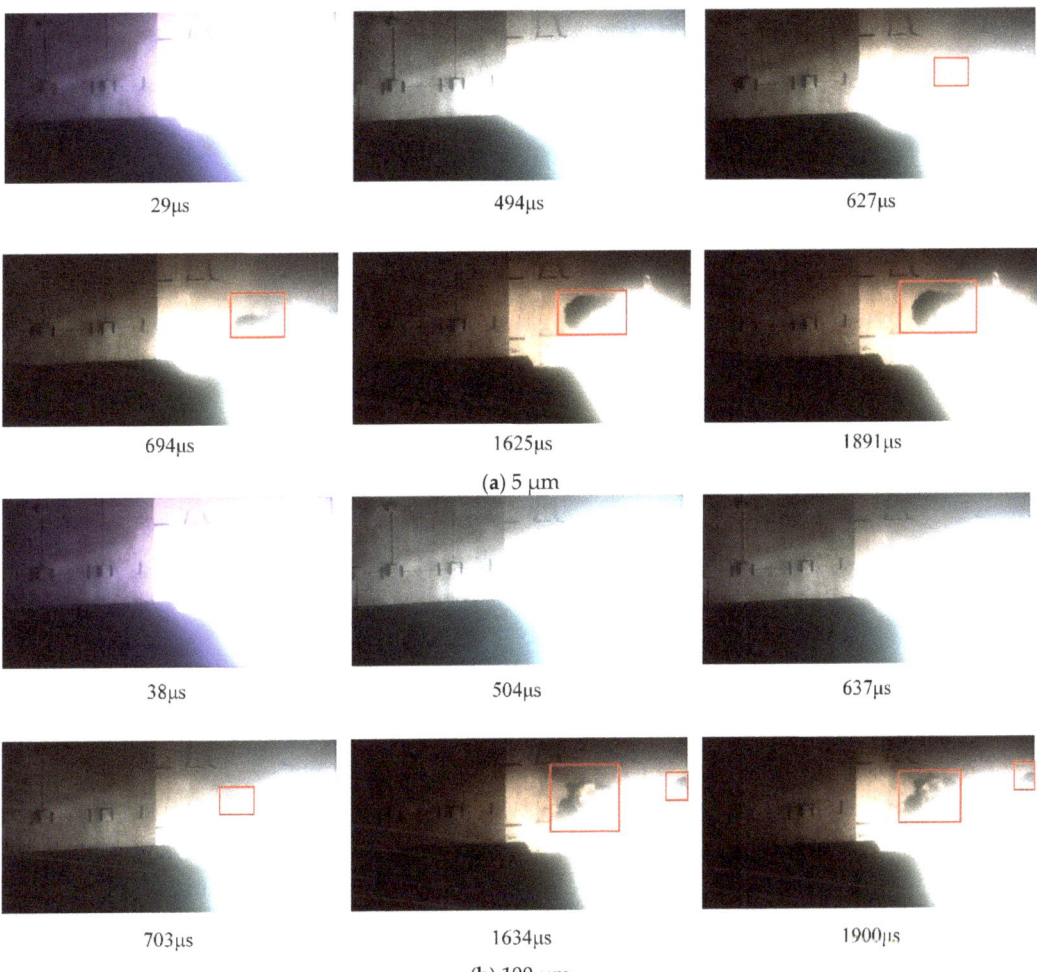

Figure 4. Energy release of RMs. (The red box refers to black smoke).

2.2.3. The Damage Effect of Witness Target

In this experiment, a witness target was placed at a height of 4.0 CD away from the liner. The diameter of the target is 150 mm and the thickness of the target is 50 mm. As shown in Figure 5, the target damage corresponding to the two groups of (a) and (b) is shown respectively.

As can be seen from the figure, the surface and both sides of the target are black. This may be caused by the deflagration products attached to the target surface. Secondly, observation can find that the witness target under the reactive liners of the two types are slightly damaged, leaving only small pits on the surface. Because the distance between the witness target and the liner is 4.0 CD, the damage ability of the jet has been greatly reduced as a result of expansion by chemical reaction before it arrives at the witness target. As can be seen from Figure 4, the number of pits on the witness target with the aluminum particle size of 100 μm is more than the number of pits on the witness target with the aluminum particle size of 5 μm, and the pits are concentrated in a certain range (damage area). When the aluminum particle size is 5 μm, the pit depth is very small, and the distribution is relatively scattered, almost distributed on the entire surface of the witness target.

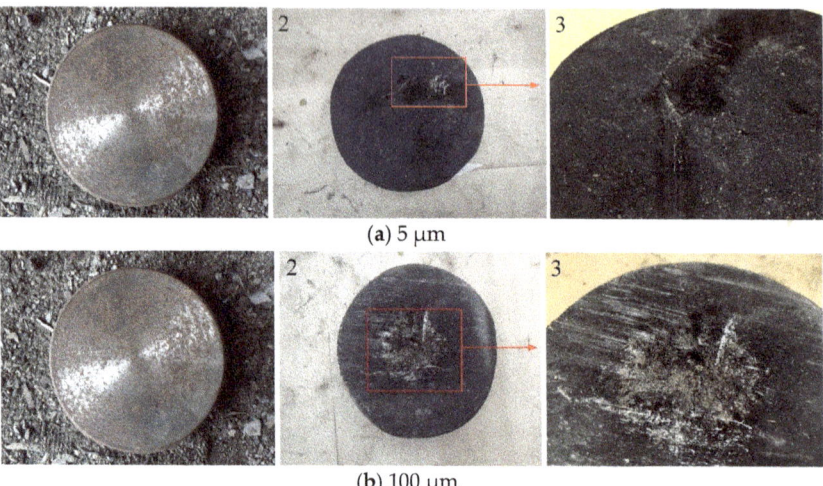

(a) 5 μm

(b) 100 μm

Figure 5. Damage to the witness target.

According to the damage of the witness target, it is speculated that most of the RMs have reacted before the reactive jet reaches the witness target, and only a few unreacted materials hit the witness target, causing slight damage to the witness target. Secondly, a large amount of RMs inside the slug undergo a chemical reaction before reaching the witness target, and the remaining unreacted part is weak due to excessive divergence. This is because if most of the slug reacted after hitting the witness target, the damage effect will be larger than that shown in Figure 5. As shown in Figure 6, through high-speed photography, it was found that the final firelight was collected above the witness target, and there was still a certain distance from the witness target. This phenomenon indicates that the slug reacts without hitting the witness target plate, and if the slug reacts after hitting the witness target plate, the final firelight should appear at the witness target plate position.

In summary, the explosive energy is sufficient to cause the RMs reaction during the jet formation. This issue should be taken into account when studying PTFE/Al reactive jet formation.

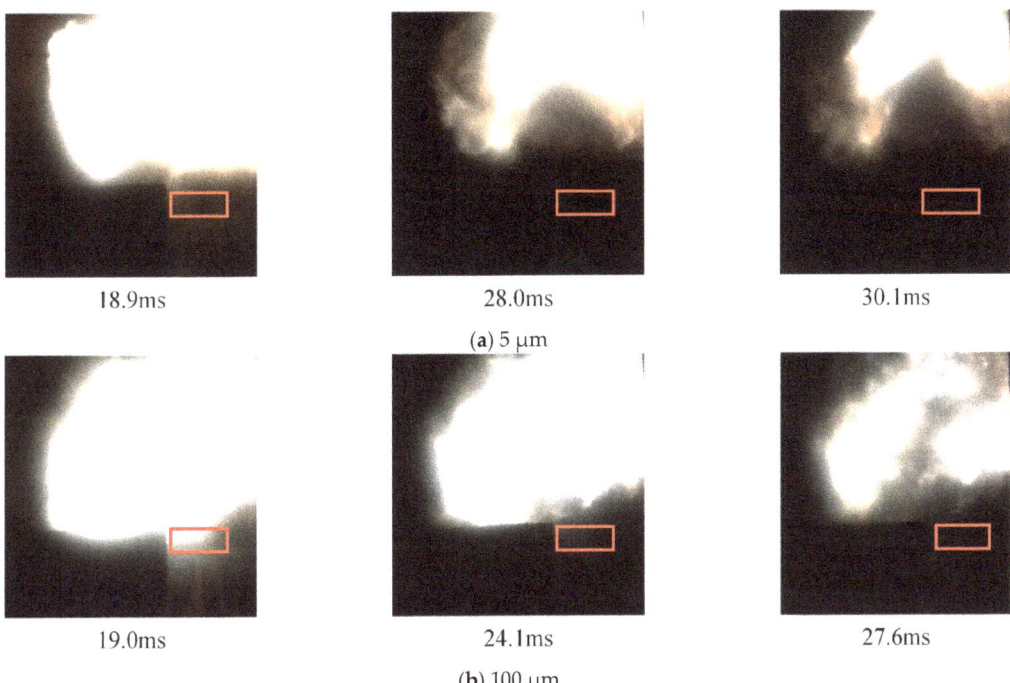

Figure 6. The position of firelight contraction. (□: The target location).

3. The Numerical Simulation

In this paper, ANSYS Autodyn-3D is used to simulate the jet formation. The real-time reaction of the reactive jet and the energy loss can be clearly seen in the results. The SPH (Smoothed Particle Hydrodynamics) method was used to simulate the jet formation. Tunable ignition threshold and relevant material model were developed to recreate the impact-induced deflagration behavior of RMs in previous work, the details of which can be found in reference [18].

3.1. Finite Element Model

In order to reduce the computational cost, a three-dimensional 1/2 simulation model is established in this paper, as shown in Figure 7. The material of the liner is PTFE/Al and 8701 explosive was used as the main charge. The material of the witness target is 45# steel. SPH was used to fill the liner, explosive, and steel target. The central point initiation mode is adopted for detonation, and the initiation point is shown as the red dot in Figure 7. The structure of the liner adopts the conical structure with equal wall thickness; the wall thickness of the liner is 5 mm. In addition, Gaussian points named 1, 2, 3, and 4 are set on the liner to examine the history variables, where 1 and 2 are located on the inner layer of the liner, and 3 and 4 are located on the outer layer of the liner.

3.2. Material Model
3.2.1. Constitutive Model

In this paper, the parameters of the Johnson–Cook constitutive model of PTFE/Al RM were determined by studying its dynamic and static mechanical properties. The Johnson–Cook constitutive model expression is as follows

$$\sigma = (A + B\varepsilon^n)\left[1 + C\ln\left(\dot{\varepsilon}/\dot{\varepsilon}_0\right)\right] \tag{1}$$

where A is the yield stress, B and n are the strain hardening rate, C is the coefficient of strain rate sensitivity, ε is the strain rate, $\dot{\varepsilon}$ is the reference strain rate. Parameter A is obtained from the quasi-static compression experiment data. The values of A, B, n, and C were obtained by combining the SHPB experimental data; see References [19,20] for details. According to the experimental data, the parameters of the Johnson–Cook constitutive model of RMs prepared with Al particles with different particle sizes can be calculated, as shown in Table 1.

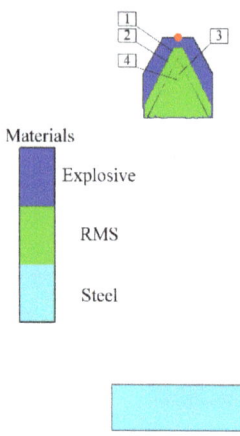

Figure 7. 3D simulation model.

Table 1. The parameters of the Johnson–Cook constitutive model.

Particle Size of Al (μm)	P (g/cm³)	A (MPa)	B (Mpa)	n	C	m
5	2.27	14.9	45.463	0.74415	0.115	1
100	2.27	13.9	49.564	0.60135	0.057	1

According to the Johnson–Cook constitutive model of the RMs obtained, the nonlinear fitting was performed with the experimental data at different strain rates as variables, and the fitting curve is high consistent with the experimental curve, this is shown in Figure 8.

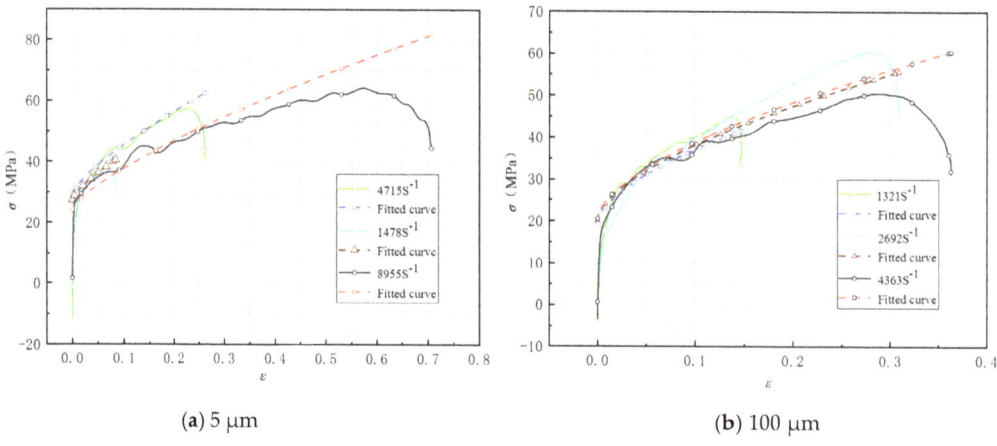

(a) 5 μm   (b) 100 μm

Figure 8. Stress–strain curves under different aluminum particle sizes.

3.2.2. EOS Model

When a material is subjected to a high impact pressure, it can be treated approximately as a fluid. The famous Rankin–Hugoniot energy equation can be obtained from the three conservation equations governing fluid motion.

$$E - E_0 = \frac{1}{2}(P - P_0)(V_0 - V) \quad (2)$$

where the subscript "0" represents the initial state, E, P, and V are the specific internal energy, pressure, and specific volume of the material, respectively.

The relationship between material energy and pressure can be obtained from thermodynamic identity

$$\left(\frac{\partial E}{\partial V}\right) = T\left(\frac{\partial P}{\partial T}\right)_V - P \quad (3)$$

If $E = C_V T$, and the specific heat C_V at constant volume is a constant or a function independent of the specific volume V, it can be obtained

$$P(V,T) = T\left(\frac{\partial P}{\partial T}\right)_V \quad (4)$$

The Grüneisen state equation of the material can be obtained by integrating Equation (4)

$$P - P_C = \frac{\gamma(V)}{V}(E - E_C) \quad (5)$$

Equations (2) and (5) can be used to obtain the equation of state expression of the compacted material

$$P(V) = \frac{\frac{V}{\gamma(V)}P_C(V) - E_C(V)}{\frac{V}{\gamma(V)} - \frac{1}{2}(V_0 - V)} \quad (6)$$

where $\gamma(V)$ is the Grüneisen coefficient, P_C is the cold pressing of the materials, and E_C is the cold energy of the materials.

The cold energy and cold pressing of materials are due to the interaction between crystals, and there are generally three potential functions used to describe them: the Born–Mayer potential, the Mie potential, and the Morse potential [21,22]. Among them, the Morse potential can better describe the cold pressing and cold energy curves of the material, and the equation is

$$P_C = A\delta^{2/3}\left[\exp\left(2B\left(1 - \delta^{-1/3}\right)\right) - \exp\left(B\left(1 - \delta^{-1/3}\right)\right)\right] \quad (7)$$

$$E_C = \frac{3AV_{0K}}{2B}\left[\exp\left(B\left(1 - \delta^{-1/3}\right)\right) - 1\right]^2 \quad (8)$$

$$\gamma(V) = \frac{B}{6\delta^{1/3}} \cdot \frac{4\exp\left[2B\left(1 - \delta^{-1/3}\right)\right] - \exp\left[B\left(1 - \delta^{-1/3}\right)\right]}{2\exp\left[2B\left(1 - \delta^{-1/3}\right)\right] - \exp\left[B\left(1 - \delta^{-1/3}\right)\right]} \quad (9)$$

where A, B are the parameters measured by the test, and $\delta = V_{0K}/V$ is the compressibility of the material at zero temperature.

On the basis of the above equations, the cold energy superposition principle is selected to calculate the cold pressing line of the mixture, and the Hugoniot relationship of the mixture is further calculated. The specific volume and specific internal energy of the mixture are

$$V(P) = \sum_{i=1}^{N} m_i V_i(P) \quad (10)$$

$$E(V,T) = \sum_{i=1}^{N} m_i E_i(V,T) \tag{11}$$

$$\sum_{i=1}^{N} m_i = 1 \tag{12}$$

where m_i is the percentage of mass of the group i, and V_i and E_i are the specific volume and specific internal energy of the group i, respectively. Similarly, the initial parameters of the mixture material can be obtained through the superposition principle

$$\frac{V}{\gamma} = \sum_{i=1}^{n} m_i \frac{V_i}{\gamma_i} \tag{13}$$

According to the three conservation equations of solid materials, the expressions of shock wave velocity U_S and particle velocity U_P with respect to P and V can be obtained

$$U_s = V_0 \left(\frac{P}{V_0 - V}\right)^{1/2} \tag{14}$$

$$U_p = [P(V_0 - V)]^{1/2} \tag{15}$$

And the relation between the shock wave velocity and the particle velocity is

$$U_s = C_0 + S U_p \tag{16}$$

where C_0 is the sound velocity of the material and S is the material proportion coefficient.

Firstly, the cold energy value of the mixture is calculated according to the cold energy superposition method, the material constants A and B are fitted by the relationship curve of E_C–V, and the A and B values of the mixture are substituted into Equations (7)–(9) to obtain the cold energy E_C, cold pressed P_C and $\gamma(V)$ of the mixture. Then, the values of these three parameters can be substituted into Formula (6) to obtain the mixture material and the P-V relationship. Finally, the P-V relationship of the material was substituted into Equations (13)–(15) to obtain the initial material parameters U_S, U_P, and $\gamma 0$ of the mixture. By fitting Equation (16), the material sound velocity C_0 and the scaling coefficient S can be obtained. The initial material parameters of the elemental material and the final calculated equation of state parameters of the PTFE/Al RMs are listed in Table 2.

Table 2. The parameters of elemental materials.

Material	ρ_0 (g/cm³)	C_0 (km/s)	S	γ_0	α_V (10^{-5}/K)
Al	2.712	5.332	1.375	2.18	6.93
PTFE	2.152	1.754	1.723	0.59	10.9
PTFE/Al	2.296	3.077	1.743	0.70	-

3.2.3. The Material Models of Explosive and Target Plates

In the paper, the explosive used was 8701 and the JWL EOS model was used in the numerical model of explosive. The model parameters are shown in Table 3. The target was made of 45# steel. The shock EOS model and Johnson–Cook strength model were used in the numerical model of the explosive. The model parameters are shown in Table 4.

Table 3. The material parameters of the explosive.

Material	A (Mbar)	B (Mbar)	R_1	R_2	W	ρ (g/cm³)	P_{CJ} (Mbar)	D (m/s)
8701	5.2423	0.7678	4.2	1.1	0.34	1.71	0.286	8315

Table 4. The material parameters of the target.

Material	ρ (g/cm^3)	C (m/s)	S	γ	A (Mbar)	B (Mbar)	n	c	m	T_m (K)
45# steel	7.83	4569	1.33	1.67	0.0057	0.0032	0.28	0.064	1.06	1811

4. Results and Discussion

4.1. The Morphology Reactive Shaped Charge Jets

Through the simulation, the morphological characteristics of reactive shaped charge jets of PTFE/Al liner prepared with aluminum powder with two particle sizes were obtained. Figure 9 shows the formation of two reactive shaped charge jets. The color isoline on the right of the figure is used to indicate the reaction degree of the RMs during the jet formation. As can be seen from the figure, during the jet formation, the jet formed by the liner prepared with 5 μm aluminum powder reacted earlier than 100 μm aluminum powder, and both jet heads reacted first. Within 0.0 CD~1.5 CD of stand-offs, the jet head with 5 μm aluminum particle size is more divergent, but the morphology of the two slugs is similar. At 1.5 CD from stand-offs, the slug with 5 μm aluminum powder has an obvious reaction. For the 100 μm aluminum powder, the internal reaction of the slug was obvious only when the jet reached 2.0 CD from stand-offs. The jet morphologies of the two are obviously different when the stand-offs are 2.0 CD. The expansion rate of the slug corresponding to 100 μm aluminum powder is smaller than that of 5 μm aluminum powder, and its jet head is more condensed. Secondly, it can be seen from Figure 9 that the slug of the reactive shaped charge jet has already started to react before it reaches the target. The reaction is violent, and the RMs have a high degree of reaction. The phenomenon is consistent with the firelight captured by high-speed photography, as the firelight finally collects above the target.

The main reason for the above phenomenon is that the smaller the size of the aluminum particles, the more easily the RMs react [15,16]. During the jet formation, the liner with small Al particles has the largest specific surface area. With the same mass, the larger the specific surface area, the greater the friction area, thus generating more heat. In addition, the strain rate constant of the RMs prepared with the two kinds of Al particles is significantly different. The strain rate constant of the RMs prepared with the 5 μm Al particles is twice that the strain rate constant of the RMs prepared with the 100 μm Al particles. The higher the strain rate constant is, the more energy is absorbed per unit volume during the jet formation, and the faster the RMs reach the reaction condition. Secondly, Al particle size has an influence on the destruction mode of the RMs, and during the reaction of the RMs, gases were generated. Therefore, the reactive shaped charge jet formed by the 5 μm aluminum liner reacts earlier, and the jet head and the slug are thicker.

During the jet formation, the top of the liner is first subjected to the blast pressure to form a jet, so the jet head begins to react first. Due to the convergence of the inner layer of the liner during the jet formation, the chemical reaction of the jet is from inside to outside. Figure 10 shows the pressure change at Gauss points during the jet formation. Gauss points 1 and 2 are located in the inner layer of the liner, and Gauss points 3 and 4 are located in the outer layer of the liner. It can be seen from Figure 10 that during the jet formation, the pressure on the inner layer of the liner rose sharply, reached a peak value, and then began to decline gradually. After reaching the first peak, the inner layer pressure of the liner decreased rapidly, and then increased rapidly after about 2.5 μs, reaching the second peak, and the second peak value was higher than the first peak value. The appearance of a secondary peak indicates that the RMs in the inner layer of the liner have indeed undergone a secondary collision.

As mentioned above, in this paper, the morphology of reactive shaped charge jet formed by the liner prepared with two aluminum particle sizes were obtained by X-ray photography technology, namely, group (a) aluminum particle size was 5 μm, and the measured time was 19.1 μs and 29.8 μs; group (b) aluminum particle size was 100 μm, measured time 18.9 μs and 29.9 μs. Figure 11 shows the comparison between X-ray images and simulated images of the reactive jet.

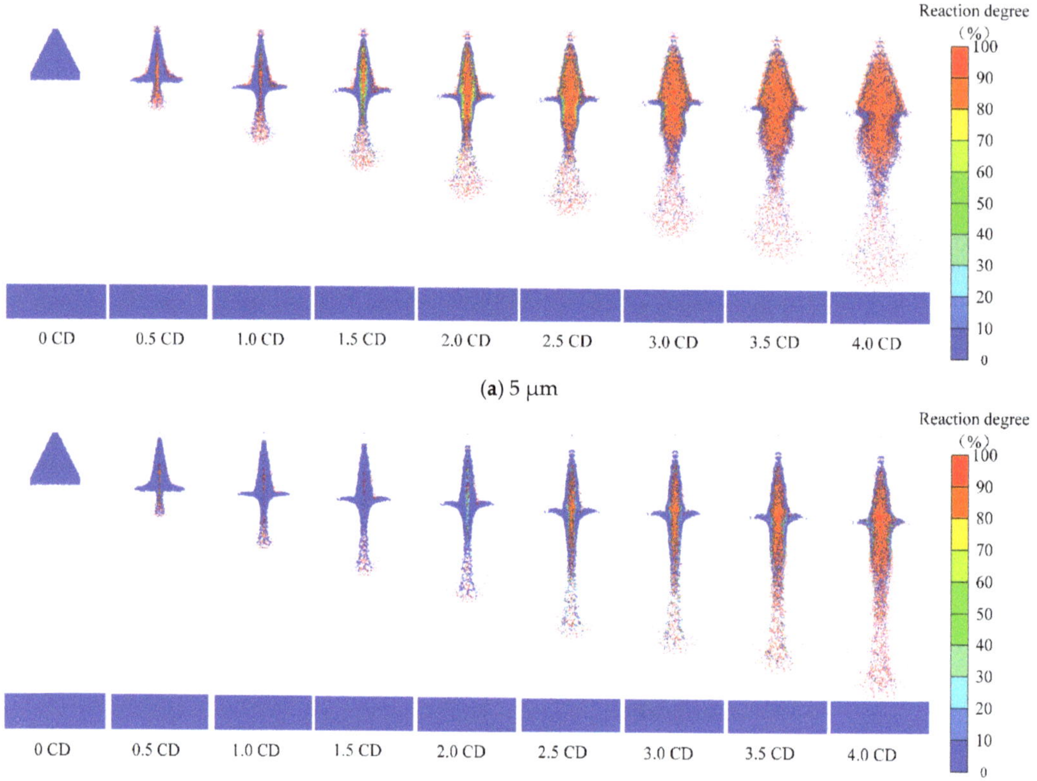

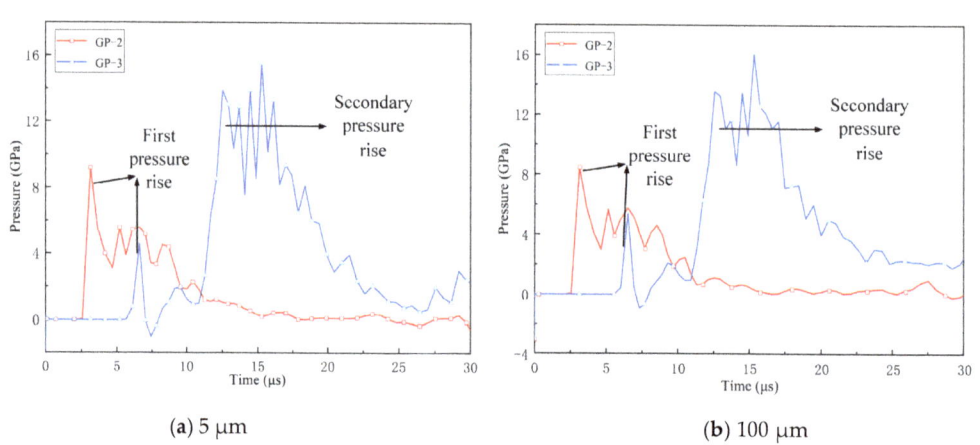

(a) 5 μm

(b) 100 μm

Figure 9. The formation of reactive shaped charge jet.

(a) 5 μm

(b) 100 μm

Figure 10. Gauss point pressure curve.

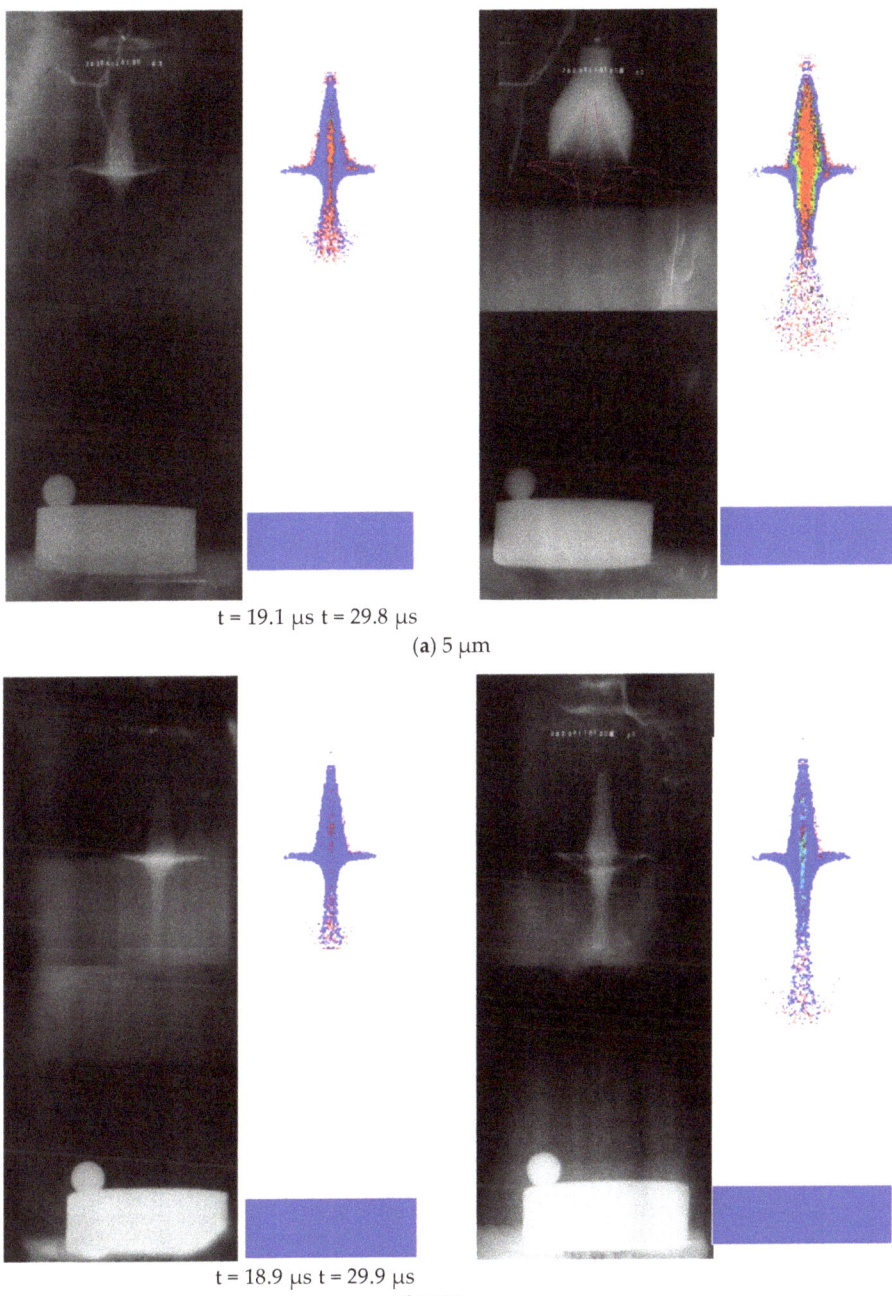

Figure 11. Shape contrast of shaped charge penetration.

According to the comparison of Figure 11, it can be seen that there is a high degree of similarity between the simulation results and experimental results. Table 5 shows the comparison between the simulation jet size and experimental jet size.

Table 5. The size of reactive jet.

Particle Size of Al (μm)	Test of Time (μs)	Diameter of Slug (mm)			Diameter of Jet (mm)		
		Experiment	Simulation	Error (%)	Experiment	Simulation	Error (%)
5	19.1	27.7	29.5	6.5	13.8	13.1	5.1
	29.8	35.0	32.8	6.3	12.5	11.7	6.4
100	18.9	25.4	27.5	7.6	9.2	9.7	5.2
	29.9	30.0	31.5	5	11.5	12.0	4.2

As can be seen from the data in Table 5, the errors between the simulation results and those measured in the experiment are within the range of 4%~8%, and all the errors are within a reasonable range. This data indicates that the simulation results in this paper are highly reliable.

4.2. Reaction Degree and Jet Velocity during the Jet Formation

Reaction degree and jet velocity are two important factors affecting the damage ability of shaped charge jets. In this paper, the energy loss was qualified through the reaction degree of the RMs during the jet formation. The reaction degree of the RMs prepared with two aluminum particle sizes was obtained through numerical simulation, as shown in Figure 12. Points 1~8 in the figure indicate the reaction degree of the jet when the jet head reached 0.5 CD, 1.0 CD, 1.5 CD, 2.0 CD, 2.5 CD, 3.0 CD, 3.5 CD, and 4.0 CD.

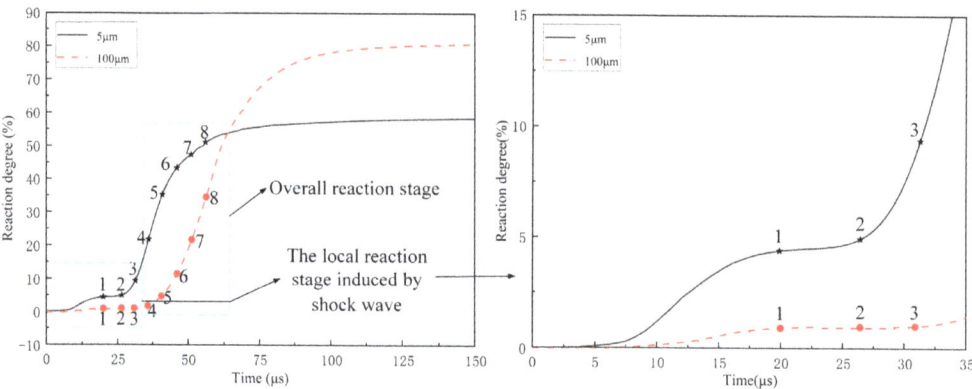

Figure 12. Reaction degree of reactive materials.

In order to obtain the reaction degree of the RMs during the jet formation, a subroutine that executed at the end of every calculation circulation is compiled. The reaction degree of RMs is defined as

$$F = \frac{\sum m_i \alpha_i}{\sum m_i} \quad (17)$$

where m_i denotes the mass of every particle contained in RMs and α_i is the corresponding reaction ratio of the particles. Then, the reaction degree of the reactive jet can be calculated.

According to Figures 9 and 12, the chemical reaction of the RMs during the jet formation can be divided into two stages: (1) The local reaction stage induced by shock wave, in which the reaction is mainly concentrated in the jet head; (2) Overall reaction stage, in which the reaction is concentrated in the slug. In the local reaction stage, the reaction of the RMs is relatively mild, but in the overall reaction stage, the deflagration of RMs is violent, and the reaction degree increases extremely. It can be found that when the aluminum particle size is 5 μm, the RMs enter the local reaction stage at about 3 μs, while when the aluminum particle size is 10 0 μm, the RMs enter the local reaction stage at about 8 μs. Correspondingly, the time for the RMs to enter the overall reaction stage with a particle size of 5 µm is also about 10 µs faster than 100 µm. Secondly, in the local reaction stage, the highest reaction degree of 5 µm aluminum particle size was about 4.5%, while the reaction degree of 100 µm aluminum particle size was less than 1%.

As shown in the figure, the final reaction degree of 5 µm aluminum particle size is about 60%, while that of 100 µm aluminum particle size is about 80%. The reason may be that the rapid reaction of the inner RMs leads to the rapid expansion of the shaped charge jet volume, which makes the jet too divergent, and the stress on the outside fails to meet the ignition conditions of the RMs. Secondly, when the aluminum particle size is 5 µm, more AlF3 is generated in the initial reaction, and the gasification of AlF$_3$ will absorb a lot of heat, so that the PTFE can't reach the decomposition temperature. Therefore, the final reaction degree of the RMs prepared with 5 µm aluminum particle size is less than that of the RMs prepared with 100 µm aluminum particle size.

According to Figure 12, the reaction degree and reaction rate of the two types of shaped charge jets can be deduced as shown in Table 6.

Table 6. Reaction rate of shaped charge projectiles during molding.

Particle Size of Al (µm)	Local Reaction Stage		Overall Reaction Stage	
	Reaction Degree (%)	Reaction Rate (g/µs)	Reaction Degree (%)	Reaction Rate (g/µs)
5	4.5	0.362	60	1.67
100	0.8	0.067	80	1.52

Figure 13 shows the velocity contour at 4.0 CD during the jet formation and the velocity curve of the main part of the jet. The velocity curves of the jet head, the middle of jet, and slug positions in Figure 13b,d correspond to 1, 2, and 3 in Figure 13a,c, respectively.

As can be seen from the figure, when the jet head reaches 0.5 CD, the jet head velocity under the two aluminum powder particle sizes reaches about 7800 m/s at about 20 µs. When the jet head reaches 1.5 CD, the velocity of the middle of jet reaches its maximum velocity at around 30 µs, about 4800 m/s.

It can be seen from the above that during the jet formation the reaction firstly concentrates on the jet head. Therefore, although the jet head has a high penetration speed, the penetration ability of the head on the target will be greatly weakened due to its excessive reaction. This is also the reason why the penetration depth of reactive shaped charge jets is smaller than that of inert shaped charge jets. For reactive shaped charge jet, the penetration depth mainly depends on the middle of the jet.

Considering the reaction degree and jet velocity, combined with Figures 12 and 13, it can be concluded that when the aluminum particle size is 5 µm, the optimal stand-off is between 0.5 CD and 1.0 CD. When the aluminum particle size is 100 µm, the optimal stand-off is between 1.0 CD and 1.5 CD.

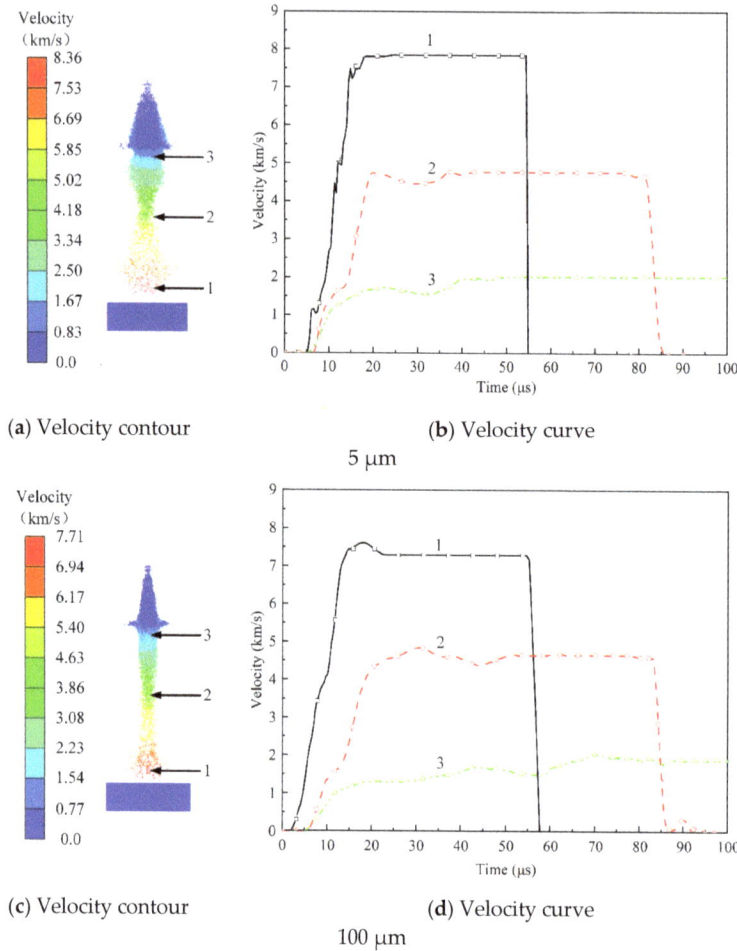

(a) Velocity contour (b) Velocity curve
5 μm

(c) Velocity contour (d) Velocity curve
100 μm

Figure 13. Velocity of shaped charge projectiles.

5. Conclusions

In this paper, the X-ray photography experiment of reactive shaped charge jet formation was performed on PTFE/Al liners prepared with two aluminum particle sizes (5 μm and 100 μm). The secondary development technology was used to simulate the formation of PTFE/Al reactive shaped charge jets and compared with the experimental results. By analyzing the experimental phenomenon and simulation results, the following conclusions were obtained:

(1) The RMs reacted during the PTFE/Al reactive shaped charge jet formation, which can be divided into local reaction stage and overall reaction stage. In the local reaction stage, the reaction is relatively mild and mainly concentrated in the jet head. In the whole reaction stage, the RMs deflagrate violently, and the reaction mainly concentrates in the slug.

(2) Secondary collision occurs in the inner layer of the liner during the jet formation, and the pressure generated by the secondary collision is higher than that given by the explosive. Therefore, the reaction of PTFE/Al reactive shaped charge jet is from inside to outside during the jet formation.

(3) The effect of Al particle size on the mechanical properties and reaction rate of PTFE/Al RMs are the main reason for the formation of the difference during the jet formation. Compared with the 5 μm aluminum powder, the PTFE/Al reactive liner prepared with 100 μm aluminum powder reacted slowly and the morphology of jet is more condensed, which is conducive to generating greater penetration depth.

Author Contributions: Conceptualization, H.W.; Data curation, M.G.; Formalanalysis, Y.W. and J.X.; Funding acquisition, H.W. and J.X.; Investigation, Y.C.; Methodology, H.W.; Project administration, J.X.; Resouces, H.W.; Software, M.G., Y.W. and J.X.; Supervision, H.W. and J.X.; Validation, Y.C. and H.W.; Visualization, Y.C.; Writing—original draft, M.G. and Y.W.; Writing—review & editing, M.G. and Y.W. All authors have read and agreed to the published version of the manuscript.

Funding: The research was funded by the National Natural Science Foundation of China (Grant No.11702256), Natural Science Foundation of Shanxi Province (Grant No.20210302124214).

Data Availability Statement: The data that support the findings of this study are available from the corresponding author upon reasonable request.

Conflicts of Interest: The authors declare that they have no known competing financial interests or personal relationships that could have appeared to influence the work reported in this paper.

References

1. Yi, J.; Wang, Z.; Yin, J.; Zhang, Z. Simulation Study on Expansive Jet Formation Characteristics of Polymer Liner. *Materials* **2019**, *12*, 744. [CrossRef] [PubMed]
2. Zheng, Y.; Su, C.; Guo, H.; Yu, Q.; Wang, H. Behind-Target Rupturing Effects of Sandwich-like Plates by Reactive Liner Shaped Charge Jet. *Propellants Explos. Pyrotech.* **2019**, *44*, 1400–1409. [CrossRef]
3. Herbold, E.B.; Nesterenko, V.F.; Benson, D.J.; Cai, J.; Vecchio, K.S.; Jiang, F.; Addiss, J.W.; Walley, S.M.; Proud, W.G. Particle size effect on strength, failure and shock behavior in Polytetrafluoroethylene-Al-W granular composites. *J. Appl. Phys.* **2008**, *104*, 1007. [CrossRef]
4. Ames, R.G. Vented chamber calorimetry for impact-initiated energetic materials. In Proceedings of the 43rd AIAA Aerospace Sciences Meeting and Exhibit, Reno, NV, USA, 10–13 January 2005; pp. 15391–15403. [CrossRef]
5. Rosencrantz, S.D. Characterization and Modeling Methodology of Polytetrafluoroethylene Based Reactive Materials for the Development of Parametric Models. Master's Thesis, Wright State University, Dayton, OH, USA, 2007.
6. Baker, E.L.; Daniels, A.S.; Ng, K.W.; Martin, V.; Orosz, J. Barnie: A unitary demolition warhead. In Proceedings of the 19th International Symposium on Ballistics, Interlaken, Switzerland, 7–11 May 2001; pp. 569–574.
7. Daniels, A.S.; Baker, E.L.; Defisher, S.E.; Ng, K.W.; Pham, J. BAM: Large scale unitary demolition warheads. In Proceedings of the 23rd International Symposium on Ballistics, Tarragona, Spain, 16–20 April 2007; pp. 239–246.
8. Xiao, J. Research on Damage Effects of Multi-layered Concrete Targets Impacted by Explosively Formed Penetrato. Ph.D. Thesis, Beijing Institute of Technology, Beijing, China, 2016.
9. Zhang, X.-P.; Xiao, J.; Yu, Q.-B.; Zheng, Y.-F. Armor penetration aftereffect overpressure produced by reactive material liner shaped charge. *Acta Armamentarii* **2016**, *37*, 1388–1394.
10. Zhang, X.; Xiao, J.; Yu, Q.; Wang, H. Penetration and blast combined damage effects of reactive material jet against steel target. *Trans. Beijing Inst. Technol.* **2017**, *37*, 789–793.
11. Wang, Y.; Yu, Q.; Zheng, Y.; Wang, H. Formation and Penetration of Jets by Shaped Charges with Reactive Material Liners. *Propellants Explos. Pyrotech.* **2016**, *41*, 618–622. [CrossRef]
12. Guo, H.; Zheng, Y.; Yu, Q.; Chao, G.; Wang, H. Penetration behavior of reactive liner shaped charge jet impacting thick steel plates. *Int. J. Impact Eng.* **2018**, *126*, 76–84. [CrossRef]
13. Guo, H.; Lu, G.; He, S.; Wang, H.; Xiao, Y.; Zheng, Y. Penetration enhancement behavior of reactive material double-layered liner shaped charge. *Trans. Beijing Inst. Technol.* **2020**, *40*, 1259–1266.
14. Li, Y.; Wang, W.; Zhang, L.; Jiang, C. Research on jet coherency of PTFE-based energetic liner. *Acta Armamentarii* **2019**, *40*, 2433–2439.
15. Liu, Y.; Ren, H.; Li, W.; Ning, J. Influence of Particle Size of Aluminum Powder and Molding Pressure on Impact-Initiation of Al/PTFE. *Chin. J. High Press. Phys.* **2019**, *33*, 123–130.
16. Mao, L.; Wei, C.; Hu, R.; Hu, W.; Luo, P.; Qi, Y.; Jiang, Y. Effects of Al Particle Size on the Impact Energy Release of Al-Rich PTFE/Al Composites under Different Strain Rates. *Materials* **2021**, *14*, 1911. [CrossRef] [PubMed]
17. Zhou, J. Study on the Impact-Induced Reaction Characteristics of Typical Fluoropolymer-Matrix Reactive Materials. Ph.D. Thesis, Nanjing University of Science and Technology, Nanjing, China, 2018.
18. Xiao, J.; Wang, Y.; Zhou, D.; He, C.; Li, X. Research on the Impact-Induced Deflagration Behavior by Aluminum/Teflon Projectile. *Crystals* **2022**, *12*, 471. [CrossRef]
19. Xiao, J.; Wang, Z.; Nie, Z.; Tang, E.; Zhang, X. Evaluation of Hugoniot parameters for unreacted Al/PTFE RMs by modified SHPB test. *AIP Adv.* **2020**, *10*, 045211. [CrossRef]

20. Wang, Z. Parameter Calibration of PTFE/Al RMs and Numerical Simulation Research of Its Impact-Induced Deflagration Behavior. Master's Thesis, North University of China, Taiyuan, China, 2021.
21. Tang, W.; Zhang, R. *Introduction to Theory and Computation of Equations of State*; Higher Education Press: Beijing, China, 2008.
22. Queen, G.M.; Marsh, S.P.; Taylor, J.W.; Cable, A.J.; Dienes, J.K.; Walsh, J.M.; Carter, W.J.; Fritz, J.N.; Gehring, J.W.; Glass, C.M.; et al. *High Velocity Impact Phenomena*; Academic Press: New York, NY, USA, 1970.

Article

Dynamic Behavior of Kinetic Projectile Impact on Honeycomb Sandwich Panels and Multi-Layer Plates

Shuai Yue [1,*], Yushuai Bai [2,*], Zhonghua Du [1], Huaiwu Zou [3], Wenhui Shi [1] and Guang Zheng [4]

1. School of Mechanical Engineering, Nanjing University of Science and Technology, Nanjing 210094, China; duzhonghua@aliyun.com (Z.D.); shiwenhui@njust.edu.cn (W.S.)
2. Academy of Opto-Electronic, China Electronic Technology Group Corporation, Tianjin 300308, China
3. Aerospace System Engineering Shanghai, Shanghai 201109, China; wzh15706038026@njust.edu.cn
4. Key Laboratory of Impact and Safety Engineering of Ministry of Education, Ningbo University, Ningbo 315211, China; zhengguang@nbu.edu.cn
* Correspondence: yueshuai@njust.edu.cn (S.Y.); baiyushuai@njust.edu.cn (Y.B.)

Abstract: In order to study the dynamic response associated with the impact of a kinetic projectile on the internal structure of an artificial satellite, we propose a kinetic projectile configuration with non-metallic material wrapping and multiple damage elements. The artificial satellite is simplified as a honeycomb sandwich panel and multi-layer plates. We carried out a ground damage test and finite element dynamic simulation, and we determined the lateral effect and penetration performance of the projectile. Then, we studied the dynamic behavior of the projectile penetrating the honeycomb sandwich panel using a theoretical model. We found that its ballistic limit velocity was 150 m/s, and the deformation of the opening had little relationship with the projectile velocity. Finally, we studied the dynamic response of the kinetic projectile impacting the multi-layer plates under various launch parameters. We found that the launch velocity required to meet the damage requirements was within 325 ± 25 m/s. Projectiles with a higher initial velocity had a stronger ability to penetrate the plates, but initial velocities higher than 325 m/s led to a reduced damage area. The kinetic projectile could adapt to incident angles less than 5° when damaging the plates. With the increasing incident angle, the penetration ability was reduced, and the damage area increased.

Keywords: dynamic modeling; impact; kinetic projectile; honeycomb; multi-layer plates

Citation: Yue, S.; Bai, Y.; Du, Z.; Zou, H.; Shi, W.; Zheng, G. Dynamic Behavior of Kinetic Projectile Impact on Honeycomb Sandwich Panels and Multi-Layer Plates. *Crystals* **2022**, *12*, 572. https://doi.org/10.3390/cryst12050572

Academic Editor: George D. Verros

Received: 23 March 2022
Accepted: 18 April 2022
Published: 20 April 2022

Publisher's Note: MDPI stays neutral with regard to jurisdictional claims in published maps and institutional affiliations.

Copyright: © 2022 by the authors. Licensee MDPI, Basel, Switzerland. This article is an open access article distributed under the terms and conditions of the Creative Commons Attribution (CC BY) license (https://creativecommons.org/licenses/by/4.0/).

1. Introduction

With the continuous advancement of the human exploration of space, more than 10,000 spacecrafts have been launched into orbit in the past 60 years. Most of these have become space junk due to a loss of function [1]. There have been more than 260 explosions and impact events involving orbiting spacecrafts, resulting in a large amount of space debris, which substantially reduces the safety of human space activities [2]. According to the detection and research of the recovered spacecraft, impacts of space debris with spacecraft are quite common and damage both the surface and internal structure of spacecrafts [3].

In order to enhance the passive protection capability of spacecraft against space debris, the American astrophysicist Fred Whipple proposed the Whipple shield in 1947. By placing a shield at a certain distance outside the spacecraft, a debris cloud would be formed when the shield was impacted by a projectile, dispersing the kinetic energy of the projectile and further reducing the damage to the spacecraft [4]. Subsequently, researchers conducted a great deal of research on the high-speed impact of projectiles on Whipple shields [5,6], which made great contributions to space debris protection technology. Poniaev et al. conducted an experimental study on the hypervelocity impact of plastic projectiles on thin aluminum plates and gave a numerical simulation of the process by the SPH (smoothed particle hydrodynamics) method [7]. Hofmann et al. tested the protective effect of a metallic glass-stuffed Whipple shield using hypervelocity impacts and explored its potential for spacecraft debris shields [8]. Duan, Han and Zhang studied the dynamic behavior of multiple projectiles, multi-particle fragments and packaged structure projectile impacting the spacecraft protective structure [9–11]. Many studies have focused on the high-speed impact of single projectiles on Whipple shields.

A honeycomb sandwich panel can be regarded as an aluminum honeycomb core bounded by a Whipple shield. Aluminum honeycomb sandwich panels are widely used in aerospace due to their advantages of light weight, high specific strength, sound insulation, heat insulation, and vibration resistance. Therefore, they are often used as load-bearing structures and outer walls of some spacecrafts [12]. Fatt established a theoretical model of a projectile impacting a honeycomb sandwich panel in three stages and obtained an analytical solution of its ballistic limit velocity [13]. Yahaya carried out an experimental study on the impact of metal foam projectiles on aluminum honeycomb sandwich panels and analyzed the dynamic response of different types of aluminum honeycomb cores under the impact load [14]. Ebrahimi et al. used a finite element simulation to study the dynamic response and failure mode of honeycomb sandwich panels under a combined impact load from an explosion and projectile; they discussed the effect of the incident angle on skin deformation and honeycomb core crushing [15].

The above studies mainly focused on the dynamics of the impact of projectiles with spacecraft protective materials or structures. However, the damage to the inner structure of a spacecraft caused by kinetic projectiles after they pass through the protective structure has not been thoroughly analyzed in simulation and experimental studies. In addition, there have been few studies about the dynamic characteristics of the penetration of honeycomb sandwich panels by projectiles with internal damage elements. Therefore, there is a need for analytical works that can provide a reference for spacecraft damage assessments and protective structure designs.

The main structure of this paper is as follows: Section 2 describes the ground damage test. Section 3 shows the finite element simulation of the damage process based on the ABAQUS program. Section 4 establishes the theoretical model of the penetration of the honeycomb sandwich material by the projectile. Section 5 analyzes the dynamic responses with different velocities and incident angles, and Section 6 concludes the paper. The purpose of this study was to analyze the characteristics of the damage caused by a kinetic projectile to a satellite's internal structure through both simulation and experimentation. This paper could provide guidance for a design in consideration of the protection and damage mitigation of satellite structures.

2. Ground Damage Test

2.1. Configuration of the Ground Damage Test

In order to study the characteristics of the damage caused by kinetic projectiles impacting upon a spacecraft, the structure of the spacecraft needs to be simplified. Taking artificial satellites as an example, many honeycomb sandwich panels are currently used in satellite structures. Even the main bearing structures of some satellites are composed of honeycomb sandwich panels, such as the A2100 satellite platform in the United States, as well as small Chinese satellite platforms [16]. Due to the harsh electromagnetic environment in space, the shell of electronic equipment in spacecrafts is generally shielded with an aluminum alloy shell [17,18], which can effectively prevent the intrusion of external interference.

In this study, the outer wall of the spacecraft was simplified as an aluminum honeycomb sandwich panel (A). The outer wall of the electronic device was simplified as an aluminum alloy plate (B), and the inner wall was plate F. Three circuit boards (C–E) were arranged between plates B and F to simulate the electronic device. The thickness of honeycomb panel A was 20 mm and was composed of 2A12 aluminum alloy. Plates B and F each had a thickness of 2 mm and were composed of 2A12 aluminum alloy. Plates C–E were three printed circuit boards with thicknesses of 1.6 mm. Three lights were connected to the printed circuit boards to simulate electronic components.

To analyze the damage mechanism, we carried out a ground damage effect evaluation test. The layout of the test device is shown in Figure 1. The shell of the projectile included a nylon cartridge case and an aluminum bottom. Forty-eight steel balls with 4 mm diameters and 0.263 g weight filled the projectile. The projectile mass was 21 g, and the diameter was 16.6 mm. The launching speed was 325 m/s. A high-speed camera was used for recording the experiment.

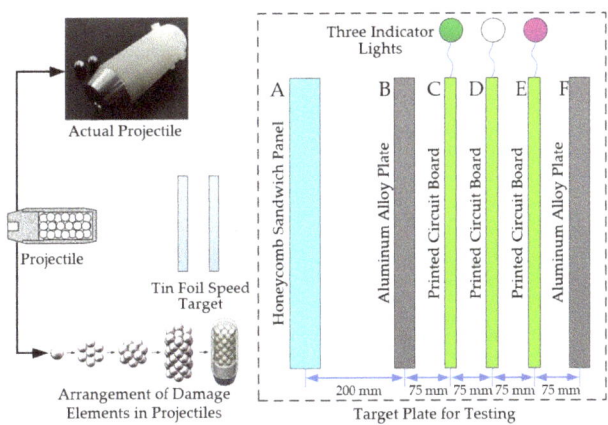

Figure 1. Schematic diagram of the layout of the ground test.

2.2. Experimental Observations

As is shown in Figure 2, the projectile hit the target plates with an initial velocity of 325 m/s. After penetrating targets A and B, the nylon cartridge case was destroyed, and the damage elements spread out, causing regional damage to the three circuit boards. Three lights were connected to the circuit boards to simulate electronic components. The extinguishment of lights indicated that the electronic components were damaged and lost their function. The fragments formed by the damage elements and the target were completely blocked by the last plate, F. The test showed that the kinetic projectile could cause regional damage to the internal structure after penetrating the outer wall of the spacecraft without generating additional space debris, demonstrating the good feasibility of the damage effect. However, the test could only validate the damage effect. It was necessary to carry out a finite element simulation to further analyze the damage performance of the kinetic projectile.

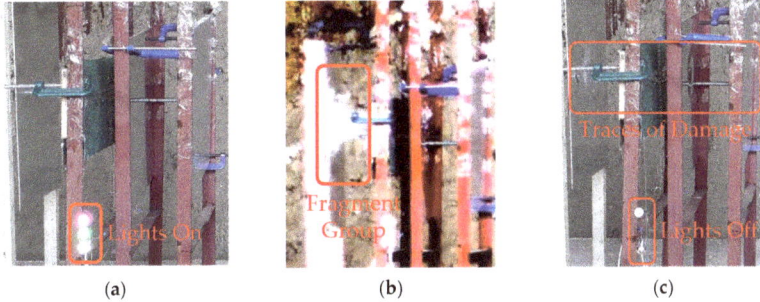

Figure 2. Test configuration. (**a**) Before the damage, the lights were on. (**b**) During the penetration, the damage elements and the plate formed fragments. (**c**) After the damage, the lights were off.

2.3. Damage Mechanism

The damage mechanism of the kinetic projectile is shown in Figure 1. The kinetic projectile needed to meet the damage requirements, which meant that the kinetic projectile must penetrate the surface of the satellite (target A) and the aluminum shell (target B); then, the projectile spread the damaged elements and generated regional damage to the internal electronic components (targets C–E); however, the projectile and the damage elements could not penetrate the surface of the satellite on the other side (target F), meaning that the generation of space debris was avoided.

The projectile consisted of a nylon shell and internal metal elements and was launched by a pyrotechnic launcher. The projectile hit and penetrated the honeycomb sandwich panel on the surface of the satellite at a speed of 325 m/s, as indicated by ① in Figure 3. When the projectile hit the higher-strength aluminum alloy shell of the electronic device, the nylon shell shattered, and the internal damage elements spread out, as shown in ② of Figure 3. In addition, the projectile was axially compressed upon impact, which was further converted to radial expansion potential energy. The damage elements spread horizontally and axially, causing regional damage to various electronic components inside the device.

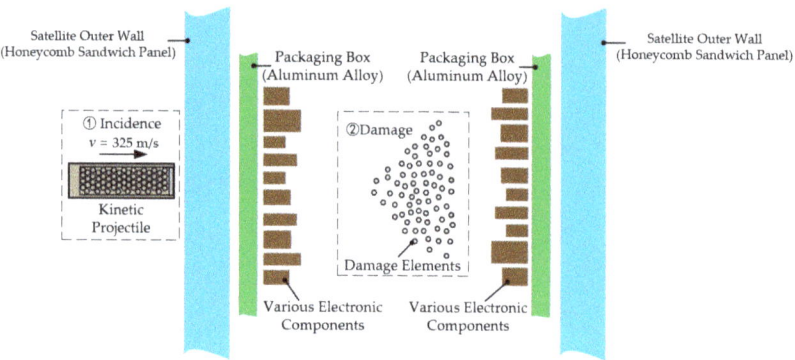

Figure 3. Mechanism of kinetic projectile damage to orbiting spacecrafts.

The kinetic projectile generated two kinds of damage effects: a lateral effect and a minor penetration effect [11]. When the projectile hit target B, it was axially compressed by the internal damage elements, causing damage and expansion of the top cover. Then, the internal damage elements spread laterally, resulting in the so-called lateral effect. Moreover, with the dispersion of the damage elements, the kinetic energy of the projectile was dispersed. The reduced kinetic energy led to a reduced penetration ability, thus the minor penetration effect.

3. Finite Element Simulation

3.1. Finite Element Simulation Model

A simulation model of the projectile damaging the multi-layer plates was established in ABAQUS; the visualization of the model is shown in Figure 4. The case and bottom of the projectile were solid parts, with element type C3D8R. According to our observations, the damage elements did not experience visible deformation during the experiment, so, in the numerical model, they were modeled as rigid bodies, and their element type was set as R3D4. The layout of the target plates was the same as in the test conditions, and the area of each target plate was 150 mm × 150 mm. In order to capture the physics of damage elements penetrating the plate, it was necessary to refine the mesh in the central area of the target plate. The honeycomb sandwich panel was composed of two skins and a honeycomb core. The thicknesses of the skin and core were 0.5 mm and 19 mm, respectively, and the wall thickness of the honeycomb cell was 0.08 mm. Based on [19], the honeycomb core was modeled using shell elements with element type S4R due to the extremely low thickness. The honeycomb skins were modeled as solid elements with element type C3D8R. The thicknesses of targets B and F were 2 mm, and the thicknesses of targets C, D and E were all 1.6 mm. These target plates were solid elements of type C3D8R. The initial velocity of the projectile was set as 325 m/s, and the boundaries of the target plates were fixed. General contact was employed for interactions between different parts.

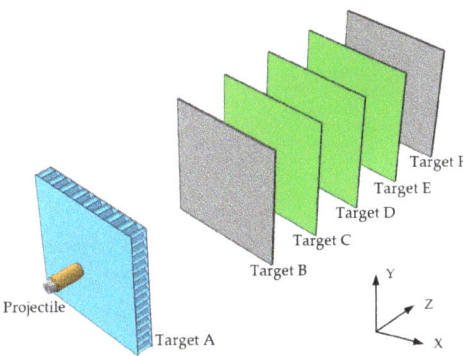

Figure 4. Visualization of the finite element simulation model.

3.2. Material Modeling

The simulation model included three materials: 2A12 aluminum alloy, nylon and printed circuit boards. In this paper, the material model of 2A12 aluminum alloy used the Johnson–Cook constitutive model. The material parameters are summarized in Table 1 based on [20]. In the table, ρ is the density, E is Young's modulus and μ is Poisson's ratio, which are the basic mechanical parameters of the material. Additionally, A is the dynamic yield strength, B is the strain rate hardening coefficient, n is the strain rate hardening index, C is the strain rate hardening coefficient and m is the temperature rise softening index, which are the parameters of the Johnson–Cook constitutive model. $D_1 - D_5$ are the Johnson–Cook failure criterion parameters.

Table 1. Material parameters of 2A12 aluminum alloy [20].

$\rho/(kg \cdot m^{-3})$	E/GPa	μ	A/MPa	B/MPa	n	C
2770	71.7	0.33	400	424	0.350	0.001
m	D_1	D_2	D_3	D_4	D_5	-
1.426	0.116	0.211	−2.172	0.012	−0.01256	-

The printed circuit boards were mainly composed of glass–fiber-reinforced material, which is a kind of anisotropic composite material. The orthotropic material model and the Hashin damage model were used to describe the composite material [21]. Since ABAQUS does not provide the Hashin criterion for 3D solid elements, it was necessary to write a VUMAT subroutine through Fortran language and implement it in the numerical solver [22]. The thickness of each layer was 0.2 mm, and the layer angles were arranged as [45/0/−45/90]. The parameters of the orthotropic material model for the circuit boards are shown in Table 2 [23]. In the table, ρ is the density, $E_1 - E_3$ are Young's moduli in the three directions, $\gamma_{12} \sim \gamma_{23}$ are Poisson's ratios in the three directions, and $G_{12} \sim G_{23}$ are the shear moduli in the three directions. The Hashin damage parameters are shown in Table 3. In the table, X_T and X_C are the tensile strength and compressive strength of the single-layer board in the fiber direction, respectively; Y_T and Y_C are the tensile strength and compressive strength, respectively, perpendicular to the fiber direction and $S_{12} \sim S_{23}$ are the shear strengths in the three directions.

Table 2. Parameters of the orthotropic material model for the circuit boards [23].

$\rho/(kg \cdot m^{-3})$	E_1/GPa	E_2/GPa	E_3/GPa	γ_{12}	γ_{13}	γ_{23}	G_{12}/GPa	G_{13}/GPa	G_{23}/GPa
1570	112	8.2	8.2	0.3	0.3	0.4	4.5	4.5	3

Table 3. Hashin damage parameters of the circuit boards [23].

X_T/MPa	X_C/MPa	Y_T/MPa	Y_C/MPa	S_{12}/MPa	S_{13}/MPa	S_{23}/MPa
2000	1100	85	250	80	60	60

Based on [24], the stress–strain data were used in the plastic deformation stage of the nylon material. The parameters of nylon are shown in Table 4. A combination of flexible damage and shear damage was used in the failure model. The element was automatically deleted when the principal strain of the nylon element was greater than 0.65.

Table 4. Material parameters of nylon [24].

$\rho/(kg \cdot m^{-3})$	E/GPa	μ
1140	3.30	0.40

3.3. Simulation Results

The curve of the total kinetic energy of the model over time is shown in Figure 5. The stress nephograms at four moments were selected to show the damage process of the projectile hitting the targets. When the projectile passed through the honeycomb sandwich plate (target A), it was compressed and deformed slightly, but the cartridge case was not broken. However, when the projectile hit target B, the cartridge case broke, and the internal damage elements were scattered both laterally and axially. At this time, the sudden dissipation of the total kinetic energy was observed. Afterwards, the damage elements hit the circuit boards regionally before falling to the bottom of projectile. All damage elements penetrated target C, while four and nine damage elements were unable to penetrate targets D and E, respectively. After penetrating the three circuit boards, the damage elements and the target fragments formed a large area fragment group. Then, the fragment group hit target F but did not penetrate. During the entire process, the fragment group remained between target A and target F, so target F was not penetrated.

A comparison of the damaged targets between the simulation and test is shown in Figure 6. The target numbers are A–F from left to right. The damage characteristics of the targets recovered from the experiment and from the simulation showed numerical consistency. When the projectile penetrated target A, the shape of the opening was affected by the honeycomb structure, and it consisted of the sides of the honeycomb cells. When the projectile hit target B, the damage elements played a key role in the penetration, so the opening was irregularly circular. Due to the lateral effect, the damage range of target B was slightly larger than that of target A. At the same time, there was no impact trace of damage elements around the opening on target B, meaning that the cartridge case had not been broken before. There was a strong randomness in the damage processes of targets C, D and E. The lateral damage ranges increased from 49.55 mm to 73.65 mm, and the local point perforations due to the penetration of individual debris also increased. Their lateral damage range increased from 49.55 mm to 73.65 mm, and the single-point damage around the damage center increased in turn. Finally, when the projectile hit target F, damage marks were produced by the impact of the fragment group; the bottom of the projectile made the most apparent damage marks but did not penetrate the target.

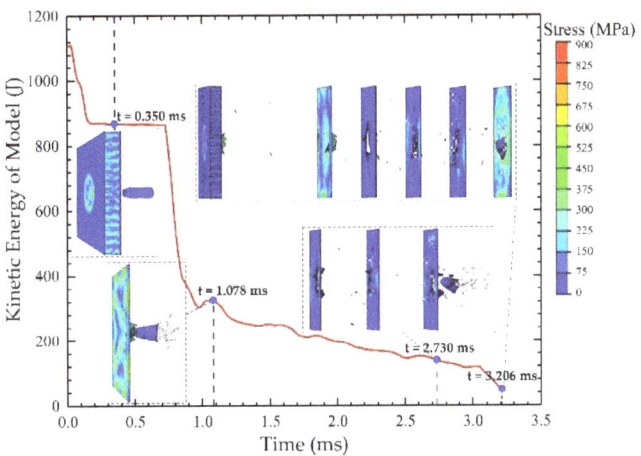

Figure 5. Simulation results of the projectile damage.

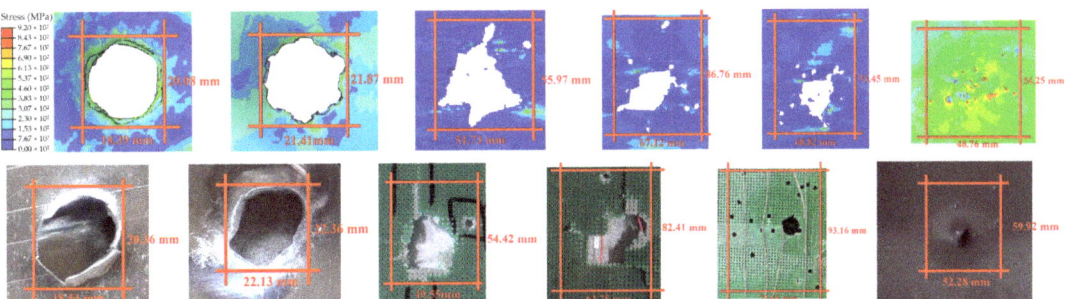

Figure 6. The comparison of damaged targets between the simulation and test.

To compare the consistency between the simulation model and the test more intuitively, we carried out a quantitative analysis on the lateral, longitudinal and oblique damage ranges, as shown in Table 5.

Table 5. Comparison of the lateral, longitudinal and oblique ranges.

Target Number	Lateral Damage Range (mm)			Longitudinal Damage Range (mm)			Oblique Damage Range (mm)		
	Simulation	Test	Error (%)	Simulation	Test	Error (%)	Simulation	Test	Error (%)
A	18.39	18.54	0.81%	20.08	20.36	1.36%	27.23	27.54	1.11%
B	21.41	22.13	3.25%	21.87	22.36	2.19%	30.60	31.46	2.72%
C	51.73	49.55	4.40%	55.97	54.42	2.86%	76.21	73.60	3.56%
D	67.12	63.75	5.29%	86.76	82.41	5.29%	109.70	104.19	5.29%
E	68.82	73.65	6.56%	93.45	93.16	0.31%	116.06	118.76	2.28%
F	48.76	52.28	6.73%	54.25	59.92	9.46%	72.95	79.52	8.27%

The qualitative analysis of the damage process showed that the damage characteristics of the test and simulation were similar; the quantitative analysis of the damage ranges of the targets showed that the deviations were within 10%. Therefore, the simulation results were consistent with the experimental results, and the simulation model can be used for further research.

4. Theoretical Model of Projectile Impact on Honeycomb Sandwich Panel

4.1. Derivation of Theoretical Model

The kinetic projectile must penetrate the outer wall of the satellite—that is, the honeycomb sandwich panel. The most commonly used parameter for describing a target's ability to withstand a projectile impact is the ballistic limit. It is defined as the maximum projectile velocity a target can withstand without being penetrated. By establishing the theoretical model of the kinetic projectile impacting the honeycomb sandwich panel, the projectile's ballistic limit velocity and residual velocity could be obtained, and the damage characteristics can be further explained.

4.1.1. Structural Model

The schematic diagram of the projectile hitting the honeycomb sandwich panel is shown in Figure 7a. The honeycomb sandwich panel was composed of a front skin, a back skin and a honeycomb core, where h is the thickness of the skins, H is the thickness of the honeycomb core, h_1 is the wall thickness of the honeycomb cell and D is the side length of the honeycomb cell. The projectile has a blunt tip, r_p is the projectile radius, d_p is the projectile diameter, M_0 is the projectile mass and V_0 is the initial velocity of the projectile. According to the study of Fatt [13], the penetration of the honeycomb sandwich panel could be decomposed into three stages: the perforation of the front skin, the honeycomb core and the back skin.

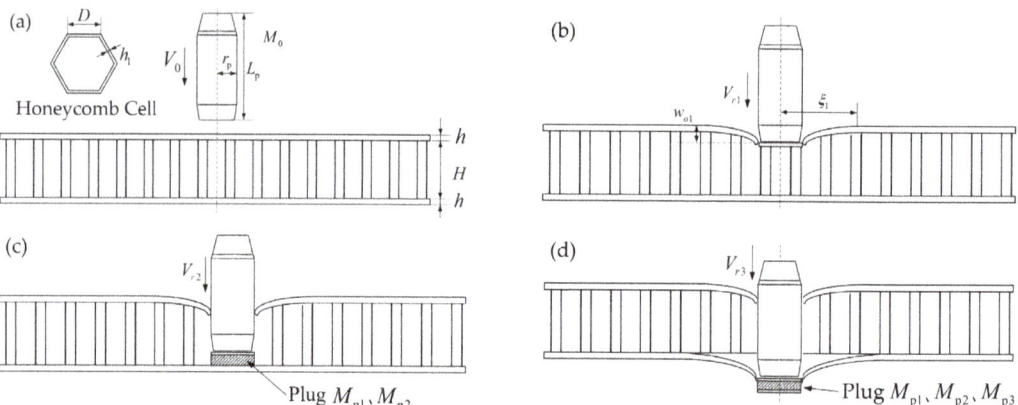

Figure 7. Theoretical model of the kinetic projectile hitting honeycomb sandwich panel: (**a**) schematic diagram; (**b**) first stage; (**c**) second stage; (**d**) third stage.

The theoretical model is based on the following assumptions:

1. The honeycomb thickness is much greater than the skin thickness, $H \gg h$. However, this assumption has some consequences (i.e., that the front skin will deflect and perforate, while the honeycomb core steadily crushes without compaction). The back skin is undeformed during the opening of the front skin.
2. The diameter of the projectile is larger than the length of the side of the honeycomb cell, $d_P > D$. Therefore, honeycomb crushing and shearing will occur under the projectile. If the projectile diameter is less than the cell side length and the projectile falls within one cell, the projectile can penetrate the cell without crushing it.

The first stage is shown in Figure 7b; the projectile penetrates the front skin, the honeycomb core and the back skin support the front skin, the honeycomb core is stably crushed and the back skin is not deformed. ξ_1 is the radial extent of the front skin deformation, w_{o1} is the transverse deflection of the front skin deformation and V_{r1} is the residual velocity after the projectile penetrates the front skin.

The second stage is shown in Figure 7c. This stage starts when the projectile completely penetrates the front skin. The projectile acts on the honeycomb core together with a plug from the front skin. According to Goldsmith's experimental research [25], Fatt obtained the ballistic limit equation of bare honeycomb [13]. In the paper, the failure mechanism of honeycomb was replaced by this ballistic limit [13]. Considering that part of the honeycomb core was crushed in the first stage, the reduced ballistic limit of bare honeycomb was used [13].

The third stage is shown in Figure 7d. It is assumed that the honeycomb core has been compacted into a plug before this stage; the plug hits the back skin together with the projectile. The energy lost upon impact with the back skin is predicted using the perforation equation developed for thin plates by Woodward [26].

4.1.2. First Stage: Perforation of Front Skin

The energy conservation equation when the projectile penetrates the front skin is [13]:

$$\frac{1}{2}M_0V_0^2 = \Gamma_1 + W_{f1} + \frac{1}{2}(M_0 + M_{p1})V_{r1}^2 \qquad (1)$$

In the formula, Γ_1 is the plastic work dissipated during membrane stretching and honeycomb deformation under the deformed skin, W_{f1} is the fracture work and M_{p1} is the mass of the plug cut out from the front skin; it is assumed that the plug diameter is equal to the projectile diameter. V_{r1} is the residual velocity of the projectile after penetrating the front skin.

The plastic work Γ_1 is dissipated in the stretching of the front skin and the crushing of the honeycomb; it is found by considering the equivalent plastic work dissipated in the static indentation of a membrane bonded to a rigid plastic foundation [13]. According to the relevant derivation of Fatt [13], the expression of the plastic work consumed by plastic deformation in the first stage is:

$$\Gamma_1 = \frac{\pi(q_c+q_s)^2 r_p^4}{16\sigma_o h}\left[\left(\frac{\xi_1}{r_p}\right)^4 \left[\ln\left(\frac{\xi_1}{r_p}\right)^4 - 1\right] + 1\right] - \frac{\pi r_p^4 q_s(q_c+q_s)}{4\sigma_o h}\left[\left(\frac{\xi_1}{r_p}\right)^2 \left[\ln\left(\frac{\xi_1}{r_p}\right)^2 - 1\right] + 1\right] \qquad (2)$$

In the formula, q_c is the dynamic crush strength of the aluminum honeycomb core, q_s is the dynamic shear strength of the honeycomb core, σ_o is the flow stress of the skin material, r_p is the projectile radius, h is the skin thickness and ξ_1 is the radial extent of the front skin deformation at the end of the first stage.

According to [13], when the radial strain reaches a critical value, the front skin experiences tensile necking. The radial extent of the front skin deformation ξ_1 is:

$$\xi_1 = r_p\sqrt{1 + 2\sqrt{2\varepsilon_{cr}}\frac{\sigma_o h}{(q_c+q_s)r_p}} \qquad (3)$$

In the formula, ε_{cr} is the critical fracture strain; Wang obtained $\varepsilon_{cr} = 1.5$ from a uniaxial tensile test [27].

The transverse deflection w_{o1} of the front skin deformation at the end of the first stage (see Figure 7b) is:

$$w_{o1} = \frac{(q_c+q_s)\xi_1^2}{2\sigma_o h}\ln\frac{\xi_1}{r_p} - \frac{(q_c+q_s)}{4\sigma_o h}\left(\xi_1^2 - r_p^2\right) \qquad (4)$$

W_{f1} is expressed as:

$$W_{f1} = \frac{\pi h^2 r_p \sigma_o}{\sqrt{3}} \qquad (5)$$

According to the impact test of Goldsmith [25], the dynamic crush strength of the honeycomb core is about 30% higher than the static crush strength. Therefore, the dynamic crush strength q_c and dynamic shear strength q_s of the honeycomb core are 1.3 times larger than the static crush strength q_{cr} and static shear strength q_{sh}.

q_{cr} is the static crush strength of the honeycomb core. Wierzbicki [28] established the relationship between the honeycomb parameters and q_{cr} and applied it in a projectile impact test. The results showed that the error was within 5% during low-velocity impact and within 16% during high-velocity impact [28]. The expression for q_{cr} is:

$$q_{cr} = 16.55\sigma_o \left(\frac{h_1}{D}\right)^{5/3} \tag{6}$$

In the formula, σ_o is the flow stress of the honeycomb core, and h_1 is the wall thickness of the honeycomb cell, as shown in Figure 7a.

When the projectile hits the honeycomb sandwich panel, the honeycomb core under the front skin is not only crushed but also experiences shear deformation, which means that the honeycomb core provides shear resistance in the deformation area from r_p to ζ_1. q_{sh} is the static shear strength of the honeycomb core obtained by McFarland [29]; its expression is:

$$q_{sh} = 1.155 \frac{\sigma_o}{\sqrt{3}} \frac{h_1}{D} \tag{7}$$

4.1.3. Second Stage: Perforation of Honeycomb Core

The energy conservation equation when the projectile penetrates the honeycomb core is [13]:

$$\frac{1}{2}(M_0 + M_{p1})V_0^2 = \Gamma_2 + \frac{1}{2}(M_0 + M_{p1} + M_{p2})V_{r2}^2 \tag{8}$$

In the formula, Γ_2 is the work involved in perforating the honeycomb, V_{r2} is the residual velocity of the projectile after penetrating the honeycomb core, M_{p2} is the mass of the plug cut out from the honeycomb core and $M_{p2} = \pi \rho H r_m^2$. r_m is the radius of the honeycomb plug; the experiments showed that r_m is 1~2 times the radius of the projectile r_p [25], so $r_m = 1.5 r_p$ is used at the end of the second stage.

When the projectile penetrates the honeycomb core, the energy loss is mainly due to the collapse and shear of the honeycomb core. Although the damage model for honeycomb penetration is very complex, the honeycomb failure mechanism can be replaced by the ballistic limit of bare honeycomb [13]. Due to the deformation of the front skin in the first stage, the honeycomb core is already crushed, and the reduced ballistic limit of the bare honeycomb is used [13]. The expression of the plastic work Γ_2 is:

$$\Gamma_2 = \frac{1}{2}(M_0 + M_{p1})V'^2_{b1} \tag{9}$$

In the formula, V'_{b1} is the reduced ballistic limit.

In the second stage, the honeycomb core is crushed from its original thickness H to H_1:

$$H_1 = H - w_{o1} \tag{10}$$

In the formula, w_{o1} is the transverse deflection of the front skin (see Figure 7b).

The reduced ballistic limit V'_{b1} can also be estimated from the original ballistic limit $(V_{b1})_H$:

$$V'_{b1} = \sqrt{\frac{H_1}{H}}(V_{b1})_H \tag{11}$$

When the projectile penetrates the bare honeycomb,

$$\frac{1}{2}M_0(V_{b1})_H^2 = (P_c + P_s)H \tag{12}$$

In the formula, P_c is the average crushing force of the honeycomb core, P_s is the average shearing force of the honeycomb core and the experiments in [25] show that P_c and P_s are constant.

According to [30], P_c and P_s can be obtained from q_c and q_s:

$$P_c = q_c \pi r_m^2 \tag{13}$$

$$P_s = 2\pi r_m q_s H \tag{14}$$

Combining the above equations, we get:

$$\Gamma_2 = \frac{M_0 + M_{p1}}{M_0}\left(q_c \pi r_m^2 + 2\pi r_m q_s H\right)(H - w_{o1}) \tag{15}$$

4.1.4. Third Stage: Perforation of Back Skin

The energy conservation equation when the projectile penetrates the back skin is:

$$\frac{1}{2}(M_0 + M_{p1} + M_{p2})V_{r2}^2 = \Gamma_3 + W_{f2} + \frac{1}{2}(M_0 + M_{p1} + M_{p2} + M_{p3})V_r^2 \tag{16}$$

In the formula, Γ_3 the plastic work due to membrane stretching, W_{f2} is the fracture work, M_{p3} is the mass of the plug cut out from the back skin and $M_{p3} = \pi \rho h r_m^2$. The plug radius $r_m = 1.5 r_p$. V_r is the residual velocity after the projectile penetrates the sandwich panel.

Woodward [26] presented the expression for the work consumed due to the plastic deformation of the back skin:

$$\Gamma_3 = \frac{\pi r_m \sigma_o h}{4}(2r_m + 2\pi h) \tag{17}$$

In the formula, σ_o is the flow stress of the skin material.

The expression for the fracture work consumed W_{f2} is [13]:

$$W_{f2} = \frac{\pi h^2 r_m \sigma_o}{\sqrt{3}} \tag{18}$$

4.2. Theoretical Calculation

4.2.1. Calculation Parameters

The modeled honeycomb sandwich panel consisted of the skins and honeycomb core. The skin material was 2A12 aluminum, and the honeycomb core material was 3003H18 aluminum. The material properties are shown in Table 6 [31], where σ_y is the yield strength, σ_u is the ultimate strength and $\sigma_o = (\sigma_y + \sigma_u)/2$ is the flow stress.

Table 6. Material properties of the honeycomb sandwich panels [31].

Part	Material	σ_y (MPa)	σ_u (MPa)	σ_o (MPa)
Skin	2A12 aluminum	325	427	376
Honeycomb core	3003H18 aluminum	170	190	180

The geometric parameters of the projectile and the honeycomb sandwich panel are shown in Table 7; these were used as the input parameters.

Table 7. Geometry parameters of the projectiles and honeycomb sandwich panel.

Name of Parameter	Symbol of Parameter	Value	Unit
Mass of projectile	M_0	21.17	g
Radius of projectile	r_p	8	mm
Length of projectile	L_p	42	mm
Thickness of skin	h	0.5	mm
Thickness of honeycomb core	H	19	mm
Wall thickness of honeycomb cell	h_1	0.08	mm
Side length of honeycomb cell	D	6	mm

4.2.2. Results

Theoretical calculations were carried out for different initial projectile velocities. The residual velocities after each stage are summarized in Table 8.

Table 8. Residual velocity after each perforation stage.

Velocity	V_0=250 m/s	V_0=300 m/s	V_0=325 m/s	V_0=350 m/s	V_0=400 m/s
V_{r1} (m/s)	216.5	272.1	299.1	325.7	378.3
V_{r2} (m/s)	210.6	266.6	293.7	320.4	372.9
V_{r3} (m/s)	196.6	254.2	181.7	308.7	361.6

4.2.3. Comparison between Simulation and Theory

Comparing the dynamic simulation with the theoretical analysis, the calculation results are listed in Table 9. The differences between the residual velocity and residual kinetic energy between simulation and theory were within 5%.

Table 9. Comparison of the results between the theoretical calculations and numerical simulation.

Initial Velocity (m/s)	Residual Velocity (m/s)			Residual Kinetic Energy (m/s)		
	Simulation	Theory	Difference	Simulation	Theory	Difference
250	199.3	196.6	1.35%	420.5	409.2	2.69%
300	258.5	254.2	1.66%	707.4	684.1	3.30%
325	286.7	281.7	1.74%	870.2	840.1	3.46%
350	314.6	308.7	1.88%	1047.8	1008.9	3.72%
400	368.5	361.6	1.87%	1437.6	1384.3	3.71%

In order to see the consistency more clearly between the theoretical model and the simulation model, Figure 8 shows the relationship between the initial velocity and the residual velocity. It can be seen that the simulation datapoints basically coincide on the theoretical curve, indicating that the simulation results support the theoretical model.

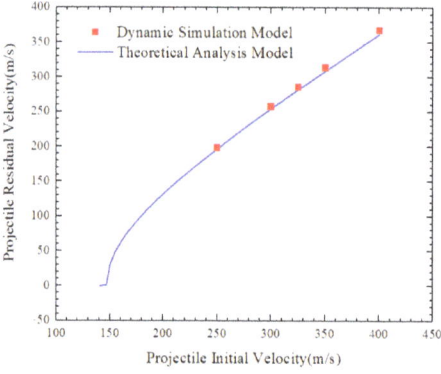

Figure 8. Comparison of the initial and residual velocity between the theory and simulation.

4.2.4. Ballistic Limit Prediction

V_{b1} is the ballistic limit velocity of the honeycomb sandwich panel impacted by the kinetic projectile. V_{b1} is obtained by gaining $V_r = 0$ using the above theoretical model. The theoretical model could obtain the residual velocity V_r by inputting the initial velocity V_0. Different values of V_0 were attempted through iterations until $V_r = 0$ was satisfied, and the ultimate obtained V_0 was V_{b1}. The theoretical calculations found the value of $V_{b1} = 146.9$ m/s [13], and the simulation results indicated that the projectile penetrated target A when the initial velocity was 150 m/s. Therefore, we considered that the ballistic limit velocity was about 150 m/s. Note that the theoretical model can not only be used to calculate the ballistic limit with the same projectile and targets but can also be used to predict the ballistic limit under different conditions, which is convenient for application in both missions and experiments.

4.2.5. Deformation Prediction

According to Equations (3) and (4), it can be calculated that $\xi_1 = 32.5$ mm and $w_{o1} = 14.5$ mm. Since the velocity variable does not appear in the expressions of ξ_1 and w_{o1}, Fatt holds that ξ_1 and w_{o1} are not related to the projectile velocity; this could be observed in the simulation. Figure 9 shows the simulated deformation of the honeycomb sandwich panel after being impacted with the projectile at different velocities. The deformation results are very similar, and the differences between the values of ξ_1 and w_{o1} at different velocities are not large. This supports the assertion that "the deformation of the front skin has little relationship with the projectile velocity", as mentioned the analytical model given in [13]. The simulation results of ξ_1 and w_{o1} are similar to the calculated results, which further illustrates the correctness of the analysis.

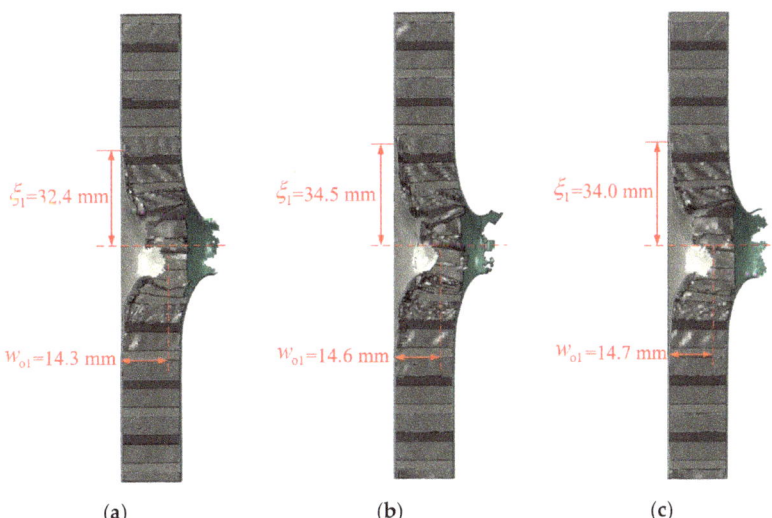

Figure 9. Deformation of the honeycomb sandwich panels after penetration, (a) $V_0 = 250$ m/s, (b) $V_0 = 325$ m/s, (c) $V_0 = 400$ m/s.

5. Influence of Launch Parameters on Damage Characteristics

5.1. Influence of Initial Velocity

Figure 10 shows the damage effects of the kinetic projectile impact under different initial velocities. When the initial velocity was 250 m/s, the projectile did not penetrate target B in the simulation. When the initial velocity was 300 m/s, the damage elements penetrated target B and spread inward, but the projectile bottom did not penetrate. When the initial velocity was in the range of 325–350 m/s, the projectile bottom penetrated target B. When the initial velocity was greater than or equal to 400 m/s, some fragments began to penetrate target F. The damage requirements for the simulation were that the entire projectile must penetrate the honeycomb sandwich panel, and the damage elements must penetrate target panel B and disperse to the inside, and none of the fragments can penetrate target F. Our results indicate that the initial velocity range that should be applied to meet these requirements is 325–350 m/s.

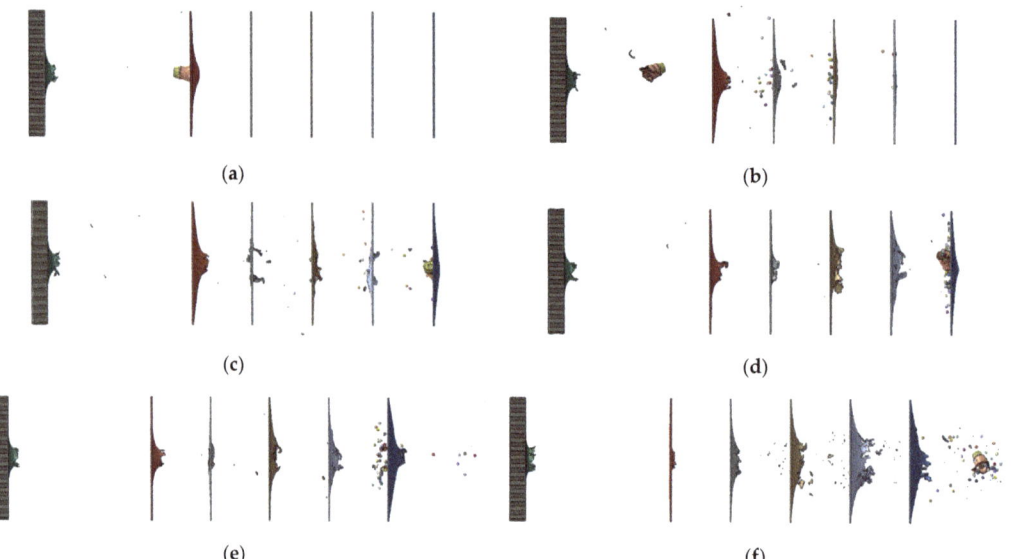

Figure 10. Damage effect of the kinetic projectile impact under different initial velocities, (**a**) $V_0 = 250$ m/s, (**b**) $V_0 = 300$ m/s, (**c**) $V_0 = 325$ m/s, (**d**) $V_0 = 350$ m/s, (**e**) $V_0 = 400$ m/s, (**f**) $V_0 = 600$ m/s.

The ability of the kinetic projectile to damage the internal structure mainly came from the lateral effect when the projectile impacted target B. Projectiles with a higher initial velocity had a more successful lateral effect, but compared to the increase in the lateral velocity, a higher initial velocity had a greater effect on the increase in the longitudinal velocity. It can be seen that projectiles with a higher initial velocity had a stronger ability to penetrate plates, but the damage area was reduced. Therefore, our results suggest that the most appropriate initial velocity is 325 m/s. At this speed, the most effective damage to the circuit boards could be achieved without the penetration of target F.

5.2. Influence of Incident Angle

Figure 11 shows the damage effect of the kinetic projectile impact under different incident angles. Due to the launch conditions, there was generally an incident angle when the projectile impacted the target—that is, the angle between the projectile axis and the normal target direction. In order to investigate the influence of the incident angle on the damage characteristics, simulations were carried out under different incident angles. Two moments were selected to analyze the damage process, as shown in Figure 11. The first (images on the left) was the partial display when the damage elements were scattered, and the second (right side) was the whole picture when the damage process was complete.

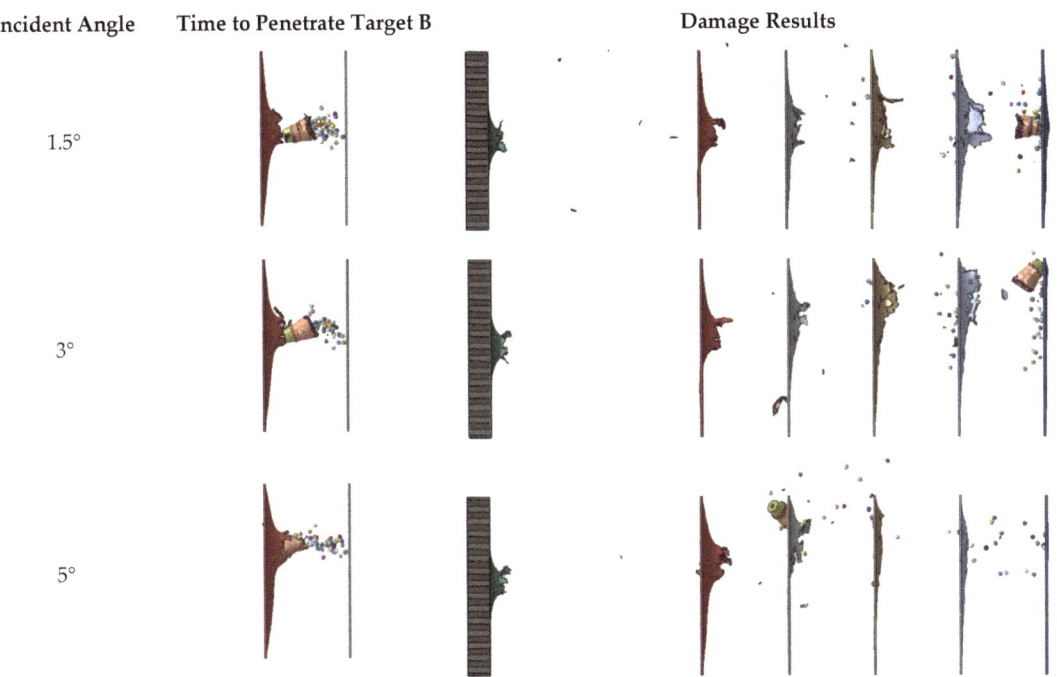

Figure 11. Damage effect of the kinetic projectile impact under different incident angles.

When the incident angles were 1.5°, 3° and 5°, the kinetic projectile could complete the effective damage to the circuit boards, indicating that the projectile could adapt to an incident angle less than 5° when the projectile velocity was 325 m/s. With the increase of the incident angle, the deflection of the projectile bottom was more significant, and more damage elements remained in front of each target plate, indicating that the penetration ability decreased with the increasing incident angle. It is worth noting that, due to the lateral effect, the front damage elements moved forward in the horizontal direction after impact, while the rear damage elements deflected with the projectile bottom. This means that higher incident angles reduced the penetration ability but increased the damage area.

6. Conclusions

In this paper, an artificial satellite was simplified as multi-layer target plates based on its structural characteristics. The damage process of a kinetic projectile hitting the multi-layer target plates was analyzed by integrating the experiments and finite element method. Then, the impact characteristics of the kinetic projectile on the honeycomb sandwich plate were analyzed. Finally, the influence of the initial velocities and incident angles on the damage characteristics were studied based on the simulation model.

The damage characteristics between the simulation and the test were consistent in terms of the damage range of the targets, validating the simulation model. The analytical results of the kinetic projectile impact on the honeycomb sandwich panel correlated well with the simulation; the ballistic limit velocity was determined as 150 m/s. We also found that the opening deformation had little relationship with the projectile velocity.

The projectile velocity necessary to achieve effective damage to the circuit boards without penetration of target F was between 300 m/s and 350 m/s, and we identified the most appropriate initial velocity as 325 m/s. Projectiles with a higher initial velocity had a stronger ability to penetrate the plates, but the damage area was reduced. The kinetic projectile could adapt to an incident angle less than 5°. Due to the lateral effect, the front damage elements still moved in the vertical direction, increasing the damage area but reducing the penetration ability.

Author Contributions: Conceptualization, S.Y., H.Z. and Z.D.; methodology, S.Y. and Y.B.; validation, S.Y., Y.B. and W.S.; formal analysis, S.Y. and Y.B.; investigation, S.Y.; resources, Z.D. and H.Z.; data curation, Y.B.; writing—original draft preparation, Y.B.; writing—review and editing, S.Y. and G.Z.; visualization, S.Y. and Y.B.; supervision, Z.D.; project administration, Z.D. and funding acquisition, H.Z. All authors have read and agreed to the published version of the manuscript.

Funding: This work was supported by the National Natural Science Foundation of China (52102436), the Fundamental Research Funds for the Central Universities (30920021109), the Natural Science Foundation of Jiangsu Province (BK20200496), the China Postdoctoral Science Foundation (2020M681615), the project of Key Laboratory of Impact and Safety Engineering (Ningbo University), the Ministry of Education (CJ202107) and the State Key Laboratory of Mechanics and Control of Mechanical Structures (Nanjing University of Aeronautics and astronautics) (Grant No. MCMS-E-0221Y01).

Institutional Review Board Statement: Not applicable.

Informed Consent Statement: Not applicable.

Data Availability Statement: The data are contained within the article. The data presented in this study can be seen in the contents above.

Acknowledgments: We wish to express our gratitude to the members of our research team, Guang-fa Gao, Zhaojun Pang, Qing Lin, Weiliang Zhu, Mengsheng Li and Chunbo Wu.

Conflicts of Interest: The authors declare no conflict of interest.

References

1. Maclay, T.; Mcknight, D. Space environment management: Framing the objective and setting priorities for controlling orbital debris risk. *J. Space Saf. Eng.* **2021**, *8*, 93–97. [CrossRef]
2. Li, M.; Gong, Z.Z.; Liu, G.Q. Frontier technology and system development of space debris surveillance and active removal. *Chin. Sci. Bull.* **2018**, *63*, 2570–2591. [CrossRef]
3. Muelhaupt, T.J.; Sorge, M.E.; Morin, J.; Wilson, S.R. Space traffic management in the new space era. *J. Space Saf. Eng.* **2019**, *6*, 80–87. [CrossRef]
4. Whipple, F.L. Meteorites and space travel. *Astron. J.* **1947**, *52*, 131. [CrossRef]
5. Schonberg, W.P. Concise history of ballistic limit equations for multi-wall spacecraft shielding. *REACH* **2016**, *1*, 46–54. [CrossRef]
6. Wen, K.; Chen, X.W.; Lu, Y.G. Research and development on hypervelocity impact protection using Whipple shield: An overview. *Def. Technol.* **2021**, *17*, 1864–1886. [CrossRef]
7. Poniaev, S.A.; Kurakin, R.O.; Sedov, A.I.; Bobashev, S.V.; Zhukov, B.G.; Nechunaev, A.F. Hypervelocity impact of mm-size plastic projectile on thin aluminum plate. *Acta Astronaut.* **2017**, *135*, 26–33. [CrossRef]
8. Hofmann, D.C.; Hamill, L.; Christiansen, E.; Nutt, S. Hypervelocity Impact Testing of a Metallic Glass-Stuffed Whipple Shield. *Adv. Eng. Mater.* **2015**, *17*, 1313–1322. [CrossRef]
9. Duan, M. Research on Damage Characteristics of the Space Debris Shield Impacted by Hypervelocity Multiple Projectiles. Master's Thesis, Harbin Institute of Technology, Harbin, China, 2018.
10. Han, X.F. Study on Damage Characteristics of High-Speed Impact on Shield Structure of Wrap Multi-Particle Fragments in Space Temperature Environment. Master's Thesis, Harbin Institute of Technology, Harbin, China, 2019.
11. Zhang, D. Research on Fragmentation Characteristics of Packaged Structure Projectile in High-Velocity Impact. Master's Thesis, Harbin Institute of Technology, Harbin, China, 2020.
12. Aborehab, A.; Kamel, M.; Nemnem, A.F.; Kassem, M. Finite Element Model Updating of a Satellite Honeycomb Sandwich Plate in Structural Dynamics. *Int. J. Space Struct.* **2021**, *36*, 105–116. [CrossRef]
13. Fatt, M.S.H.; Park, K.S. Perforation of honeycomb sandwich plates by projectiles. *Compos. Part A Appl. Sci. Manuf.* **2000**, *31*, 889–899. [CrossRef]
14. Yahaya, M.A.; Ruan, D.; Lu, G.; Dargusch, M.S. Response of aluminium honeycomb sandwich panels subjected to foam projectile impact–An experimental study. *Int. J. Impact Eng.* **2015**, *75*, 100–109. [CrossRef]
15. Ebrahimi, H.; Ghosh, R.; Mahdi, E.; Nayeb-Hashemi, H.; Vaziri, A. Honeycomb sandwich panels subjected to combined shock and projectile impact. *Int. J. Impact Eng.* **2016**, *95*, 1–11. [CrossRef]
16. Guo, J.S.; Yue, C.F.; Wang, F. Lightweight structure design method for satellites with high payload ratios. *J. Harbin Eng. Univ.* **2021**, *42*, 1535–1542.

17. Man, X. The Design and Dynamics Simulation of Satellite Structure and Mechanism with Complex Constraints. Ph.D. Thesis, Center for Space Science and Applied Research Chinese Academy of Sciences, Beijing, China, 2015.
18. Liu, S.; Lan, S.W.; Ma, Z.X.; Li, Y.; Huang, J. Experimental study on characteristics of satellite breakup debris. *J. Astronaut.* **2012**, *33*, 1347–1353.
19. Zhang, Y.; Yan, L.; Zhang, C.; Guo, S. Low-velocity impact response of tube-reinforced honeycomb sandwich structure. *Thin-Walled Struct.* **2021**, *158*, 107188. [CrossRef]
20. Zhang, W.; Wei, G.; Xiao, X.K. Constitutive relation and fracture criterion of 2A12 aluminum alloy. *Acta Armamentarii* **2013**, *34*, 276–282.
21. Zhang, Y.; Chen, P. An improved methodology of constructing inter-fiber failure criteria for unidirectional fiber-reinforced composites. *Compos. Part A Appl. Sci. Manuf.* **2021**, *145*, 106369. [CrossRef]
22. Feito, N.; Díaz-Álvarez, J.; López-Puente, J.; Miguelez, M.H. Experimental and numerical analysis of step drill bit performance when drilling woven CFRPs. *Compos. Struct.* **2018**, *184*, 1147–1155. [CrossRef]
23. Yan, Q.X. Experimental and Simulation Research on High Speed Drilling of Printed Circuit Board. Master's Thesis, Shenzhen University, Shenzhen, China, 2017.
24. Cheng, B.; Wang, H.Y. Finite element analysis of nylon rotating bands design for gas gun. *J. Ordnance Equip. Eng.* **2019**, *40*, 43–46, 51.
25. Goldsmith, W.; Sackman, J.L. An experimental study of energy absorption in impact on sandwich plates. *Int. J. Impact Eng.* **1992**, *12*, 241–262. [CrossRef]
26. Woodward, R.L.; Cimpoeru, S.J. A study of the perforation of aluminium laminate targets. *Int. J. Impact Eng.* **1998**, *21*, 117–131. [CrossRef]
27. Wang, L.; Li, Y.L.; Suo, T.; Guo, Y.Z. Mechanical behavior of commonly used aeronautical aluminum alloys under dynamic tension. *J. Aeronaut. Mater.* **2013**, *33*, 71–77.
28. Wierzbicki, T. Crushing analysis of metal honeycombs. *Int. J. Impact Eng.* **1983**, *1*, 157–174. [CrossRef]
29. McFarland, R.K. Hexagonal cell structures under post-buckling axial load. *AIAA J.* **1963**, *1*, 1380–1385. [CrossRef]
30. Zhang, Q.N. Study on the Ballistic Impact Effect of Sandwich Panels with Aluminum Honeycomb Cores. Master's Thesis, Beijing Institute of Technology, Beijing, China, 2014.
31. Liu, M.F.; Liu, S.X. Mechanical properties of aluminum and aluminum alloys. In *Handbook of Mechanical Properties of Metallic Materials*, 1st ed.; China Machine Press: Beijing, China, 2011; Volume 1, pp. 527–530.

Article

Dynamic Simulation and Parameter Analysis of Contact Mechanics for Mimicking Geckos' Foot Setae Array

Qing Lin [1,*], Chunbo Wu [2,*], Shuai Yue [2], Zhonghui Jiang [3], Zhonghua Du [2] and Mengsheng Li [2]

[1] Aerospace System Engineering Shanghai, Shanghai 201109, China
[2] School of Mechanical Engineering, Nanjing University of Science and Technology, Nanjing 210094, China; yueshuai@njust.edu.cn (S.Y.); duzhonghua@aliyun.com (Z.D.); lms@njust.edu.cn (M.L.)
[3] Xi'an North Qinghua Electrical Co., Ltd., Xi'an 710025, China; shiwenhui@njust.edu.cn
* Correspondence: linqing@nuaa.edu.cn (Q.L.); wcb@njust.edu.cn (C.W.)

Abstract: According to the dynamic characteristics of the adhesion desorption process between gecko-like polyurethane setae and the contact surface, the microcontact principle of an elastic sphere and plane is established based on the Johnson–Kendall–Robert model. On this basis, combined with the cantilever beam model, microscale adhesive contact models in the case of a single and an array of setae are obtained. The contact process is numerically simulated and verified by the adhesion desorption test. After that, the effects of external preload, the elastic modulus of setae material, the surface energy, and the surface roughness on the contact force and depth during the dynamic contact process of setae are studied. The results show that the error between the simulation and test is 15.9%, and the simulation model could reflect the real contact procedure. With the increase in preload, the push-off force of the setae array would grow and remain basically constant after reaching saturation. Increasing the elastic modulus of setae material would reduce the contact depth, but have little effect on the maximum push-off force; with the increase in the surface energy of the contact object, both the push-off force between the objects and the contact depth during desorption would increase. With the increase in wall roughness, the push-off force curve of the setae array becomes smoother, but the maximum push-off force would decrease. By exploring the dynamic mechanical characteristics of the micro angle of setae, the corresponding theoretical basis is provided for the numerical simulation of the adsorption force of macro materials.

Keywords: Johnson–Kendall–Robert model; polyurethane setae array; surface energy; modeling; simulation

1. Introduction

In recent years, substantial progress has been achieved in gecko-inspired adhesive technology since the discovery of the uniquely layered footpad structure of geckos [1,2]. Geckos can climb on almost any surface, or even stand upright or upside down, because their layered structure is composed of inclined villi, and the end of each villus is composed of many thin setae [3]. Researchers have been able to understand the adhesion mechanism of gecko feet and have promoted the development of biomimetic adhesion materials [4–6]. Based on the van der Waals force contact principle, biomimetic adhesive materials may exhibit stable contact performance in complex environments [7,8]. The van der Waals force exists in the molecules on the surface of various objects without any requirements regarding the environment, so it is employed by animals such as geckos to adhere to the surfaces of various objects [9].

A lot of research has been carried out on the design and production of gecko-like adhesive materials and structures at home and abroad. In 2000, Autumn et al. [10,11] studied the adhesion mechanism of gecko setae. In the United States, there has been an upsurge in research on biomimetic dry adhesive materials and structures. The adhesion mechanism of gecko setae dominated by the van der Waals force and supplemented by the capillary force has been widely accepted. However, the elastic deformation of the setae was rarely involved, leading to the overestimation of the adhesion to the setae. In 2003, Geim et al. proposed a "gecko tape" material, which is fabricated with an array of geometrically shaped flexible plastic posts. The structure has a repeatable attachment function and self-cleaning capability [12]. Persson proposed a simple model to study the effect of surface roughness on the adhesion of gecko foot setae. This model simplifies the tongue depressor at the end of the setae and regards the tongue depressor as an equivalent plane. Assuming that the surface roughness follows a normal distribution, the simulation results show whether the adhesion depends on the magnitude of the surface roughness [13,14]. The literature carries out experimental and theoretical analyses on adhesion to rough surfaces, but does not emphasize the change in the adhesion force of microscopic setae during adhesion. In 2005, the Max Planck Institute for Metal Research in Germany simulated setae found on the foot of a gecko, then optimized and simulated the pulling process of the gecko setae, but lacked research on the adhesion properties of array materials [15,16]. Dai Liming et al., of the University of Dayton in the United States, recently used chemical vapor deposition to obtain high-density, large aspect ratio carbon nanotube-adhered arrays [17]. The literature did not further elaborate on the adhesion principle of the array material. Guo et al. studied the dry adhesion of VACNTs in different temperature ranges and explained the phenomenon of adhesion with temperature [18]. The microscopic morphology of the material was optimized by Mark Cutkosky et al. at Stanford University to develop a millimeter-scale array with sharp ends. The adhesion strength of the experimentally obtained adhesion array is about 0.24 N/cm^2, with obvious anisotropy, but the connection between the microscopic setae adhesion theory and the macroscopic adhesion has not been explained further [19,20].

Research on the adhesion characteristics of geckos has good references regarding the development of bionic adhesion materials. Therefore, we could explore the principle of the microstructure of gecko feet to study the adhesion characteristic of polyurethane setae array materials. The traditional Hertz contact theory is mainly used for the contact between two elastomers. When the research scale is further reduced and the surface energy is introduced, the Hertz contact theory cannot be explained. Based on the Hertz theory, the Johnson–Kendall–Robert (JKR) contact theoretical model is established by introducing surface adhesion. The dynamic mechanical properties of setae in microcontact are further studied by integrating the JKR contact theory model with the cantilever beam model. Through the established mathematical model, the influence mechanism of roughness, surface energy, and elastic modulus on the microcontact dynamic characteristics of polyurethane, the setae array is analyzed.

The structure of this paper is as follows: Section 1 summarizes the relevant research and the main contributions of this work. In Section 2, the microscale adhesion theory and cantilever beam theory are studied and integrated to form the setae quasistatic contact theory. Descriptions of experiments, simulations, and the model analysis are included in Section 3. Section 4 concludes the whole research.

2. Micro Contact Theory of Setae

2.1. Microscale Adhesive Contact Theory

In the macroscopic theory, when objects are in contact with each other, the elastic force is much larger than the surface force of the objects. Thus, the surface force is often ignored. However, when the characteristic scale of the research object is reduced to a certain range, there can be many phenomena that cannot be explained by the traditional macroscopic contact theory. The reason is that the role of the surface force and surface energy between two objects is the key factor in determining the adhesion, contact, and deformation behavior of solid surfaces. In areas where surface forces are dominated, traditional continuum mechanics methods are no longer applicable. The concept of surface force was introduced in classical mechanics; thus, forming the theory of the microscale adhesive contact.

In the Hertz contact model, due to the lack of an adhesion force, when the applied external load is P_h, a contact circular surface with a radius a_h would be generated, and its contact depth δ_h and pressure distribution $P_h(r)$ in the contact area could be given by the following formula:

$$a_h^3 = \frac{3RP_h}{4E^*} \tag{1}$$

$$a_h^3 = \frac{3RP_h}{4E^*} \tag{2}$$

$$p_h(r) = \frac{3P_h}{2\pi a_h^2}\left(1 - \frac{r^2}{a_h^2}\right)^{1/2} \tag{3}$$

where E^* is the equivalent elastic modulus, $E^* = \left(\frac{1-v_1^2}{E_1} + \frac{1-v_2^2}{E_2}\right)^{-1}$ and R are the equivalent radius $\frac{1}{R} = \frac{1}{R_1} + \frac{1}{R_2}$, $R_1 R_2$ are the radii of two contact spheres, respectively, and r is the distance between the contact point and the center of the contact area.

Based on the Hertz contact model, the adsorption force P_a between the contact surfaces is introduced, and its pressure distribution $P_h(r)$ in the contact area could be expressed by the following formula:

$$P_a(r) = \frac{P_a}{2\pi a^2}\left(1 - \frac{r^2}{a_h^2}\right)^{-1/2} \tag{4}$$

The adsorption term P_a has a crack propagation singularity at the contact boundary, and the corresponding pressure increase coefficient is $K_I = \frac{P_a}{2a\sqrt{\pi a}}$. Using the Irwin formula, the relationship between the adsorption energy and the adsorption energy could be obtained as:

$$G = \frac{K_I^2}{2E^*} = w \tag{5}$$

Among them, G is the elastic energy release rate, w is the adhesion energy, and $w = 2\sqrt{\gamma_1 \gamma_2}$, γ_1, and γ_2 are the surface free energy of the two objects in contact with each other, respectively. The expression of the adsorption term could be obtained from the above formula:

$$P_a = \sqrt{8\pi w E^* a^3} \tag{6}$$

The final JKR theoretical contact pressure distribution can be obtained:

$$P_{JKR} = P_h - P_a = \frac{4E^* a^3}{3R} - \sqrt{8\pi w E^* a^3} \tag{7}$$

The expression of the contact circle radius a_{JKR} is:

$$a_{JKR}^3 = \frac{3R}{4E^*}\left(P + 3\pi wR + \sqrt{6\pi wRP + (3\pi wR)^2}\right) \tag{8}$$

The corresponding normal displacement is δ_{JKR} and could be written as:

$$\delta_{JKR} = \delta_h + \delta_a \tag{9}$$

where δ_h is the Hertz compression depth, δ_a is the adsorption compression depth, and $\delta_a = \frac{P_a}{2\pi E^*} = \frac{\sqrt{8\pi E^* w a^3}}{2\pi E^*}$; finally, the normal displacement could be written as:

$$\delta_{JKR} = \frac{a_{JKR}^2}{R} - \sqrt{\frac{2\pi w a_{JKR}}{E^*}} \tag{10}$$

To facilitate the analysis of the essence of the JKR contact model, the obtained dimensionless external load P^*, contact circle radius a^*, and contact depth δ^* are:

$$P^* = \frac{P_{JKR}}{\pi R w} \tag{11}$$

$$a^* = a_{JKR} \left(\frac{4E^*}{3\pi R^2 w}\right)^{1/3} \tag{12}$$

$$\delta^* = \delta_{JKR} \left(\frac{9\pi^2 R w^2}{16E^{*2}}\right)^{-1/3} \tag{13}$$

Then, according to the formula (7), the expression of dimensionless load P^* could be obtained as:

$$P^* = P_h^* - (6P_h^*)^{1/2} = a^{*3} - (6a^{*3})^{1/2} \tag{14}$$

Based on Equation (10), so that the contact depth δ^*:

$$\delta^* = a^{*2} - 2\left(2a^*/3\right)^{1/2} \tag{15}$$

When the external load is 0, the contact radius generated by the van der Waals force is $a_0 = \left(6\pi R^2 w / K\right)^{1/3}$, and the following could be seen from Equation (8): the maximum adhesion force: $F_{ad} = -1.5\pi R w$; the separation radius: $a_s = 0.63 a_0$; the normal displacement during separation: $\delta_s = -0.21 a_0^2 R^{-1}$.

2.2. Cantilever Beam Theory

Since the JKR theory could only describe the micro-scale adhesion properties and cannot characterize the elastic deformation of the setae rods, it may lead to the overestimation of the adhesion strength. When analyzing the microscopic contact of setae, a single seta rod could be regarded as a large flexible beam, and the end could be simplified as a viscous elastic sphere. This simplified model could deal with issues related to setae contact.

By applying vertical pressure on the setae, they can gradually achieve close contact with the surface. The setae would then be in the adsorption state. After that, a push-off force could be applied on the setae to separate them from the contact surface. Based on Figure 1, the mechanical analysis of a single polyurethane seta was carried out.

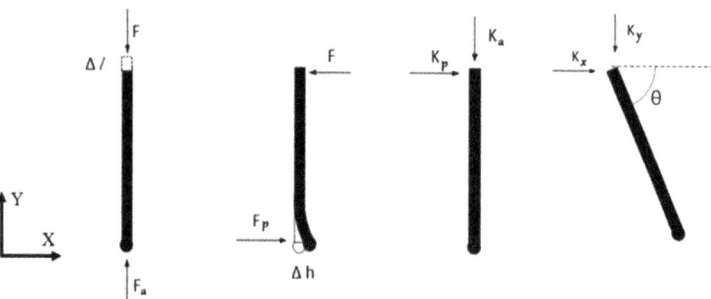

Figure 1. Schematic diagram of the stiffness of a single seta.

Neglecting the instability, when the setae were compressed in their axial direction, the ends of the setae were also subjected to a reaction force in the axial direction. According to the deformation theory of large flexible beams in the mechanics of materials, the deformation of the seta in their axial direction was:

$$\Delta l = \frac{Fl}{\pi r^2 E_1} \tag{16}$$

where l is the length of the setae rod, E_1 is the elastic modulus of the setae material, r is the diameter of the setae rod, and F is the axial load on the setae. The above Equation (16) could be transformed into a function $F = K_a \Delta l$, where K_a is the stiffness of the setae along the axial direction, with the expression:

$$K_a = \frac{\pi r^2 E_1}{l} \tag{17}$$

when the end of the setae is subjected to an external force along its radial direction, the bending deformation of the material would cause a relative displacement Δh in the radial direction. Therefore:

$$\Delta h = \frac{Fl^3}{3E_1 I} \tag{18}$$

The above formula could be transformed into a function $F = K_p \Delta h$, where K_p is the stiffness of the setae along the radial direction, and its expression is:

$$K_p = \frac{F}{\Delta h} = \frac{3\pi r^4 E}{4l^3} \tag{19}$$

The longitudinal stiffness K_y of the setae at the inclination angle θ could be expressed as:

$$K_y = \frac{F_y}{\delta_y} = \frac{K_P \cdot K_a}{K_P \sin^2 \theta + K_a \cos^2 \theta} \tag{20}$$

In the same way, the expression of the lateral stiffness K_X of the setae could be expressed as:

$$K_x = \frac{F_x}{\delta_x} = \frac{K_P \cdot K_a}{K_P \cos^2 \theta + K_a \sin^2 \theta} \tag{21}$$

To facilitate the study of the plastic deformation of the seta rod, the axial stiffness and the radial stiffness of the inclined seta rod were converted into the longitudinal stiffness K_y and the transverse stiffness K_X. The comprehensive deformation of the seta rod could be decomposed into the X-direction deformation and the Y-direction deformation according to the coordinate system.

2.3. Quasi-Static Contact Theory

When the setae were pressed in or pulled out vertically at a low speed, the contact state of a single polyurethane seta with the surface could be approximately regarded as the quasistatic contact. As shown in Figure 2, the distance between the base of the setae and the contact surface was Z, the length of the setae rod was L, and the radius of the end ball was R. The seta was displaced as y in the Y direction. The initial distance of the setae from the contact plane was h_0. The distance between the end of the seta and the contact surface during the movement was h.

$$h = Z - L \sin \theta - R - y \tag{22}$$

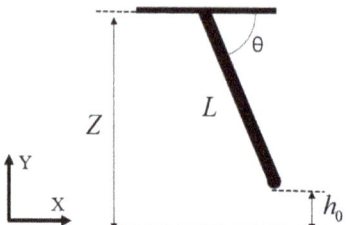

Figure 2. Schematic diagram of single polyurethane setae and surface.

The setae contact state was judged by analyzing the size of h. Since the interaction force of the setae in the process of contact and desorption was different, it was necessary to analyze the mechanical state of the pressing and desorption of the polyurethane setae.

2.3.1. Quasistatic Indentation Force Analysis of Setae

The polyurethane setae were pressed vertically in the Y direction. Due to the van der Waals forces acting at a small distance, the polyurethane setae were just in contact with the surface without deformation at that time $h = 0$. The interaction force between the polyurethane setae and the surface was 0 at this time. When $h < 0$, the polyurethane setae were squeezed, and the interaction force between them was expressed as a repulsive force $F > 0$.

The displacement of the base of the setae was y, and the contact depth between the ball at the end of the setae was y_1. When the rod of the setae was compressed, its elastic deformation was y_2. As shown in Figure 3a, the relationship between their displacements could be obtained:

$$y = y_1 + y_2 \tag{23}$$

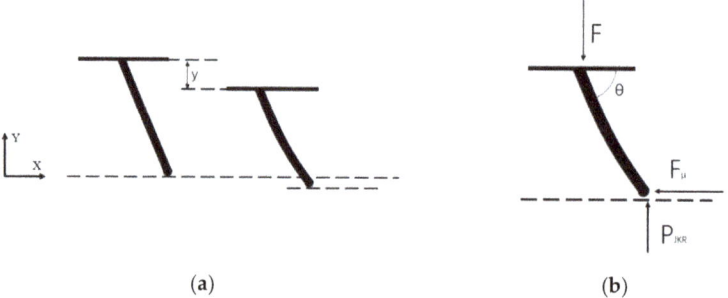

(a)   (b)

Figure 3. Static relationship between seta and contact surface;(a) variation of a single seta press-in displacement;(b) force analysis of a single seta in contact with the surface.

As the setae moved down, the displacement y_1 for the end ball was the JKR theoretical depth of contact:

$$y_1 = \delta_{JKR} \tag{24}$$

Based on the force balance theory, as shown in Figure 3b, we could obtain:

$$F = P_{JKR} = y_2 \times K_y \tag{25}$$

By combining Equations. (7), (10), (20), (23), and (25), the relationship between the contact force and the displacement of the setae in the Y direction could be obtained.

During the pressing process, the seta would produce a lateral deformation x, which could be approximated according to the deformation coordination relationship:

$$x = (y - y_1) \times \tan\theta \tag{26}$$

The coefficient of friction between the setae and the surface was μ, and the setae experienced a leftward frictional force F_μ during the press-in process:

$$F_\mu = \mu \times F \tag{27}$$

If the force generated by the lateral deformation of the setae was less than the friction force, that is $xK_x \leq \mu F$, then the setae would not produce a lateral displacement. Additionally, the friction force value was equal to the static friction. If the force generated by the lateral deformation of the setae was greater than the friction force, that is $xK_x > \mu F$, then the setae would produce a lateral displacement. Additionally, the friction force value at this time was equal to the size of the kinetic friction force.

2.3.2. Quasistatic Push-off Force Analysis of Setae

The polyurethane setae were perpendicular to the surface in the Y direction. When $h < 0$, the polyurethane setae were squeezed, and the interaction force between them was expressed as the repulsive force $F > 0$. In the Y direction, the contact force between the setae and the surface was the same as that in the pressed state. The friction force was opposite to that in the pressed state in the X direction.

When $\delta_s \geq h > 0$, the polyurethane setae end beads adhered to the surface, and the interaction force between them was an attractive force $F < 0$, as shown in Figure 4. Equation (25) was also applicable to the solution of the contact force in the desorption state, but the force state was opposite to that in the press-in state.

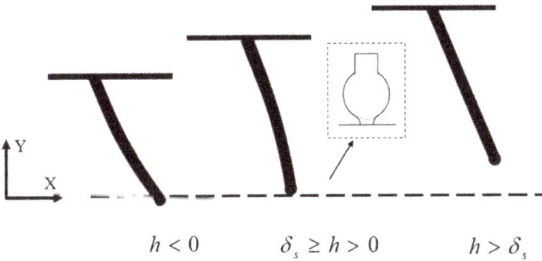

Figure 4. Schematic diagram of desorption of single polyurethane setae.

When $h > \delta_s$ the polyurethane setae were separated from the surface, and the interaction force between them was $F = 0$.

3. Results and Discussion

3.1. JKR Model Analysis

By solving the normalized JKR model of the setae contact, the relationship between the contact circle radius, contact force, and contact depth was obtained. Through the analysis of the curve characteristics, the dynamic mechanical relationship between the two objects in the mutual adsorption and desorption process was obtained.

The relationship between the force and the radius when the setae were in contact with the surface is shown in Figure 5a. It could be seen from the figure that the maximum adsorption force was at point C. The corresponding adsorption force was -1.5 μN at this point, and the radius of the contact circle was 1.24 μm. This point was the critical point where the ball end of the setae and the contact surface were separated from each other.

When the ball at the end of the setae was 0 from the surface, the adsorption force changed from point O (0, 0) to point A (0, −1.35 µN) in Figure 5b. The increase in the contact depth led to a growth in the adsorption force, which changed in the AB direction from point A. When the ball surface and the contact surface were separated from each other, the force would gradually decrease along with BA. The positive pressure gradually became the adsorption force, and then changed from point A to point D (−0.8, −0.75). When it was further detached from point D, the adsorption force would change suddenly from point D (−0.8, −0.75) to point E (−0.8, 0), which also meant that the polyurethane setae were completely detached from the contact surface. Point D was the critical point, where the polyurethane setae and the contact surface were separated from each other.

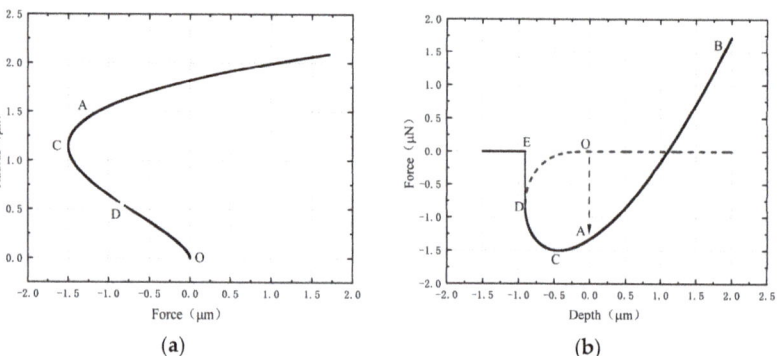

Figure 5. The dynamic mechanical relationship between the sphere and the contact surface: (**a**) the relationship between the radius and the force; (**b**) the relationship between the contact force and the contact depth.

3.1.1. Influence by Elastic Modulus

Based on the above theory, the effect of the material's elastic modulus and surface energy on the interaction force model was further analyzed. The dynamic mechanical simulation analysis was carried out by selecting real polyurethane seta material parameters. The sphere and the plane were selected to be in contact with each other, the radius of the sphere at the end of the polyurethane setae was 4 µm, the elastic modulus was 1.413 Mpa, the Poisson's ratio was 0.3 mJ/m², and the adhesion energy was 40 mJ/m². The elastic modulus of the contact plane was 55 Gpa, the Poisson's ratio was 0.25, and the adhesion energy was 170 mJ/m².

Taking the elastic modulus as 0.1 Mpa, 1 Mpa, and 10 Mpa for the dynamic simulation, we could obtain the relationship curve between the circle radius and the external load, and the external load and the depth. It could be seen from Figure 6a that with the increase in the elastic modulus, the contact circle radius under the same contact force decreased significantly from 3.69 µm at 0.1 Mpa to 0.74 µm at 10 Mpa. An increase in the elastic modulus resulted in less deformation under the same contact force. Under the same contact force, the deformation became smaller as the elastic modulus increased. It was also found that the maximum push-off force of spheres with different elastic moduli experienced little change during the separation process.

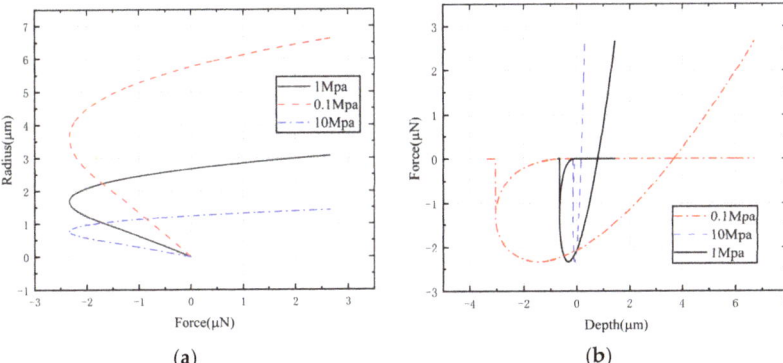

Figure 6. Dynamic analysis results under different elastic moduli of 0.1 Mpa, 1 Mpa, and 10 Mpa: (**a**) the relationship between the contact radius and the contact force under different elastic moduli; (**b**) the relationship between the contact force and the contact depth under different elastic moduli.

3.1.2. Influence by Material Surface Energy

Taking the sphere adhesion energy as 4 mJ/m^2, 40 mJ/m^2, and 100 mJ/m^2, the dynamic simulation was carried out. The relationship between the radius of the circle and the load, and the load and the depth were obtained. It could be seen from Figure 7a that with the increase in the surface adhesion energy, the maximum adsorption force when the setae were desorbed increased significantly. The maximum adsorption force increased from 0.71 µN at 4 mJ/m^2 to 3.55 µN at 100 mJ/m^2. With the increase in the surface adhesion energy of the contacting object, the corresponding contact depth also increased gradually during desorption, from 0.23 µm at 4 mJ/m^2 to 0.71 µm at 100 mJ/m^2.

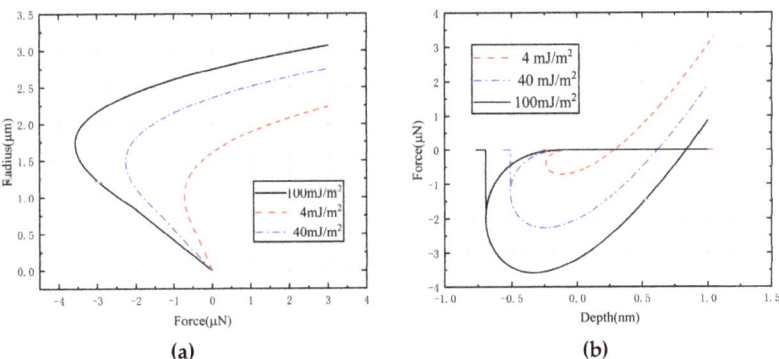

Figure 7. Dynamic analysis results under the conditions of different surface energies of 4 mJ/m^2, 40 mJ/m^2, and 100 mJ/m^2: (**a**) the relationship between the contact radius and force under different surface energies; (**b**) the relationship between the contact force and depth under different surface energies.

To sum up, the contact depth would increase with the decrease in the elastic modulus under the same contact load, but had little effect on the maximum push-off force. The depth of the contact and maximum push-off force would increase as the surface energy of the contacting object increased. Therefore, the main factor affecting the setae push-off force was the surface energy of the material. When the material for creating the setae was determined, no matter how the morphology and size of the setae changed, the value of the push-off force was constant.

3.2. Simulation of the Single Seta

The interaction force between a single seta and the surface adhesion or detachment had different forms, so the micro-adhesion state of the setae needed to be considered and analyzed separately. When $\delta > 0$, the seta would be in contact with the surface and elastically deform. When $\delta < 0$, the polyurethane seta would undergo a plastic deformation due to the effect of adhesion and remained in contact with the surface. Until the elastic force overcame the adhesion $\delta > \delta_0$, the setae detached from the surface.

The dynamic desorption process of a single polyurethane seta was simulated and analyzed, and the parameters were selected as shown in Table 1. The dynamic desorption process of setae was solved by Matrix Laboratory (MATLAB) programming.

Table 1. Selection of polyurethane setae parameters.

Material Name	Slope	Setae Diameter	Length	Poisson's Ratio	Elastic Modulus	Adhesive Energy
Polyurethane	30°	3 μm	20 μm	0.3	1.413 Mpa	40 mJ/m²

The relationship curve between the contact force and the contact depth was obtained through a simulation, as shown in Figure 8. The setae were slowly pulled out at a certain speed, and the contact force decreased with the contact depth and changes toward BA. When the setae were pulled apart to −7.2 μm, the contact was suddenly disconnected. From point D to point E, the adsorption force suddenly changed from 4.1 μN to 0 N. The EF segment setae were separated from the contact surface, and there was no interaction force detected.

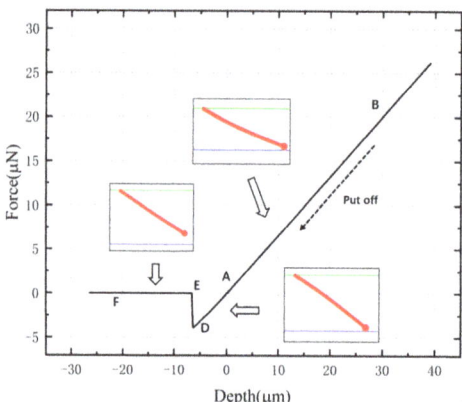

Figure 8. Simulation curve of contact force and contact depth for setae array.

3.3. Setae Array Simulation and Experiment
3.3.1. Rough Surface Model

An important assumption of the classical contact theory is that the contact surface is geometrically smooth, but the real surface in engineering is rough. Therefore, it was necessary to establish a roughness model that could approximate the actual surface and could be easily calculated. As shown in Figure 9, assuming that the base surface of the setae was parallel to the reference plane of the contact surface, the spacing was Z, the length of each seta was L, the setae were parallel to each other, and the angle between the setae and the base surface was θ.

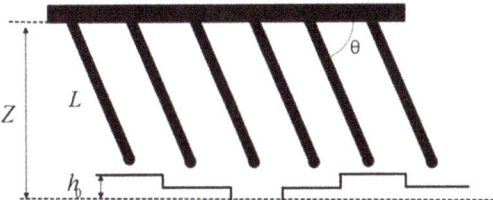

Figure 9. Roughness model.

It was assumed that the peak height distribution of the contact surface conformed to a normal distribution, and the reference surface of the rough surface was a plane. The distance was h between the contact surface profile and the reference plane, and the height satisfied the following Gaussian distribution:

$$g(h) = \frac{1}{\sqrt{2\pi}\sigma} e^{-\frac{h^2}{2\sigma^2}} \tag{28}$$

3.3.2. Dynamic Simulation and Experimental Verification

The dynamic adhesion process of polyurethane setae arrays was simulated and analyzed, and the setae were pulled out vertically from the bottom to the top. When the material area was 1 cm² and the seta spacing was 4 μm, there was 1×10^7 setae in total, and the inclination angle of the seta was 60°. The elastic modulus was taken as 1.413 Mpa, the Poisson's ratio was 0.3, the surface adhesion energy was 40 mJ/m², and the friction coefficient between the end of the setae and the surfaces was 0.2. The elastic modulus of the contact surface was taken as 55 Gpa, the Poisson's ratio was 0.25, and the adhesion energy was taken as 170 mJ/m². The mean square error of the surface roughness was 1. Using MATLAB programming to solve the dynamic desorption process of setae, the results obtained by dynamic calculation are shown in the following Figure 10.

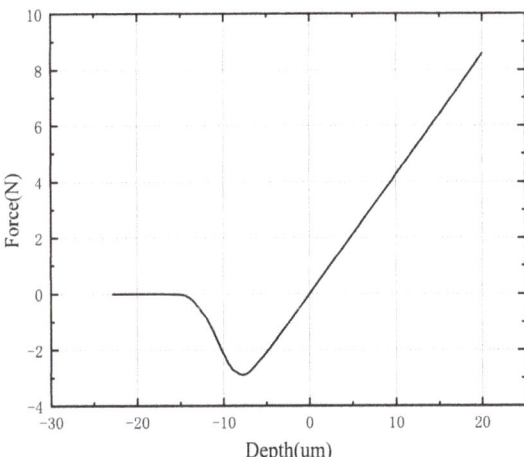

Figure 10. Simulation curve of the contact force with different contact depths.

The relationship between the push-off force and the depth was obtained through a simulation, as shown in Figure 10. With the continuous pulling out of the setae array, the contact force decreased gradually from 8.6 N. When the contact depth was less than 0 μm, the direction of the contact force changed from pressure to adsorption. The curve had an extreme point as the adsorption force reached a maximum value of −2.85 N. As the setae pulled further apart, the adsorption force gradually decreased to 0 N, and all the setae were separated from the surface.

Figure 11a shows the relationship between the adsorption force and the preload on the setae array when the mean square error of the roughness was 1. We could see from the simulation curve that the adsorption force of the setae array increased rapidly with the increase in pressure. After the pre-pressure reached 1 N, the adsorption force stabilized at about 2.3 N. The experimental results matched well and the deviation error of the data was 7.81%.

As the preload force increased, the number of setae arrays in contact with the surface increased, so the preload force determined the distribution of adhesion between the array and the surface. Different adhesion distributions would produce different interaction force relationship curves during the detachment process. There was an extreme point of adhesion force in the desorption curve under a certain pre-pressure, as shown in Figure 10. By collecting the maximum desorption force under different pre-pressures, the relationship curve between the pre-pressure and the maximum desorption force was obtained. During dynamic detachment, the maximum adhesion force produced by the setae array increased with an increasing preload, as shown in Figure 11a. The preload was further increased, and the maximum adhesion attraction tended to reach a stable value. Because the area of the material was constant in both experiments and simulations, there was a saturation value for the number of setae contacts. The mean square error of surface roughness was selected as 0.5, 1, and 1.5, respectively. The experimental and simulation results of setae in contact with surfaces of different roughness are shown in Figure 11b. The root mean square (RMS) was used to describe the dispersion between the experimental data and the simulation data, and the calculation result was 15.9% [21]. The peak adsorption force increased continuously at different preloads as the root mean square deviation of the surface decreased. The contact of the polyurethane setae array with the rough surface became better as the root mean square deviation decreased, so the smooth surface was more favorable to the adsorption of the setae array.

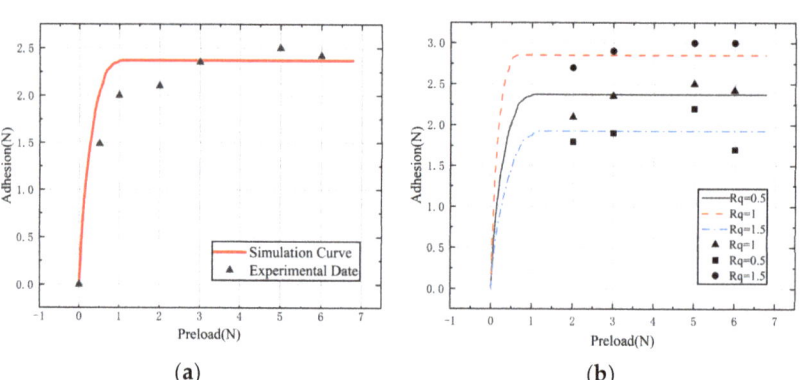

Figure 11. The relationship curves of the adsorption force between experiment and simulation: (**a**) results data with different preloads; (**b**) results data with different surface roughness.

4. Conclusions

This paper established a complete contact model for polyurethane setae arrays, which considered the Hertz contact theory, the surface energy theory, the cantilever beam theory, and the rough surface mode comprehensively. When analyzing the contact of a single seta, the setae rod was simplified as a cantilever beam, with its axial and radial stiffnesses converted into longitudinal and transverse stiffnesses. Integrating the single setae quasistatic contact theory and the rough surface model, the setae array contact model was further obtained. After that, a parameter analysis was conducted and verified by the corresponding experimental data. The main conclusions were as follows:

(1) By comparing the simulation with the experiment, the average discrepancy of the data was 15.9%, which demonstrated the feasibility of the polyurethane setae arrays contact model.
(2) Through the analysis of the *JKR* model for the ball end of the setae, it was found that reducing the elastic modulus of the object would increase the depth of contact under the same external load. The elastic modulus had little effect on the maximum push-off force. Increasing the surface energy of the contacting object would grow the distance between the objects, the push-off force, and the depth of contact during desorption.
(3) When the polyurethane setae array was in contact with the rough surface, it was found that with the growth of the preload applied to the setae array, the number of setae in contact rose. Its maximum adsorption force would also increase, but it would reach saturation.
(4) Different properties of the contact surface material had different effects on adhesion. The greater the surface energy of the contact surface, the greater the adsorption force generated when the seta arrays were separated. The elastic modulus and Poisson's ratio of the contact surface had little effect on the adsorption force. The roughness of the contact surface had a negative effect on the adsorption force of the seta array, and the smoother the surface, the greater the adsorption force.

Author Contributions: Conceptualization, Q.L. and S.Y.; Software, C.W.; Validation, Z.J., M.L. and Q.L.; Writing—Original Draft Preparation, Q.L. and Z.J.; Writing—Review and Editing, Z.D. and C.W.; Supervision, S.Y. All authors have read and agreed to the published version of the manuscript.

Funding: This work was supported by the National Natural Science Foundation of China (52102436), the Fundamental Research Funds for the Central Universities (30920021109), the Natural Science Foundation of Jiangsu Province (BK20200496), the China Postdoctoral Science Foundation (2020M681615), and the project of Key Laboratory of Impact and Safety Engineering (Ningbo University), the Ministry of Education (CJ202107).

Institutional Review Board Statement: Not applicable.

Informed Consent Statement: Not applicable.

Data Availability Statement: Data are contained within the article. The data presented in this study can be seen in the content above.

Acknowledgments: We wish to express our gratitude to the members of our research team, Shuai Yue, Chunbo Wu, Zhonghui Jiang, Zhonghua Du, and Mengsheng Li.

Conflicts of Interest: The authors declare no conflict of interest.

References

1. Cadirov, N.; Booth, J.A.; Turner, K.L.; Israelachvili, J.N. Influence of humidity on grip and release adhesion mechanisms for gecko-inspired microfibrillar surfaces. *ACS Appl. Mater. Interfaces* **2017**, *9*, 14497–14505. [CrossRef] [PubMed]
2. Wan, Y.; Xia, Z. Self-cleaning and controlled adhesion of gecko feet and their bioinspired micromanipulators. *MRS Adv.* **2018**, *3*, 1641–1646. [CrossRef]
3. Hu, S.; Jiang, H.; Xia, Z.; Gao, X. Friction and adhesion of hierarchical carbon nanotube structures for biomimetic dry adhesives: Multiscale modeling. *ACS Appl. Mater. Interfaces* **2010**, *2*, 2570–2578. [CrossRef] [PubMed]
4. Raut, H.K.; Baji, A.; Hariri, H.H.; Parveen, H.; Soh, G.S.; Low, H.Y.; Wood, K.L. Gecko-inspired dry adhesive based on micro–nanoscale hierarchical arrays for application in climbing devices. *ACS Appl. Mater. Interfaces* **2018**, *10*, 1288–1296. [CrossRef] [PubMed]
5. Sitti, M.; Fearing, R.S. Synthetic gecko foot-hair micro/nano-structures for future wall-climbing robots. In Proceedings of the IEEE International Conference on Robotics and Automation (Cat. No. 03CH37422), Taipei, Taiwan, 14–19 September 2003; pp. 1164–1170.
6. Shah, G.J.; Sitti, M. Modeling and design of biomimetic adhesives inspired by gecko foot-hairs. In Proceedings of the IEEE International Conference on Robotics and Biomimetics, Shenyang, China, 22–26 August 2004; pp. 873–878.
7. Peng, M.; Han, X.; Xiao, G.Z.; Chen, X.; Xiong, W.; Yang, K.; Luo, H. Spherical volume elements scheme for calculating van der Waals force between irregular particles and rough surfaces. *Chin. J. Phys.* **2021**, *72*, 645–654. [CrossRef]
8. Lee, J.; Bush, B.; Maboudian, R.; Fearing, R. Gecko-Inspired Combined Lamellar and Nanofibrillar Array for Adhesion on Nonplanar Surface. *Langmuir* **2009**, *25*, 12449–12453. [CrossRef] [PubMed]

9. Li, X.; Tao, D.; Lu, H.; Bai, P.; Liu, Z.; Ma, L.; Meng, Y.; Tian, Y. Recent developments in gecko-inspired dry adhesive surfaces from fabrication to application. *Surf. Topogr. Metrol. Prop.* **2019**, *7*, 023001. [CrossRef]
10. Autumn, K.; Liang, Y.A.; Hsieh, S.T.; Zesch, W.; Chan, W.P.; Kenny, T.W.; Fearing, R.; Full, R.J. Adhesive force of a single gecko foot-hair. *Nature* **2000**, *405*, 681–685. [CrossRef] [PubMed]
11. Autumn, K.; Sitti, M.; Liang, Y.A.; Peattie, A.M.; Hansen, W.R.; Sponberg, S.; Kenny, T.W.; Fearing, R.; Israelachvili, J.N.; Full, R.J. Evidence for van der Waals adhesion in gecko setae. *Proc. Natl. Acad. Sci. USA* **2002**, *99*, 12252–12256. [CrossRef] [PubMed]
12. Geim, A.K.; Dubonos, S.V.; Grigorieva, I.V.; Novoselov, K.; Zhukov, A.A.; Shapoval, S.Y. Microfabricated adhesive mimicking gecko foot-hair. *Nat. Mater.* **2003**, *2*, 461–463. [CrossRef] [PubMed]
13. Persson, B. On the mechanism of adhesion in biological systems. *J. Chem. Phys.* **2003**, *118*, 7614–7621. [CrossRef]
14. Dai, Z.D.; Hui, C.; Stannislav, G. Effect of surface roughness on the adhesive properties of polyurethane. *Tribology* **2003**, *23*, 245–249.
15. Gao, H.; Yao, H. Shape insensitive optimal adhesion of nanoscale fibrillar structures. *Proc. Natl. Acad. Sci. USA* **2004**, *101*, 7851–7856. [CrossRef] [PubMed]
16. Gao, H.; Xiang, W.; Yao, H.; Gorb, S.; Arzt, E. Mechanics of hierarchical adhesion structures of geckos. *Mech. Mater.* **2005**, *37*, 275–285. [CrossRef]
17. Qu, L.; Dai, L.; Stone, M.; Xia, Z.; Wang, Z.L. Carbon nanotube arrays with strong shear binding-on and easy normal lifting-off. *Science* **2008**, *322*, 238–242. [CrossRef] [PubMed]
18. Guo, F.; Li, Y.; Meng, G.; Sun, C.; Dai, Z. Adhesion property of vertically aligned carbon nanotubes under high temperature. *Trans. Nanjing Univ. Aeronaut. Astronaut.* **2018**, *35*, 41–45.
19. Santos, D.; Spenko, M.; Parness, A.; Kim, S.; Cutkosky, M. Directional adhesion for climbing. theoretical and practical considerations. *J. Adhes. Sci. Technol.* **2007**, *21*, 1317–1341. [CrossRef]
20. Chen, R.; Fu, L.; Qiu, Y.; Song, R.; Jin, Y. A gecko-inspired wall-climbing robot based on vibration suction mechanism. *Proc. Inst. Mech. Eng. Part C J. Mech. Eng. Sci.* **2019**, *233*, 7132–7143. [CrossRef]
21. Yue, S.; Titurus, B.; Nie, H.; Zhang, M. Liquid spring damper for vertical landing Reusable Launch Vehicle under impact conditions. *Mech. Syst. Signal Process.* **2019**, *121*, 579–599. [CrossRef]

Article

Dynamic Simulation and Parameter Analysis of Weaved Composite Material for Unmanned Aerial Vehicle Parachute Recovery in Deployment Phase

Wenhui Shi [1], Shuai Yue [1,2,*], Zhiqian Li [1], Hao Xu [3], Zhonghua Du [1], Guangfa Gao [1], Guang Zheng [2] and Beibei Zhao [4]

1. School of Mechanical Engineering, Nanjing University of Science and Technology, Nanjing 210094, China; shiwenhui@njust.edu.cn (W.S.); lizhiqian@njust.edu.cn (Z.L.); duzhonghua@aliyun.com (Z.D.); gfgao@ustc.edu.cn (G.G.)
2. Key Laboratory of Impact and Safety Engineering of Ministry of Education, Ningbo University, Ningbo 315211, China; zhengguang@nbu.edu.cn
3. Hongguang Airborne Equipment Company, Nanjing 210022, China; xu_hao1231@163.com
4. Xi'an North Qinghua Electrical Co., Ltd., Xi'an 710025, China; fizzsk@163.com
* Correspondence: yueshuai@njust.edu.cn

Citation: Shi, W.; Yue, S.; Li, Z.; Xu, H.; Du, Z.; Gao, G.; Zheng, G.; Zhao, B. Dynamic Simulation and Parameter Analysis of Weaved Composite Material for Unmanned Aerial Vehicle Parachute Recovery in Deployment Phase. *Crystals* 2022, 12, 758. https://doi.org/10.3390/cryst12060758

Academic Editors: Yong He, Wenhui Tang, Shuhai Zhang, Yuanfeng Zheng, Chuanting Wang and Per-Lennart Larsson

Received: 28 April 2022
Accepted: 24 May 2022
Published: 25 May 2022

Publisher's Note: MDPI stays neutral with regard to jurisdictional claims in published maps and institutional affiliations.

Copyright: © 2022 by the authors. Licensee MDPI, Basel, Switzerland. This article is an open access article distributed under the terms and conditions of the Creative Commons Attribution (CC BY) license (https://creativecommons.org/licenses/by/4.0/).

Abstract: Aiming at the parachute recovery of fixed-wing unmanned aerial vehicles, a method of parachute deployment by tractor rocket is proposed. First, the tensile tests were carried out on high-strength polyethylene and brocade silk-weaved composite materials. The dynamic property parameters of the materials were obtained, which was the input for the dynamic model of the parachute deployment phase. Second, the model was verified by the experiment results. Finally, parachute weight and rocket launch temperature during the deployment phase were studied. The results showed that the dynamic model has good accuracy; as the parachute weight increases, the maximum snatch force of the extraction line and the sling decreases as the force on the suspension lines increases and the deployment effect worsens. With the temperature rise, the maximum snatch force on the extraction line, sling, and suspension lines increases and the deployment length changes slightly.

Keywords: dynamic model; weaved composite material; unmanned aerial vehicle; parachute recovery; deployment phase; parameter analysis

1. Introduction

An Unmanned Aerial Vehicle (UAV) is an unmanned aircraft operated by radio remote control equipment and its program control device [1]. It can be divided into fixed-wing UAV [2], rotor UAV [3], parawing UAV [4], and flapping-wing UAV [5] according to the flight platform configuration. The fixed-wing UAV has the advantages of fixed-wing gliding, stronger "robustness" to control technology errors, enormous carrying capacity, more extended flight, etc. [6]. It is widely used in various fields, which increases the requirements for survival and recovery. The takeoff technology of UAVs is relatively mature, whereas the landing technology is still facing many challenges, especially the more difficult emergency landings. According to statistics, 60% of UAV accidents are caused by human operation errors, among which takeoff and landing phases account for 50%, and landing accidents account for the majority [7]. Therefore, one of the critical problems currently for UAVs is how they can be recovered safely and accurately [8].

At present, the typical recovery methods of UAVs are net collision recovery [9], cable hook recovery [10], parachute recovery [11], and vertical takeoff and landing UAV [12]. Parachute recovery has been widely adopted due to its simple overall layout, low level of technical difficulty, and low dependence on other systems. In the parachute recovery of UAVs, the deployment model can be divided into deployment by a crown chute or a moving mechanism [13]. The crown chute will take a long time to open the parachute, which is not conducive to emergency release. The moving mechanism can shoot the traction device quickly to get a fast opening speed. Whether it is a normal or emergency situation, the parachute can be opened reliably and maintain the safe recovery of the UAV. In respect of parachutes, the deployment phase and the inflation phase are always research hotspots in the aviation field. Compared to the research on the parachute inflation phase, the research on the deployment phase is more complicated, so there is little research literature. However, the essence of the parachute deployment process is rope dynamics, where the research has been a hot topic in recent years [14].

Composite materials are widely used in the aviation field because of their excellent strength-weight ratio [15,16], and the weaved composite materials are widely used in flexible mechanisms such as parachutes. The parachute deployment phase is short and complicated, involving several movements. Studies on this process first appeared in the 1970s. Wolf proposed the continuous deployment model [17], which assumed that the suspension line was always straight during the deployment phase. The calculated snatch force was equal to the actual situation. Shen Guanghui et al. [18] established a 6-DOF mathematical model of the Mars deceleration system. Ortega Enrique et al. [19] accurately calculated the structural load and stress during parachute inflation based on the filling-time expansion model and Ludtke's area law. Yu Li et al. [20] analyzed the deployment phase of the parachute and established a multi-particle dynamics model, focusing on the rope's flexible and significant deformation characteristics. Zhang Qingbin, Wang Haitao, et al. [21,22] systematically studied the spring damping model of rope and the influence of wake on the trajectory of the parachute; they applied it to the calculation and analysis of the parachute deployment phase, obtaining good results.

The researchers studied theoretical models of the deployment and inflation phases, but few studies considered the initial folding of the parachute. In this paper, a method of deployment of parachutes by tractor rocket is proposed to improve the survivability of UAVs in emergencies. Based on the explicit dynamics method, the dynamics model of the parachute deployment phase is established. Experiments were designed and carried out to verify the accuracy of the dynamic model by comparing the parachute deployment time and height. The influence of parachute weight and the launch temperature on the deployment performance was further studied. The above research provides a new idea and method for the design and simulation of the parachute deployment stage in the parachute recovery process of unmanned aerial vehicles.

The structure of this paper is as follows: Section 1 summarizes the relevant research and the main contributions of this paper. In Section 2, the dynamics model of the parachute deployment phase is established. In Section 3, the accuracy of the simulation model is verified by comparing it with the experiment. Parametric analysis of parachute weight and rocket launch temperature are included in Section 4. Section 5 concludes the whole research.

2. Dynamic Model of the Parachute Deployment Phase

2.1. Overview of Parachute Ejecting Tractor Rocket

Parachute recovery has a wide range of applications [23]. The mass of what is recovered ranges from a few kilograms to dozens of tons; the opening speed ranges from low to supersonic; and the height spans from thousands of meters to dozens of kilometers, which can safely and effectively slow the recovered UAVs down to the ground or water surface. In this paper, a typical fixed-wing UAV is selected, which adopts a conventional H-type double-tail-support layout, with a width of 7020 mm, a length of 4180 mm, a height of 1470 mm, and a weight of about 250 kg. It needs to be recycled within the height range of 150–5000 m. The temperature range is $-40\ °C$–$+45\ °C$. The carrying platform of the UAV recovery system cannot be subjected to excessive recoil force. The tractor rocket is used as the traction device to keep the overload within a reasonable range.

The working process of the tractor rocket is as follows: When the igniter receives instructions to light a solid propulsive agent to produce a large amount of gas, the gas blow-out from the rocket nozzle jets makes the rocket accelerate. The rocket's move drives the extraction line, canopy, suspension lines, and sling to keep a straight direction. As the UAV descends, the parachute inflates until it lands smoothly.

2.2. Model Description

The parachute is the core part of the whole UAV parachute recovery system. The working process of the parachute includes two stages: parachute deployment, and inflation [24]. The UAV recovery system has strict weight and small swing angle requirements. It is necessary to select an appropriate parachute with a light weight to meet the specified swing angle due to its significant drag coefficient, small nominal area, and low volume. So the extended skirt parachute was chosen as the main parachute, which is shown in Figure 1.

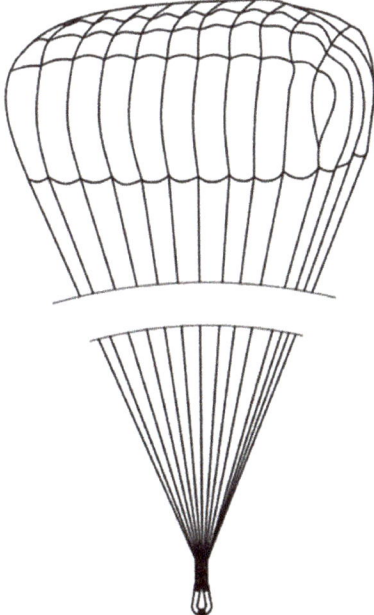

Figure 1. Parachute structure.

There is also an extraction line and sling in the parachute system. The extraction line connects the tractor rocket to the parachute canopy, and the sling connects the suspension lines to the UAV. The canopy material is 1056 Anti-burning brocade silk material, the material of the extraction line and sling are 25–1000 Arlene belt, and the tensile force measured by the test is 19,860 N. The material of suspension lines is 2–200 high-strength polyethylene rope. The tensile strength is 2000 N. The main parameters of the parachute are shown in Table 1.

Table 1. Main parameters of parachute.

Parameter	Value
Nominal radius/m	7.78
Nominal area/m^2	190
Number of gore section	40
Length of suspension lines/m	14
Drag coefficient	0.83
Length of extraction line/m	1
Length of sling/m	5

As the first action in the parachute recovery process, the deployment phase requires the parachute to complete a series of pre-set steps in a short time. As is shown in Figure 2. There are mainly two parachute deployment procedures: canopy-first deployment and lines-first deployment [25]. Under the same conditions, the snatch force of the canopy-first deployment method is larger than the lines-first deployment method, so the lines-first deployment method is adopted in this paper.

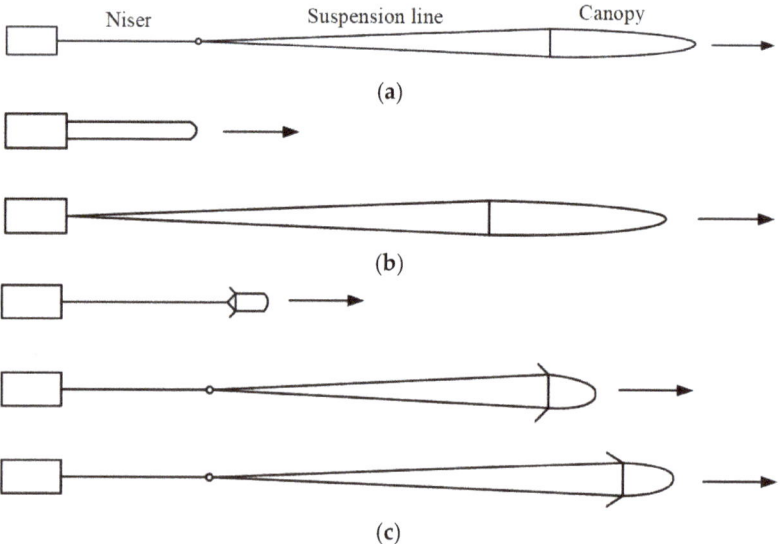

Figure 2. Deployment procedures. (**a**) The final state of a parachute deployment phase (**b**) Canopy-first deployment (**c**) Lines-first deployment.

A finite element simulation model was needed to simulate the parachute deployment phase. The parachute uses a "Z" fold in the parachute container. Firstly, a single gore section model was established according to its size. Then, the individual gore sections were stacked and placed according to the constraints of each gore section position, with a certain angle and distance between each gore section. The partially enlarged view of the gore model is shown in Figure 3.

Figure 3. Gore model.

Then the gore model was divided, and the split length was consistent with the folding size of the parachute in the parachute container. The geometric parachute folded model was established by placing the parachute from the bottom to the top. The folded model was taken as the initial state of the parachute in the deployment phase. On this basis, an extractor rocket, suspension lines, and other components were added. The parachute in the final folded state is shown in Figure 4. The main parameters of the tractor rocket are shown in Table 2.

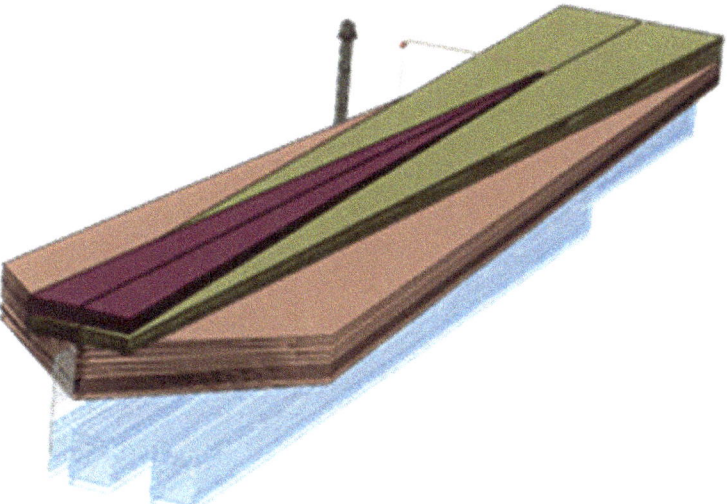

Figure 4. Simulation model.

Table 2. Main parameters of parachute.

Parameter	Value
Diameter/mm	40
Height/mm	360
Weight/g	806

2.3. Establishment of Suspension Line

The T3d2 element (two-node linear three-dimensional truss element) in ABAQUS was selected to construct the suspension lines model [26,27]. Element nodes connected the truss elements in the finite element model. When the truss elements are under compression loads, they will rotate around the nodes freely, making them unable to bear compression loads. Therefore, multiple truss elements connected by element nodes can simulate the dynamics performance of the rope. Each truss element can bear internal and external forces during the movement. The dynamic model of the suspension lines consists of a set of dynamic equations for a single truss element [28]. Figure 5 shows the force transmission of the truss element in the deployment process. The truss element is a double node element (node i and node j), and the unit axial force T_{ij} of the element can be denoted as [29]:

$$T_{ij} = k_{ij}\left(l_{ij} - l_{ij}^0\right) \quad (1)$$

where l_{ij} is the deformed length of the truss element H_{ij}, parameter l_{ij}^0 is the initial length, K_{ij} is the equivalent stiffness of the truss element, which depends upon the material of the rope. The equivalent stiffness K_{ij} is written as:

$$k_{ij} = \frac{EA_{ij}}{l_{ij}^0} \quad (2)$$

where E is the elastic modulus of the rope material, A_{ij} is the cross-section area of the rope.

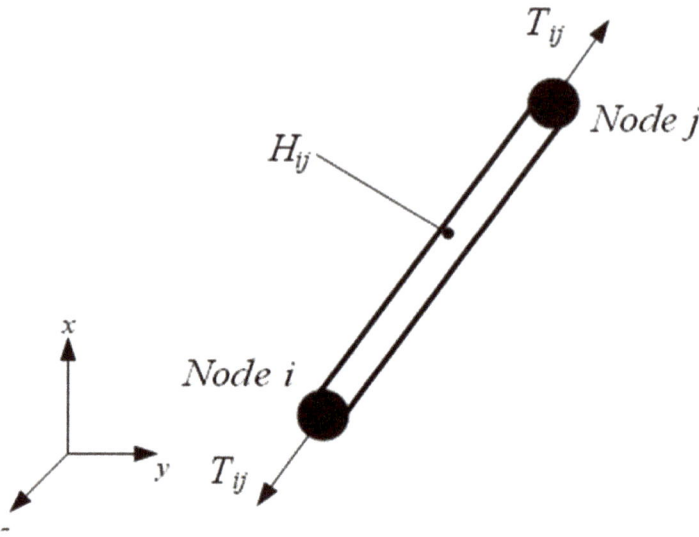

Figure 5. The stress of the truss element.

The test was carried out on the SANS testing machine [30]. The measurement range for the force was between 1 KN and 100 KN, and the precision of force indication can be achieved within ±1%. Before the test, the end of the suspension line was tied with the upper and lower gripper of the testing machine, and the suspension line was in a relaxed state. The suspension line was stretched during the test, with the displacement and force sensor recording the real-time distance and tensile force. The initial length of the suspension line was 100 mm, and the maximum strain in the test was 0.2. The tensile stiffness tests for the suspension line are shown in Figure 6.

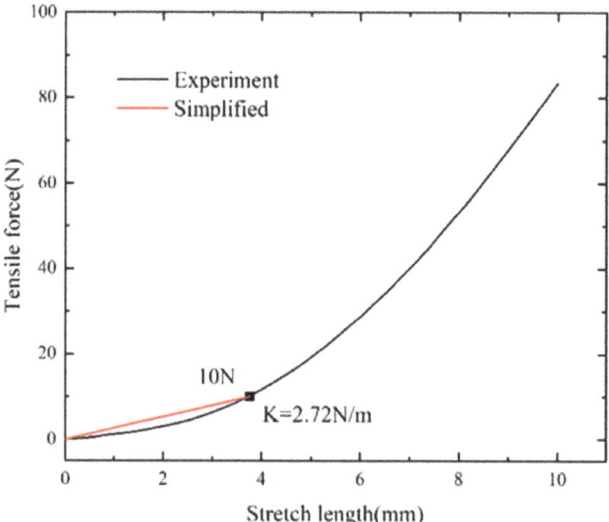

Figure 6. Tensile stiffness test of suspension line.

According to Figure 7, the line stiffness increases with the growth of external load. However, when the external load is under around 10 N, the relationship between the rope stiffness and the elastic modulus of the material can be simplified as linear:

$$K = \frac{EA}{L} \qquad (3)$$

where K is the stiffness of the rope, and E is the elastic modulus of the rope material. A is the equivalent area of the rope, which is 0.196 mm^2. L is the equal length, which is the original length of the rope (100 mm) in the tensile test.

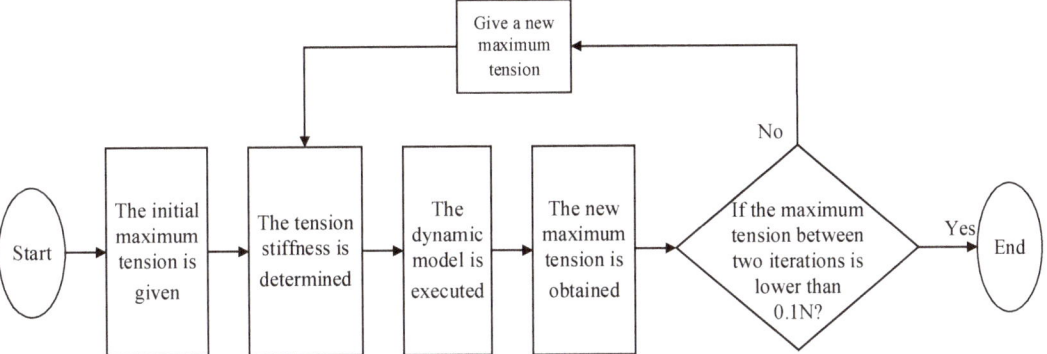

Figure 7. The flowchart to determine line stiffness.

The procedure to determine the suspension line stiffness is listed as follows. Firstly, the initial maximum tension of the line segment is given, and the constant tensile stiffness is determined, based on Figure 6. After that, the dynamic model of suspension line deployment is executed, and the new maximum tension during deployment is obtained. Based on this, the new tensile stiffness was reselected, based on Figure 6. The above steps were repeated until the maximum tension between two iterations was lower than 0.1 N. Then, the current tensile stiffness was determined as the ultimate rope stiffness for the following analysis. The iterative flowchart is shown in Figure 7.

The initial maximum tension of the rope was first given as 10 N, in which the stiffness of the suspension line is 2.72 N/mm. Based on Equation (3), the elastic modulus was calculated as 1385.3 MPa. Based on Figure 7, the iteration was performed to obtain the tensile stiffness of the suspension line. After eight iterations, the ending condition was satisfied, and the whole loop was finished. The final force curve of the suspension line during deployment is shown in Figure 8.

From Figure 8, it can be seen that the maximum tension suspension line was 4.5 N. Therefore, the stiffness of the suspension line was 1.47 N/mm. The chosen stiffness was shown in Figure 9, and the time for maximum tension to appear ranged around 1.57 s. The suspension line was in full deployment and started straight state. Equation (3) obtains the elastic modulus of the suspension line, and the value is 748.7 MPa. The elastic moduli computed are equivalent to the corresponding structure, which can be applied to the suspension line structure.

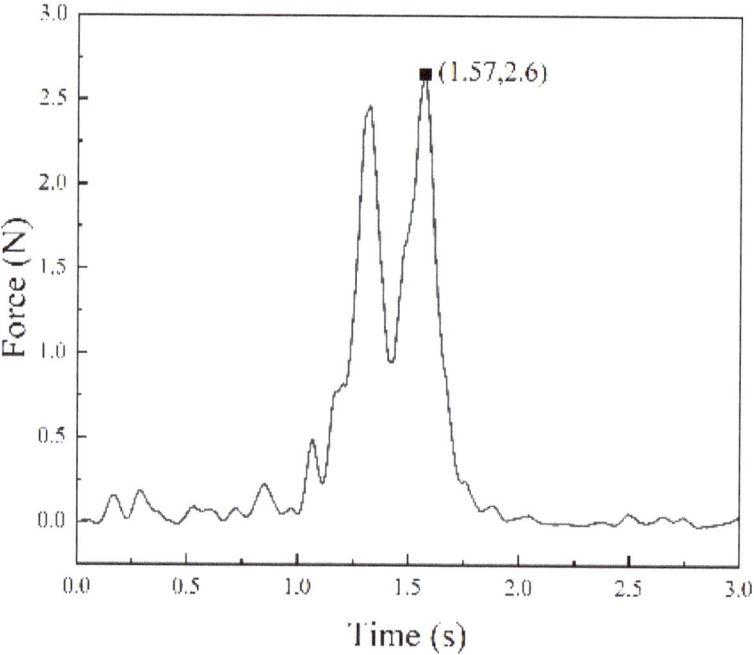

Figure 8. Line segment tension during deployment.

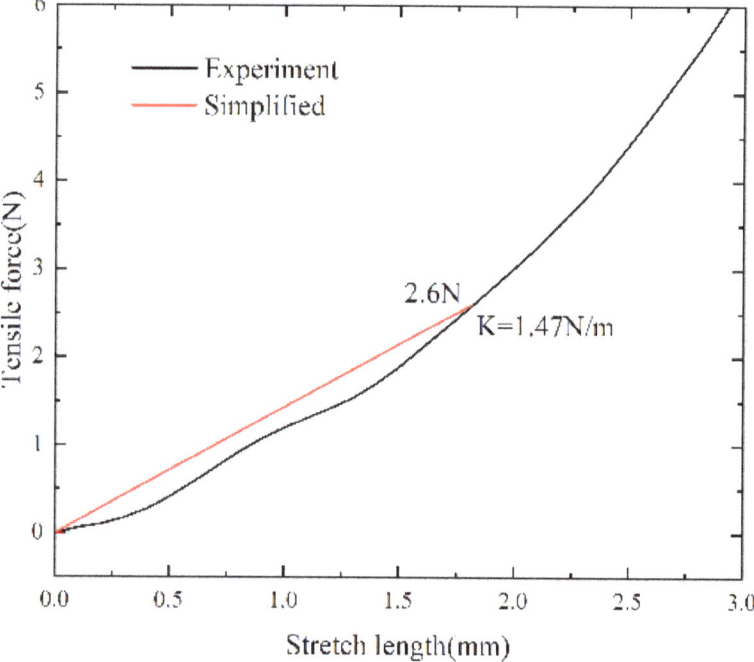

Figure 9. Result of stiffness selection.

2.4. Establishment of Canopy

The membrane element is a thin plate that can withstand the membrane force without bending or transverse shear stiffness, so the only non-zero stress components in the membrane were those parallel to the middle surface of the membrane. The membrane was in a state of plane stress. Discretization and meshing obtains a combination of quadrilateral elements with four nodes. In the local coordinate system, according to the theoretical hypothesis of the finite element method, the displacement vector of any node of the membrane element is d [31,32]:

$$\{d\} = \{u, v, w\}^T \tag{4}$$

where u, v, and w are all functions of three coordinate directions in the global coordinate system. Therefore, the displacement of this membrane element in the local coordinate system can be expressed as:

$$\{x\} = \{x_i, y_i, z_i, x_j, y_j, z_j, x_m, y_m, z_m, x_n, y_n, z_n\} \tag{5}$$

For quadrilateral elements, the form function can be expressed in local coordinates as:

$$N = [IN_i, IN_j, IN_m] \tag{6}$$

where I is the element matrix of size 4×4.

According to the finite element theory, the relation between the displacement function and the node displacement is:

$$\{d\} = N \cdot \{x\}^T \tag{7}$$

Due to the nonlinear characteristics of the membrane material, the green strain tensor was adopted, and the strain matrix of the membrane element $[B]$ could be expressed in two parts:

$$[B] = [B_0] + [B_N] \tag{8}$$

where $[B_0]$ is the linear part of the strain matrix; $[B_N]$ is the nonlinear part of the strain matrix, which is a function of the node displacement vector $\{d\}$. On this basis, the virtual displacement principle was used, and the finite element was written in the incremental form as follows:

$$[K] \cdot d\{d\} = d\{P\} = \int_V [B]^T \cdot d\{\sigma\} dv + \int_V d[B]^T \cdot \{\sigma\} dv \tag{9}$$

According to mechanical theory, $[B_0]$ has nothing to do with node displacement, so the following formula can be obtained:

$$d\{\varepsilon\} = d([B] \cdot \{d\}) = [B] \cdot d\{d\} + d[B_N] \cdot \{d\} \tag{10}$$

The final finite element incremental equation can be expressed as:

$$([K] + [K_\sigma]) \cdot d\{d\} = d\{P\} \tag{11}$$

where $[K_\sigma]$ is the geometric stiffness matrix.

The tensile test of the canopy was carried out to obtain the weft and warp tensile modulus of the canopy material. The thickness of the specimen was 0.1 mm, the width was 10 mm, and the length was 450 mm. As is shown in Figure 10, the weft modulus was calculated to be 7.29 MPa, and the warp modulus 10.75 MPa.

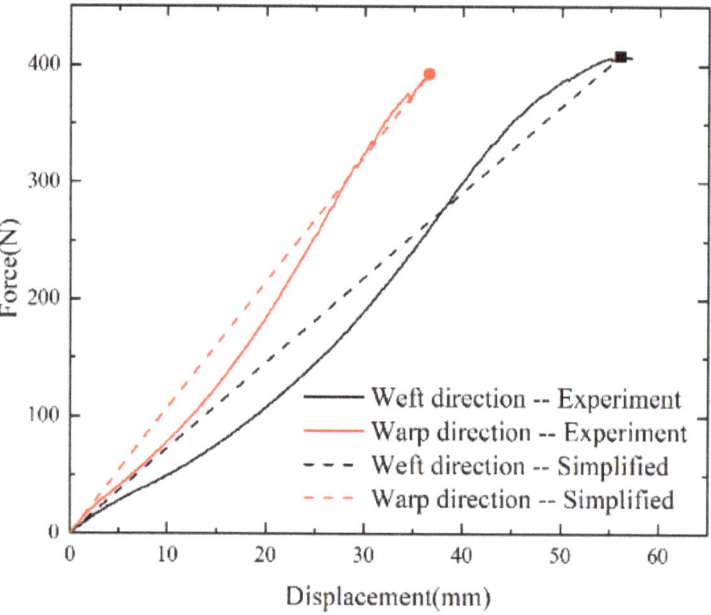

Figure 10. Canopy tensile test.

2.5. The Tractor Rocket Thrust Test

In this paper, a tractor rocket is used to deploy the parachute. The rocket thrust tests are carried out at different temperatures to obtain the thrust parameters. The test system consists of a testbed, rocket, sensor, data transmission lines, and data collector, as is shown in Figure 11. The experiment proceeded three times, one at standard temperature (+20 °C), one at low temperature (−40 °C), and one at high temperature (+45 °C). Before the test, the rocket was kept in an incubator for more than twelve hours. During the trial, the rocket was taken out and installed quickly. The ambient temperature was 20 °C, and the sampling frequency was 10 kHz.

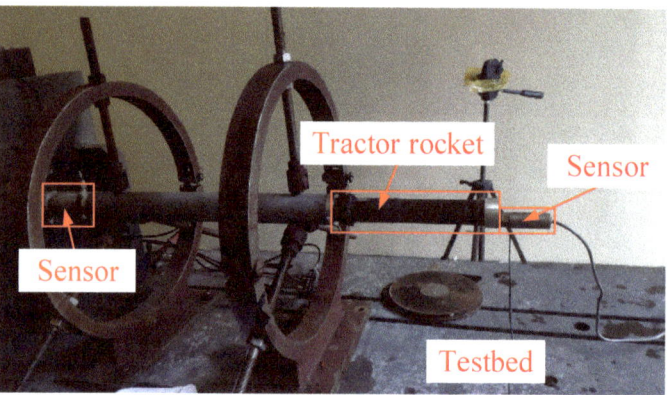

Figure 11. The tractor rocket thrust test layout.

The test results are shown in Figure 12 and Table 3. It can be seen that with the temperature increase, the peak thrust value, average thrust value, and peak time of the rocket all increase, and the working time of the rocket decreases.

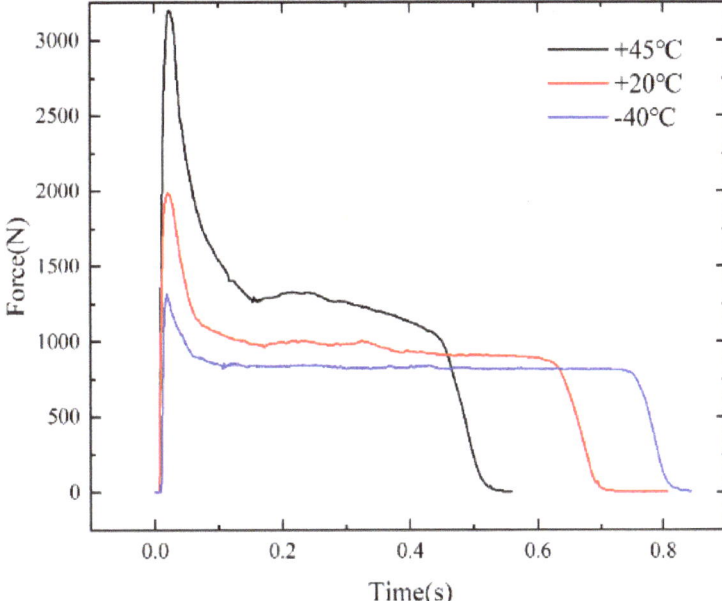

Figure 12. Curves of rocket thrust.

Table 3. Main parameters of parachute.

Temperature/°C	+45	+20	−40
Thrust peak/N	3199.6	1991.0	1318.3
Average thrust/N	1254.1	911.1	831.3
Peak time/s	0.0234	0.0212	0.0198
Work time/s	0.538	0.709	0.834

3. Experimental Validation

3.1. Experiment Design

The parachute deployment experiment was designed to verify the correctness of the simulation results. The launch temperature of the tractor rocket was 20 °C. The test was carried out on a clear and windless day to facilitate the test record and avoid the impact of environmental factors. The test layout is shown in Figure 13.

In this test, the parachute weight was 10 kg. Before the test, the rocket was kept at +20 °C for more than twelve hours. Next, test parts were trialed together and placed on the test site, and the high-speed camera was set up in front of the parachute container to record time and parachute track. The aerial drone hovered above to record the test process. When the rest of the preparation work was completed, the rocket was removed from the oven, installed in the rocket cabin, and fired quickly to ensure that the rocket temperature was around +20 °C.

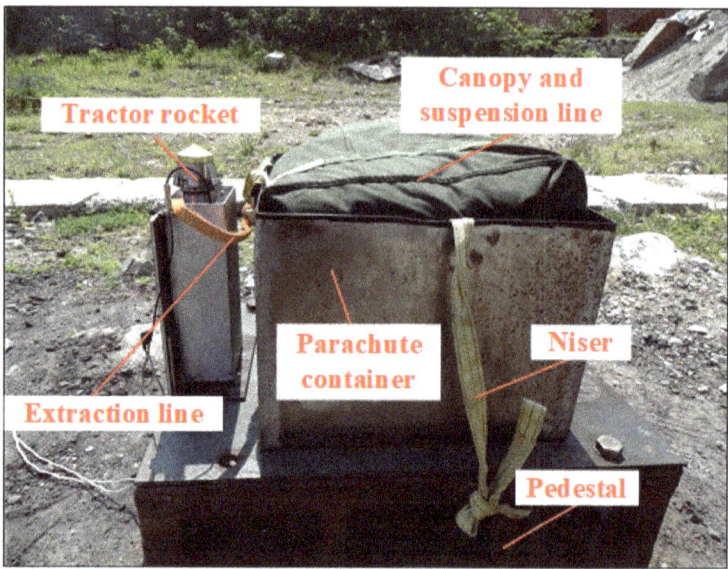

Figure 13. Experiment layout.

3.2. Comparison between Simulation and Experiment

The simulation and experiment deployment process configuration at different times is obtained, as shown in Figure 14.

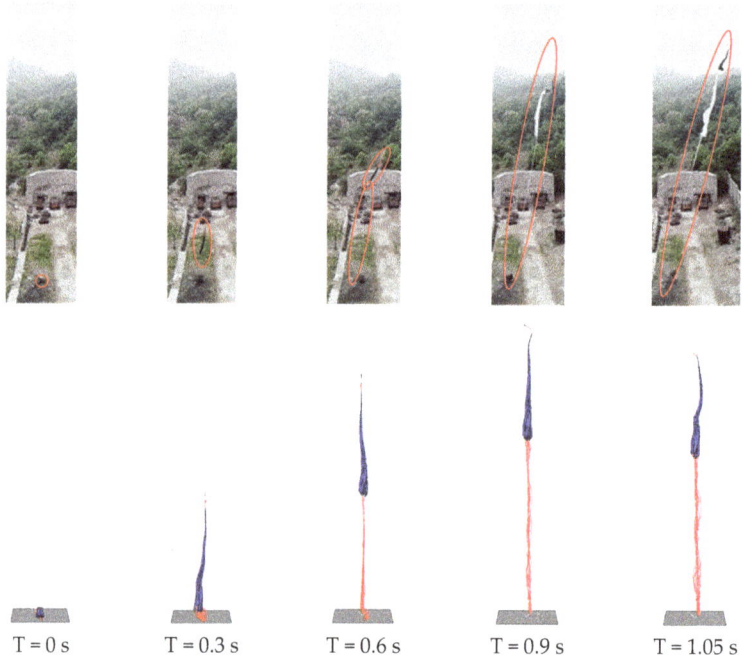

Figure 14. Comparison of deployment configuration between simulation and experiment.

The displacement of the tractor rocket directly reflected the deployment degree of canopy and suspension lines. As is shown in Figure 15, the tractor rocket reached its maximum displacement of 29.91 m in 0.8 s. Observation of the test video shows that the time for the parachute to be deployed was about 0.93 s, with a time error of 13.97%. The nominal radius of the canopy was 7.78 m; when it was deployed straight, the measured length was about 11 m. The length of the suspension lines, extraction line, and sling was 14 m, 1 m, and 5 m, respectively. Finally, the height of the rocket was 0.36 m. Therefore, the displacement of the rocket under ideal conditions was about 31.36 m, and it can be seen from the test video that both the canopy and the suspension lines were straight, so the perfect length can be used as the test length comparison. The total displacement error was 4.62%. The main reason for the error was that the friction between the canopy, suspension lines, parachute container, and the air resistance was not considered in the simulation. Figure 16 shows the change in the rocket velocity. It can be seen that the rocket began to accelerate, and with the parachute weight increase, the rocket speed decreased. Because the canopy was folded three times, there were three distinct fluctuations in the rocket's velocity. With the canopy and suspension lines fully straight, the rocket finished its work, and the gravity slowed it down until it fell back to the ground.

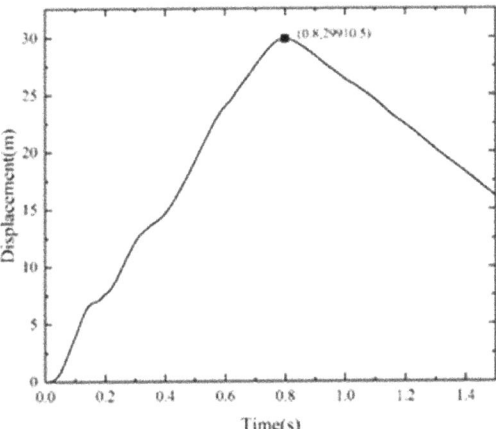

Figure 15. Displacement of tractor rocket.

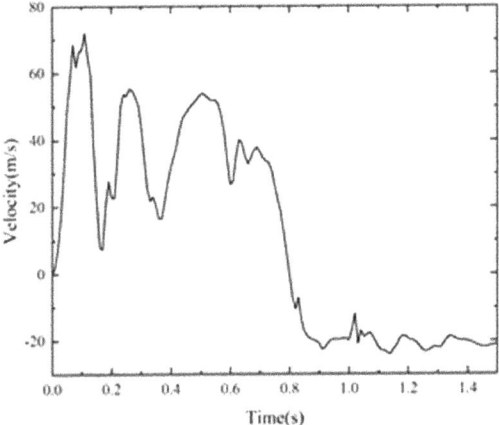

Figure 16. Velocity of tractor rocket.

In general, the simulation results matched the test results in trends and magnitudes, validating the accuracy and reliability of the dynamic model, which provided a solid foundation for the following parameter analysis of deploying performance.

4. Parameter Analysis of Parachute Deployment

4.1. Influence of Parachute Weight

This simulation researched the deployment phase of parachutes with different weights of 10 kg and 20 kg under the launch temperature of +20 °C. The simulation process is shown in Figure 17. It can be concluded that the rocket deployed parachutes successfully with a relatively smooth phase and a good result.

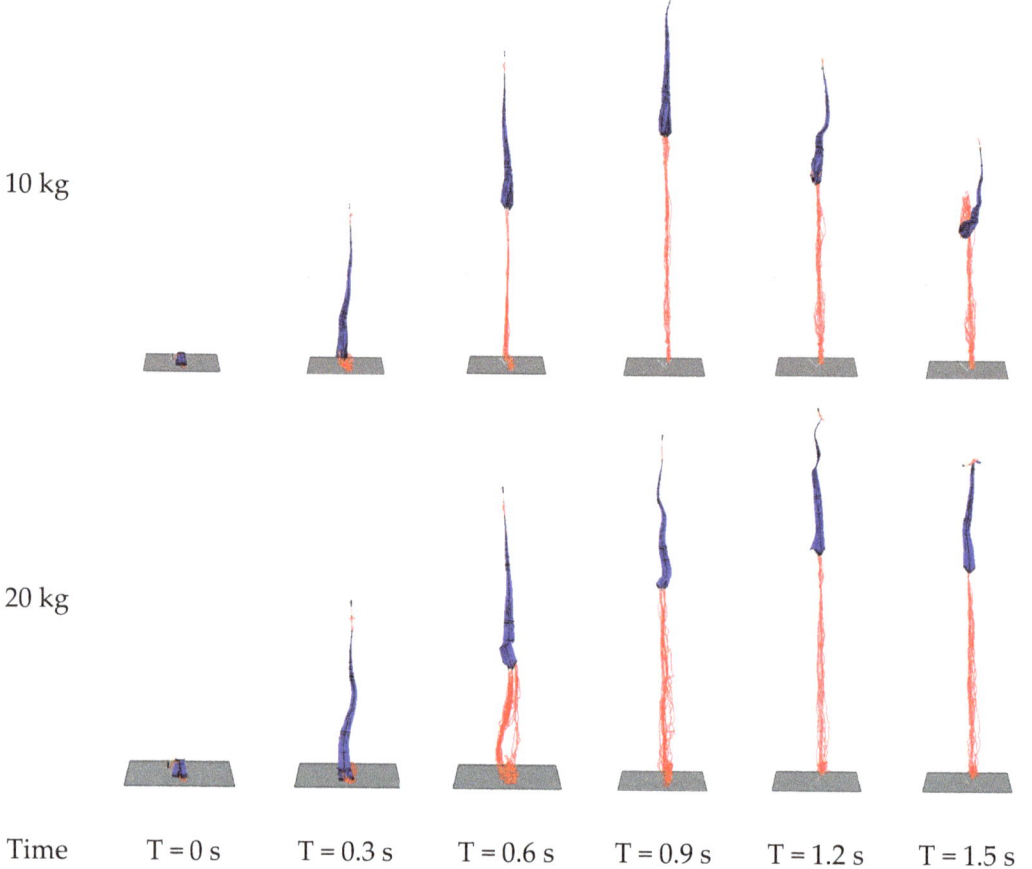

Figure 17. Different parachute weight deployment process.

The force curves of the extraction line and the sling are shown in Figure 18a,b. The force of the extraction line and the sling varied in the same way in the two conditions. The maximum snatch force of the parachute with a weight of 10 kg was greater than 20 kg, which is because the more weighted parachute could offset more rocket thrust, resulting in a smaller force acting on the extraction line and the sling. The strength of the extraction line and of the sling met the design requirements.

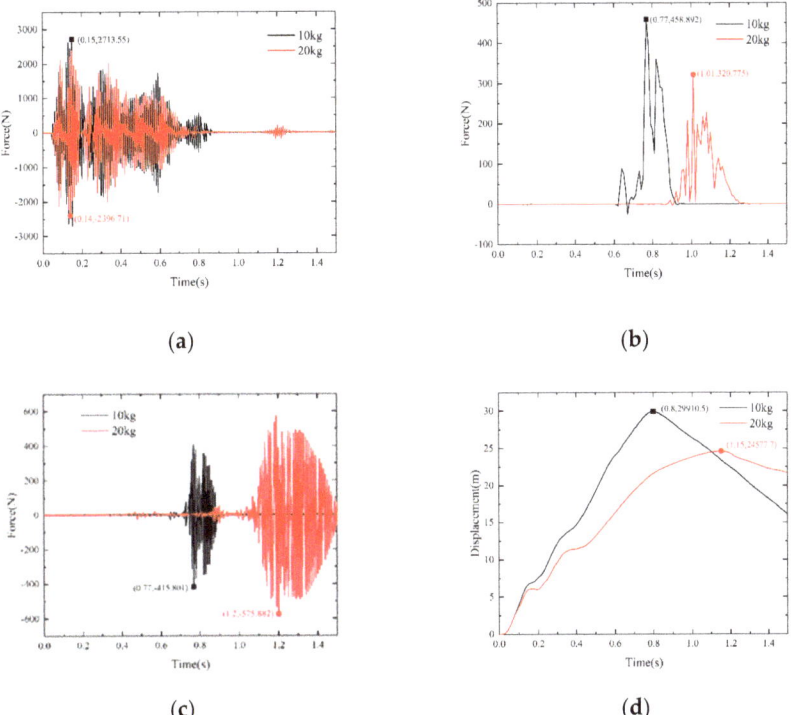

Figure 18. Deployment performance under different weight. (**a**) Extraction line force (**b**) Sling force (**c**) Suspension lines force (**d**) Tractor rocket displacement.

The force curves of the suspension lines are shown in Figure 18c. The maximum snatch force of the parachute with a weight of 20 kg was greater than 10 kg, and the lower the weight, the less the cord fluctuated. Meanwhile, the strength of the suspension lines met the design requirements.

The displacement curves of the rocket are shown in Figure 18d. The maximum displacement of the 10 kg parachute was 29.91 m at 0.8 s, and that of the 20 kg parachute was 24.58 m at 1.15 s. The error of time and displacement was 23.66% and 21.6%, respectively. Thus, with the increase in the parachute weight, the degree of deployment reduced.

To sum up, as the parachute weight increased, the maximum snatch force on the extraction line and the sling decreased, but that on the suspension lines increased, and the deployment effect reduced. Therefore, the 10 kg parachute is the most effective.

4.2. Influence of Temperature

This simulation studied the deployment phase of a rocket under −40 °C, +20 °C, and +45 °C, with a parachute weight of 10 kg. The deployment simulation phase is shown in Figure 19. The canopy and the suspension lines were deployed gradually from the initial folding state, and the whole process was relatively smooth. After the rocket was fired, a continuous thrust pulled the parachute and the suspension lines out of the container. As the rocket rose, the parachute and the suspension lines straightened. When the propellant was burned out, the parachute and rocket gradually fell to the ground due to gravity. During the test, the deployment process of the parachute was smooth and orderly, with a good parachute shape and no strong disturbance. As the temperature rose, the parachute took less time to become straight.

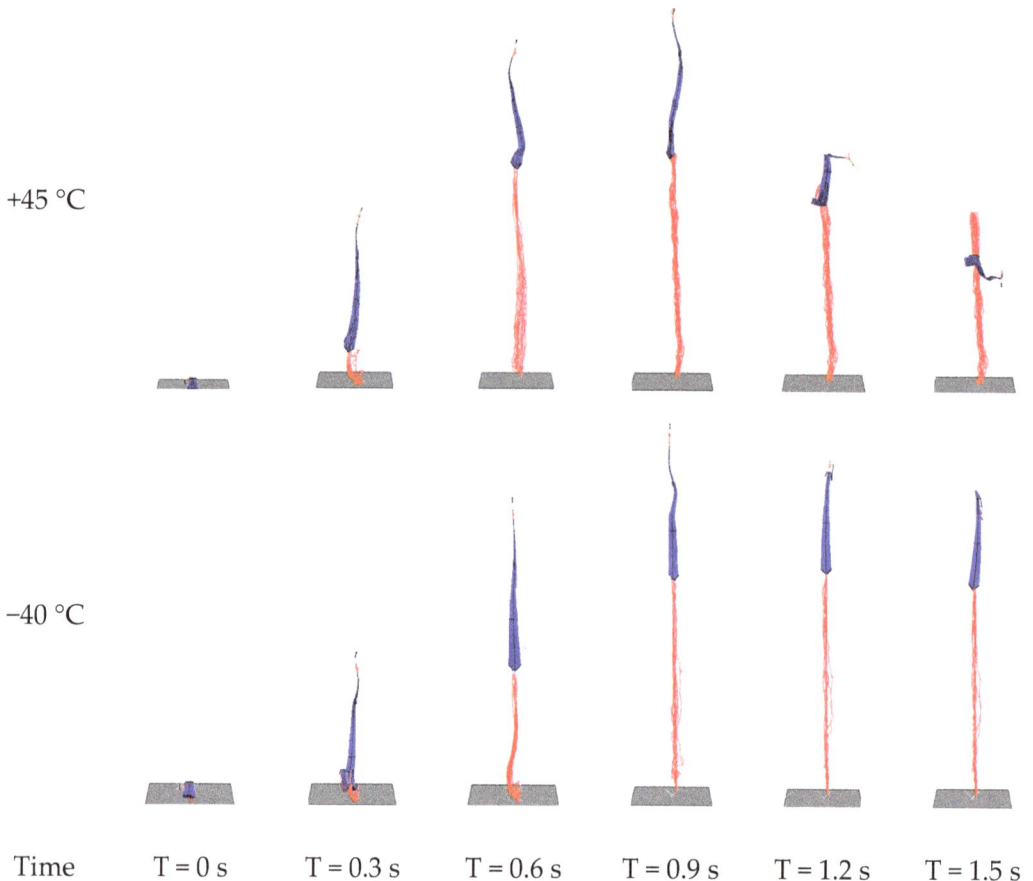

Figure 19. Different temperature parachute deployment process.

The force curves of the extraction line, the sling, and the suspension lines are shown in Figure 20a–c. The snatch forces varied in the same way at different temperatures. For the extraction line, at +45 °C, the maximum force was 3814.98 N at 0.28 s; at +20 °C, the maximum force was 713.55 N at 0.16 s; and at −40 °C, the maximum force was 2857.3 N at 0.18 s. The force at +45 °C was the largest and was little different at the other two temperatures. This is mainly because with the temperature increase, the peak thrust value of the rocket changed significantly. The peak thrust value was 3199.6 N at +45 °C, and only 1991 N and 1318.3 N at +20 °C and −40 °C, respectively, leading to a change in the extraction line.

In the sling and suspension lines, it was evident that as the temperature increased, the force increased, although the maximum force appeared earlier, with higher requirements for the strength of the material of the sling and suspension lines. The reason is mainly that the average thrust of the rocket increased as the temperature rose. The results accord with the variation trend of thrust peak value and mean value of rocket at different temperatures. At the same time, the other parameters in the parachute system remained unchanged.

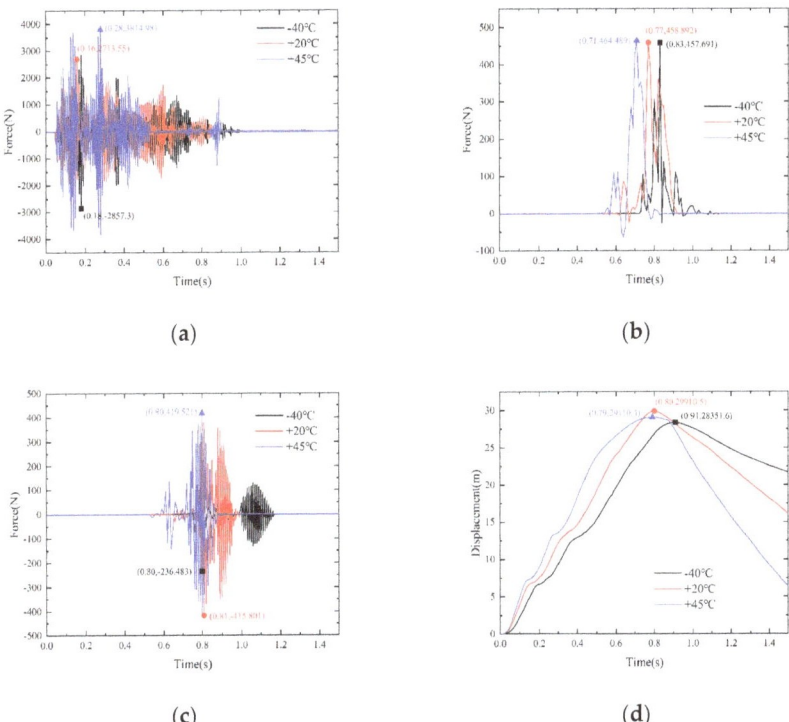

Figure 20. Deployment performance under different temperatures. (**a**) Extraction line force (**b**) Sling force (**c**) Suspension lines force (**d**) Tractor rocket displacement.

The rocket displacement directly reflected the deployment degree of the parachute. In Figure 20d, at −40 °C, the maximum displacement of the rocket reached 28.35 m at 0.91 s, and the error of displacement was 9.6%. At +45 °C, the maximum displacement of the rocket was 29.91 m at 0.79 s, and the error of displacement was 4.6%. It can be concluded that as the temperature increased, the deployment time became shorter, and the deployment length did not change by much. The results accord with the rocket's working time and thrust variation trend at different temperatures. At these three temperatures, the difference between the maximum and minimum values of rocket displacement was only 1.559 m, accounting for 4.97% of the ideal full length. Therefore, it can be found that temperature has little influence on the degree of parachute straightening, which provides a theoretical basis for the normal operation of the UAV parachute recovery system at different temperatures.

It can be seen from the above analysis that with the temperature increase, the deployment time became shorter, and the deployment length of the canopy and suspension lines only changed slightly. The maximum snatch force on the extraction line, sling, and suspension lines increased.

5. Conclusions

This paper proposed deploying the parachute by the tractor rocket to make the unmanned aerial vehicle land safely in an emergency. The deployment process of unmanned aerial vehicle parachute recovery was analyzed by combining experiment and finite element methods. Then, the dynamic characteristics of deployment process were studied and researched. Finally, the effects of parachute weight and rocket launch temperature on the deployment characteristics were discussed, based on the dynamics model. The conclusions are as follows:

(1) The elastic modulus of the high-strength polyethylene suspension line was 748.7 MPa. The brocade silk canopy weft modulus was 7.29 MPa, and the warp modulus was 10.75 MPa. The elastic models of weaved composite material could be proved by a tensile test, and the results could be used for parachute dynamic simulation.

(2) The simulation results agree with the experimental results; the time and displacement error was 13.97% and 4.62%, which verify the accuracy of the dynamic model. This provides a new idea and method for the design and simulation of parachute deployment process.

(3) With the parachute weight increase, the maximum snatch force on the extraction line and the sling decreased, as the snatch force on the suspension lines increased. The increase in parachute weight reduced the deployment effect, which is not conducive to the recovery of an unmanned aerial vehicle.

(4) With the temperature increase, the deployment time became shorter, and the deployment length changed only slightly. The maximum snatch force on the extraction line, sling, and suspension lines increased. Different launch temperatures have little influence on the parachute deployment process, ensuring the successful recovery of unmanned aerial vehicle under various conditions.

Author Contributions: Conceptualization, W.S. and S.Y.; Software, W.S.; Validation, Z.L., H.X., Z.D., G.G., G.Z. and B.Z.; Writing—Original Draft Preparation, Z.L., H.X. and W.S.; Writing—Review and Editing, W.S.; Supervision, S.Y. and G.Z. All authors have read and agreed to the published version of the manuscript.

Funding: This work was supported by the National Natural Science Foundation of China (52102436), the Fundamental Research Funds for the Central Universities (30920021109), the Natural Science Foundation of Jiangsu Province (BK20200496), the China Postdoctoral Science Foundation (2020M681615), and the project of Key Laboratory of Impact and Safety Engineering (Ningbo University), the Ministry of Education(CJ202107), and the State Key Laboratory of Mechanics and Control of Mechanical Structures (Nanjing University of Aeronautics and astronautics) (Grant No. MCMS-E-0221Y01).

Institutional Review Board Statement: Not applicable.

Informed Consent Statement: Not applicable.

Data Availability Statement: Data are contained within the article. The data presented in this study can be seen in the content above.

Acknowledgments: We wish to express our gratitude to the members of our research team, Wenhui Shi, Shuai Yue, Zhiqian Li, Hao Xu, Zhonghua Du, Guangfa Gao, Guang Zheng and Beibei Zhao.

Conflicts of Interest: The authors declare no conflict of interest.

References

1. Xie, C.; Huang, X. Energy-efficiency maximization for fixed-wing UAV-enabled relay network with circular trajectory. *Chin. J. Aeronaut.* **2021**. [CrossRef]
2. Levin, J.M.; Nahon, M.; Paranjape, A.A. Real-time motion planning with a fixed-wing UAV using an agile maneuver space. *Auton. Robot.* **2019**, *43*, 2111–2130. [CrossRef]
3. Shi, Q.; Liu, D.; Mao, H.; Shen, B.; Li, M. Wind-induced response of rice under the action of the downwash flow field of a multi-rotor UAV. *Biosyst. Eng.* **2021**, *203*, 60–69. [CrossRef]
4. Zhu, H.; Sun, Q.; Wu, W. Accurate Modeling and Control for Parawing Unmanned Aerial Vehicle. *Acta Aeronaut. Et Astronaut. Sin.* **2019**, *40*, 122593.
5. Zhu, Z.Q.; Wang, X.L.; Wu, Z.C.; Chen, Z.M. Aerodynamic characteristics of small/micro unmanned aerial vehicles and their shape design. *Acta Aeronaut. Et Astronaut. Sin.* **2006**, *27*, 353–364.
6. Wang, X.; Liu, Z.; Cong, Y.; Li, J.; Chen, H. Miniature fixed-wing UAV swarms: Review and outlook. *Acta Aeronaut. Et Astronaut. Sin.* **2020**, *41*, 023732.
7. Arrabito, G.R.; Ho, G.; Lambert, A.; Rutley, M.; Keillor, J.; Chiu, A.; Hou, M. *Human Factors Issues for Controlling Uninhabited Aerial Vehicles*; Defense Research and Development Canada: Toronto, ON, Canada, 2010.

8. White, F.M.; Wolf, D.F. A theory of three-dimensional parachute dynamic stability. *J. Aircr.* **1968**, *5*, 86–92. [CrossRef]
9. Zhen, Z. Research development in autonomous carrier-landing/ship-recovery guidance and control of unmanned aerial vehicles. *Acta Autom. Sin.* **2019**, *45*, 669–681.
10. Gajjar, B.I.; Zalewski, J. A07: On-ship landing and takeoff of Unmanned Aerial Vehicles (UAV'S). *IFAC Proc. Vol.* **2004**, *37*, 42–46. [CrossRef]
11. Wu, H.; Wang, Z.; Zhou, Z.; Wang, R. Establishment and Simulation of Twelve-Degree-of-Freedom Model for UAV Parachute Recovery System. *J. Northwest. Polytech. Univ.* **2020**, *38*, 68–74. [CrossRef]
12. Zhang, Y.; Shen, H.; Wang, B.; Liu, Y. Transition process adaptive switch control of a hybrid-wing vertical takeoff and landing UAV. *J. Harbin Eng. Univ.* **2020**, *41*, 1675–1682.
13. Wu, H.; Wang, Z.; Zhou, Z.; Wang, R. Dynamics modeling and Simulation of UAV parachute recovery based on Kane equation. *J. Beijing Univ. Aeronaut. Astronaut.* **2019**, *45*, 1256.
14. Zhang, Q.B.; Peng, Y.; Chen, W.K.; Qin, Z.Z. A Mass Spring Damper Model Parachute Deployment. *J. Ballist.* **2003**, *1*, 31–36.
15. Koklu, U.; Morkavuk, S.; Featherston, C.; Haddad, M.; Sanders, D.; Aamir, M.; Giasin, K. The effect of cryogenic machining of S2 glass fibre composite on the hole form and dimensional tolerances. *Int. J. Adv. Manuf. Technol.* **2021**, *115*, 125–140. [CrossRef]
16. Köklü, U.; Mayda, M.; Morkavuk, S.; Avcı, A.; Demir, O. Optimization and prediction of thrust force, vibration and delamination in drilling of functionally graded composite using Taguchi, ANOVA and ANN analysis. *Mater. Res. Express* **2019**, *6*, 085335. [CrossRef]
17. McVey, D.F.; Wolf, D.F. Analysis of deployment and inflation of large ribbon parachutes. *J. Aircr.* **1974**, *11*, 96–103. [CrossRef]
18. Shen, G.; Xia, Y.; Sun, H. A 6DOF mathematical model of parachute in Mars EDL. *Adv. Space Res.* **2015**, *55*, 1823–1831. [CrossRef]
19. Ortega, E.; Flores, R. Aeroelastic analysis of parachute deceleration systems with empirical aerodynamics. *J. Aerosp. Eng.* **2020**, *234*, 729–741. [CrossRef]
20. Xue, Y.; Yu, L.; Liu, M.; Pang, H. Fluid structure interaction simulation of supersonic parachute inflation by an interface tracking method. *Chin. J. Aeronaut.* **2020**, *33*, 1692–1702.
21. Zhang, Q.; Feng, Z.; Ma, Y.; Ge, J.Q.; Cao, X.L.; Cao, Q.Y. Modeling and simulation of Mars EDL process. *J. Astronaut.* **2017**, *38*, 443–449.
22. Wang, H.; Cheng, W. Research on Ejecting and Deploying Process of Parachute Considering Wake Flow Effects. *Spacecr. Recovery Remote Sens.* **2017**, *38*, 3–9.
23. Chen, B.; Wang, Y.; Zhao, C.; Sun, Y.; Ning, L. Numerical visualization of drop and opening process for parachute-payload system adopting fluid-solid coupling simulation. *J. Vis.* **2022**, *25*, 229–246. [CrossRef]
24. Zhang, Q.; Feng, Z.; Zhang, M.; Chen, Q. Multi-objective optimization of parachute triggering algorithm for Mars exploration. *Adv. Space Res.* **2020**, *65*, 1367–1374. [CrossRef]
25. Yu, L. *Aerodynamic Deceleration Technology*, 1st ed.; Science Press: Beijing, China, 2018; pp. 190–191.
26. Yu, Y.; Baoyin, H.; Li, J. Dynamic modeling and simulation of space flying net projectile deployment. *J. Astronaut.* **2010**, *31*, 1289–1296.
27. Li, J.; Yu, Y.; Baoyin, H. Projecting Parameters Optimization for Space Web Systems. *J. Astronaut.* **2012**, *33*, 823–829.
28. Liu, H.; Zhang, Q.; Yang, L.; Zhu, Y. Analysis of deployment dynamics of space tether-net system. *J. Natl. Univ. Def. Technol.* **2015**, *37*, 68–77.
29. Liu, H.T.; Zhang, Q.B.; Yang, L.P.; Zhu, Y. The deployment dynamic characteristics analysis of space web system. *J. Natl. Univ. Def. Technol.* **2015**, *37*, 68–77.
30. Barry, C.P.; Olson, B.G.; Bergeron, K.; Willis, D.J.; Sherwood, J.A. Modeling Tensile Tests of a Braided Parachute Suspension Line using a Mesomechanical Finite Element Model. In Proceedings of the AIAA Scitech 2020 Forum, Orlando, FL, USA, 6–10 January 2020; p. 1312.
31. Shi, W.H.; Chen, X.; Chen, Y.H.; Sheng, S.J. Dynamic Simulation and Test of Parachute Deployment. *Sci. Technol. Eng.* **2021**, *21*, 3379–3386.
32. Zhu, L. *Numerical Simulation Analysis on the Deformation and the Stress of Inflated Membrane Structure*; Shanghai Jiao Tong University: Shanghai, China, 2014.

Article

Flow Field and Inclusions Movement in the Cold Hearth for the Ti-0.3Mo-0.8Ni Alloy

Zhenze Zhu, Rongfeng Zhou *, Xiangming Li *, Wentao Xiong and Zulai Li

National-Local Joint Engineering Laboratory for Technology of Advanced Metallic Solidification Forming and Equipment, Faculty of Materials Science and Engineering, Kunming University of Science and Technology, Kunming 650093, China
* Correspondence: zhourfchina@hotmail.com (R.Z.); lixm@kust.edu.cn (X.L.)

Abstract: To investigate the melt flow field and inclusions movement in the cold hearth for the Ti-0.3Mo-0.8Ni alloy during electron-beam cold-hearth melting, a three-dimensional numerical model was established. By using solidification and discrete phase models, the information on the melt flow field and inclusions movement in the cold hearth were obtained. As the casting velocity increased, the melt flow velocity increased, the solid–liquid interface moved down. Inclusions with a density of 4.5 g/cm^3 were the most difficult to remove. When the density of the inclusions was 3.5 g/cm^3, the number of inclusions that escaped decreased with an increase in the inclusion diameter; these inclusions easily floated on the pool surface and remained in the cold hearth. Inclusions with a density of 5.5 g/cm^3 have a similar escaping trend to the inclusions with a density of 3.5 g/cm^3; as the diameter of these inclusions increased, gravity on these inclusions had a larger effect and caused them to sink more easily. Generally, for high and low density inclusions with a large diameter, the effect of density can be eliminated; the most effective method to remove inclusions in the metallurgical industry is to promote the polymerization and growth of the inclusions.

Keywords: movement; inclusions; cold hearth; flow field

1. Introduction

Titanium is the fourth most abundant structural metal on Earth, with a content of about 0.6 % in the crust [1]. The most common forms found in nature are rutile (TiO$_2$) and ilmenite (FeTiO$_3$). The density of titanium and its alloys is about 60% of that of nickel or iron alloys, but their tensile strength is comparable; they also have great corrosion resistance in different environments [2].

The main problem limiting the wider use of titanium and its alloys is the high cost of its production. In the metallurgical industry, electron beam cold hearth melting (EBHM) is often used to produce titanium and titanium alloy ingots; this has an excellent ability to remove high and low density inclusions [3]. This device uses high-energy electron beams to scan the feedstock. The feedstock melts and flows into a cold hearth for refining and then into a crystallizer, where it solidifies and is removed under a high vacuum (Figure 1). Refining is a very important process. Some of the inclusions floating on the surface of the molten pool are melted by the high-energy electron beams and evaporated; some flow into the crystallizer, while others settle on the solidification shell during the cold hearth.

The inclusions with their size, locations, and sources often affect the quality of titanium and its alloy products. With the development of computer technology, many scholars have gradually started using numerical simulations to study the behavior of the inclusions in the mold [4]. Li et al. [5] simulated the fractal profile of two-dimensional and three-dimensional inclusions by using the diffusion-limited aggregation (DLA) model of inclusion growth. Yu et al. [6] calculated the heat transfer, flow, solidification, and movement behavior of the inclusions in molds with different corner structures using mathematical models; they found that inclusions larger than 100 μm were removed easily. Lei et al. [7] used a solidification model and fluid flow coupled with the discrete phase model (DPM) to predict inclusions trapped in the solidification front of high-speed continuous casting slab and their distribution in the inner surface layer. Liu et al. [8,9] developed a DPM and used a Euler–Euler model to predict the quasi-four phases of argon-steel slag-air in a slab mold.

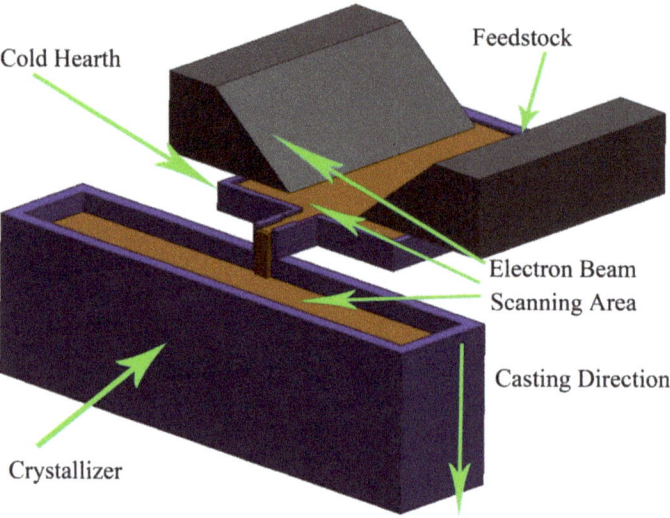

Figure 1. Schematic of electron beam cold hearth melting (EBCHM).

The Ti-0.3Mo-0.8Ni alloy is a near α alloy and has great properties, such as good welding property, crevice corrosion resistance, and excellent processing plasticity [10]. It is widely used in the chemical industry; its applications include being a crystallizer for salt production, plate brine coolers, heaters in the chlor-alkali industry, tubular reactors for the oxidizing acid tank, a reactor for the treatment of wastewater with high-chlorine content by using wet oxidation, and heat exchanger in treating strong corrosive and thermally concentrated chlorides [11]. Li et al. [12] studied the flow field motion and segregation of the Ti-0.3Mo-0.8Ni alloy in a crystallizer at different casting processes during EBCHM. Truong et al. [13] studied the effect of the electron beam scanning strategy in aluminum volatilization of Ti-6Al-4V during the cold hearth. Bellot et al. [14] used mathematical models to simulate the melting process of titanium in EBCHM and simulated the behavior of hard α inclusions in the melting process, including the dissolution trajectory and kinetics. Kroll-Rabotin et al. [15] investigated inclusion interactions in the plane shear flow, computing the fully resolved hydrodynamics at finite Reynolds numbers, using a lattice Boltzmann method with an immersed boundary method. In order to determine the collision efficiency, different initial conditions, different shear values, and different sizes of inclusions were studied.

The main research focus is the movement of inclusions during the cold hearth; the removal mechanism of different specifications of inclusions by cold hearth can be theoretically modeled and understood. A literature review found that the current study on the movement of inclusions and melt flow in cold hearth is still not involved for the Ti-0.3Mo-0.8Ni alloy during EBCHM; this study will systematically research the movement of inclusions in the cold hearth. The three-dimensional numerical model considers the melt flow field, solidification process, and inclusions movement of the Ti-0.3Mo-0.8Ni alloy; the DPM in ANSYS was used to quantitatively analyze the movement of inclusions in the cold hearth.

The purpose of this study was to analyze the melt flow field and inclusions movement in the cold hearth. Two main aspects were studied in detail: (1) the movement of different inclusions in the cold hearth and (2) the melt flow field of the cold hearth.

2. Mathematical Model

2.1. Assumptions

The conditions of the three-dimensional numerical model are simplified as follows:
(1) All inclusions were treated as spheres.
(2) The density and size of the inclusions were assumed to be constant.
(3) The agglomeration, cracking, chemical reaction, and growth of the inclusions in the cold hearth were not considered.
(4) Marangoni flow was neglected.
(5) The density of the melt is a constant of 4.5 g/cm^3.

2.2. Boundary Conditions and Computational Domain

To improve the calculation efficiency, only a half-symmetrical part of the cold hearth was selected for calculation [16]. A region of the cold hearth with a width of 200 mm and a length of 1570 mm was established. The computational domain consisted of 1.1 million hexahedron cells. In the previous work, the cell parameters of different sizes have been verified and found that the cell parameters in this work have met the requirements. For this model, the perfect 90° hexahedron cell is achieved, with an aspect ratio of 0.95, determinant value of 0.97, and cell quality value of 0.98. The boundary conditions of the model (Figure 2a) were the pool surface, sidewall, symmetry plane, inlet, and outlet. The type of inlet is "velocity inlet" with different casting velocities, and outlet is "outflow"; the value is obtained from the actual production experience of the factory and the setting of incompressible fluid in the model.

The molten Ti-0.3Mo-0.8Ni alloy with inclusions flowed into the cold hearth from the inlet and flowed out of the calculation domain from the outlet. Due to the different sizes and densities of the inclusions, they potentially floated on the surface, flowed into the crystallizer, or were captured by the solidification shell of the cold hearth. The heat transfer coefficient of the water-cooled copper mold (the cold hearth) attached to the sidewall was 2000 W/(m^{2}K) [12], and other forms of heat transfer are not considered.

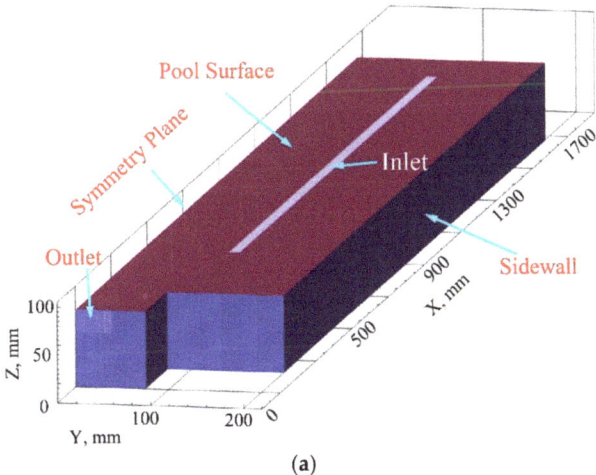

Figure 2. *Cont.*

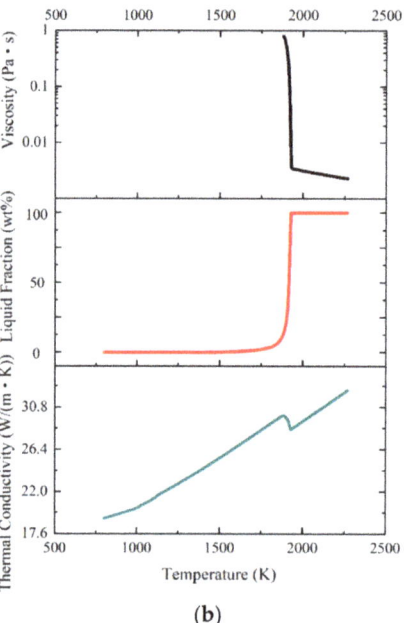

Figure 2. (a) Boundary conditions of the numerical model and (b) physical properties of the Ti-0.3Mo-0.8Ni alloy used in the model.

The solution method in the software includes these components: the SIMPLEC scheme; the pressure–velocity coupling algorithm; and the discretization model with PRESTO pressure, third-order MUSCL, and Green-Gaussian cell-based gradients.

The iteration time step was 1 s until the calculation reached the steady state; streamlines of the inclusions in the cold hearth that tended to be in a stable state were selected for this study. By default, the software used a residual value to determine the convergence of the calculations.

2.3. Governing Equations

This model was coupled with the solidification of Ti-0.3Mo-0.8Ni alloy and DPM. Previous research [17,18] explained the necessity of the governing equations of mass conservation, energy conservation, and momentum conservation in the model, as shown by Equations (1)–(6). The mushy zone flow followed the enthalpy–porosity method for simulation [19]. The exact details regarding the models are found in the literature [20–25].

2.3.1. Mass Conservation

$$\frac{\partial \rho}{\partial t} + \frac{\partial (\rho \vec{v})}{\partial X_j} = 0 \tag{1}$$

where $\vec{v}$ represents the cell velocity.

2.3.2. Momentum Conservation

$$\frac{\partial}{\partial t}(\rho \vec{v}) + \frac{\partial (\rho \vec{v} \vec{v})}{\partial X_j} = -\frac{\partial p}{\partial t} + \frac{\partial (\bar{\bar{\tau}})}{\partial X_j} + \rho \vec{g} + \vec{S}_{i,p} + \vec{S}_{i,m} \tag{2}$$

Here, $\overline{\overline{\tau}} = \mu \frac{\partial u_i}{\partial X_j} - \rho \cdot \overline{u_i' u_j'}$, and $-\rho \cdot \overline{u_i' u_j'} = \mu \left(\frac{\partial u_i}{\partial X_j} + \frac{\partial u_j}{\partial X_i} \right) - \frac{2}{3} \left(\rho K + \mu \frac{\partial u_i}{\partial X_j} \right) \delta_{ij}$ is the turbulent Reynolds stress.

Here, $\vec{S}_{i,p}$ is the thermo-solutal buoyancy, $\overline{\overline{\tau}}$ represents the stress-tensor, p is the static pressure, and $\vec{g}$ is gravity. u_i is the velocity components on three axes, X_i is the coordinate value components on three coordinate axes, and δ_{ij} is the unit tensor.

$$\vec{S}_{i,p} = \rho \vec{g} B_T \left(T - T_{liq} \right) + \sum \rho \vec{g} C_{c,i} \left(Y_{i,liq} - Y_0 \right) \tag{3}$$

Here, B_T represents the thermal-expansion coefficient, T is the temperature, $C_{c,i}$ is the solutal-expansion coefficient, $Y_{i,liq}$ is the locally averaged concentration of the solute element i in the melt, and $\vec{S}_{i,m}$ is the momentum sink in the mushy zone, which has the following form:

$$\vec{S}_{i,m} = \frac{(1-\beta)^2}{(\beta^3 + 0.01)} A_{mushy} \left(\vec{v} - \vec{v}_p \right) \tag{4}$$

where A_{mushy} is the mushy zone constant, and $\vec{v}_p$ represents the casting velocity.

2.3.3. Energy Equations

$$\frac{\partial}{\partial t}(\rho H) + \frac{\partial (\rho \vec{v} H)}{\partial X_j} = \frac{\partial (k \frac{\partial T}{\partial t} - \sum_j h_j \vec{J}_i + \overline{\overline{\tau}} \cdot \vec{v})}{\partial X_j} + \vec{Q}_r \tag{5}$$

$$H = h + \Delta H = h_{ref} + \int_{T_{ref}}^{T} C_p dT + \beta \Delta H_f \tag{6}$$

where $k \frac{\partial T}{\partial t}$ is the energy of heat conduction, $\sum_j h_j \vec{J}_i$ is the energy of species diffusion, $\overline{\overline{\tau}} \cdot \vec{v}$ is the energy transport due to viscous dissipation, H is the enthalpy, h represents the sensible enthalpy, ΔH_f is the pure solvent melting heat, C_p represents the specific heat, h_{ref} is the reference enthalpy, k represents the thermal conductivity, and $\vec{Q}_r$ is the source term.

These are the basic equations of fluid mechanics above. Fluid mechanics studies the moving and static states of the fluid itself under the action of various forces, along with the interaction and flow when there is relative movement between the fluid and the solid boundary wall.

Additionally, a realistic $k - \varepsilon$ model of Equations (7 and 8) was adopted, which was recommended in another study [26]; so far, this model has been widely used in engineering and accumulated the most experience. It has achieved basic success in many aspects. It is widely used for viscous simulations, compressible and incompressible fluid problems, and the external flow of complex geometry.

The turbulence kinetic energy K is given by:

$$\frac{\partial}{\partial t}(\rho K) + \frac{\partial}{\partial X_j}(\rho K u_j) = \frac{\partial}{\partial X_j} \left[\left(\mu + \frac{\mu_t}{\sigma_K} \right) \frac{\partial K}{\partial x_j} \right] + G_K + G_b - \rho \varepsilon - Y_M + S_K \tag{7}$$

The turbulence energy dissipation rate is given by:

$$\frac{\partial}{\partial t}(\rho \varepsilon) + \frac{\partial}{\partial X_j}(\rho \varepsilon u_j) = \frac{\partial}{\partial X_j} \left[\frac{\partial \varepsilon}{\partial X_j} \left(\mu + \frac{\mu_t}{\sigma_\varepsilon} \right) \right] + \rho C_1 S \varepsilon - \rho C_2 \frac{\varepsilon^2}{K + \sqrt{\nu \varepsilon}} + C_{1w} \frac{\varepsilon}{k} C_{3\varepsilon} G_b + S_\varepsilon \tag{8}$$

where G_K is the generation of turbulence kinetic energy, ε is the turbulent dissipation rate, G_b represents the generation of turbulence kinetic energy because of buoyancy, and Y_M is the contribution of fluctuating expansion to the total dissipation rate in the compressible turbulent flow. The turbulence model constants are:

$$C_{1\omega}= 1.44,\ C_2= 1.9,\ \sigma_K= 1.0,\ \sigma_\varepsilon = 1.2$$

2.3.4. Equations of Motion for Inclusions

The trajectory of the discrete inclusions was calculated by integrating the force balance; this is written in a Lagrangian reference frame. This kind of force balance equates the inclusion inertia with forces acting on the inclusions and is expressed as the following Equations (9)–(11):

$$\frac{d\vec{u}_p}{dt} = \frac{\vec{v} - \vec{u}_p}{\tau_r} + \frac{\vec{g}(\rho_p - \rho)}{\rho_p} + \vec{F} \tag{9}$$

where $\vec{F}$ is an additional mass-force, $\frac{\vec{v}-\vec{u}_p}{\tau_r}$ is the drag force per unit inclusion mass, and

$$\tau_r = \frac{\rho_p d_p^2}{18\mu} \frac{24}{C_d Re} \tag{10}$$

Here, τ_r is the inclusion relaxation time; $\vec{u}_p$ is the inclusion velocity; μ is the viscosity of the melt; ρ_p and ρ represent the densities of the inclusion and melt, respectively; d_p is the inclusion diameter; C_d is the drag coefficient of inclusions; and Re is the inclusion Reynolds number, which was reported in other studies [27,28].

$$Re = \frac{\rho_p d_p |\vec{u}_p - \vec{v}|}{\mu} \tag{11}$$

2.4. Physical Properties

The solidus and liquidus temperatures of the Ti-0.3Mo-0.8Ni alloy are 1860 and 1930 K. Figure 2b shows the thermophysical properties of Ti-0.3Mo-0.8Ni alloy; these include the thermal conductivity, liquid fraction, and viscosity. Combined with the actual production experience in the factory, normal inclusions have a diameter below 100 μm. For a quantitative analysis, the diameters of 10, 40, 70, and 100 μm and the densities of 3.5, 4.5, and 5.5 g/cm^3 were studied. The casting process parameters used casting velocities of 2×10^{-4}, 2.5×10^{-4}, 3×10^{-4}, and 3.5×10^{-4} m/s with a pouring temperature of 2273 K.

3. Results and Discussion

To study the movement of the inclusions in the cold hearth, the molten pool morphology was calculated. As shown in Figure 3a, the left side is the surface of the molten pool, and the right side is the internal slice of the molten pool, which shows the solid–liquid interface and the streamline of the melt. The streamline entered the cold hearth from the inlet, extended around, and then flowed out of the outlet. The farthest melt from the outlet had fewer streamlines and poor melt fluidity. Combined with Figure 3b, it can best reflect the characteristics of a molten pool and can also visually compare the differences of molten pool morphology under different casting process parameters, as can be seen from the cross-section of the molten pool, two parts below the inlet were the deepest, and the symmetry plane was relatively shallow; this caused the molten pool to appear W-shaped. The velocity field of the melt is analyzed in detail below. These factors affected the movement of the inclusions.

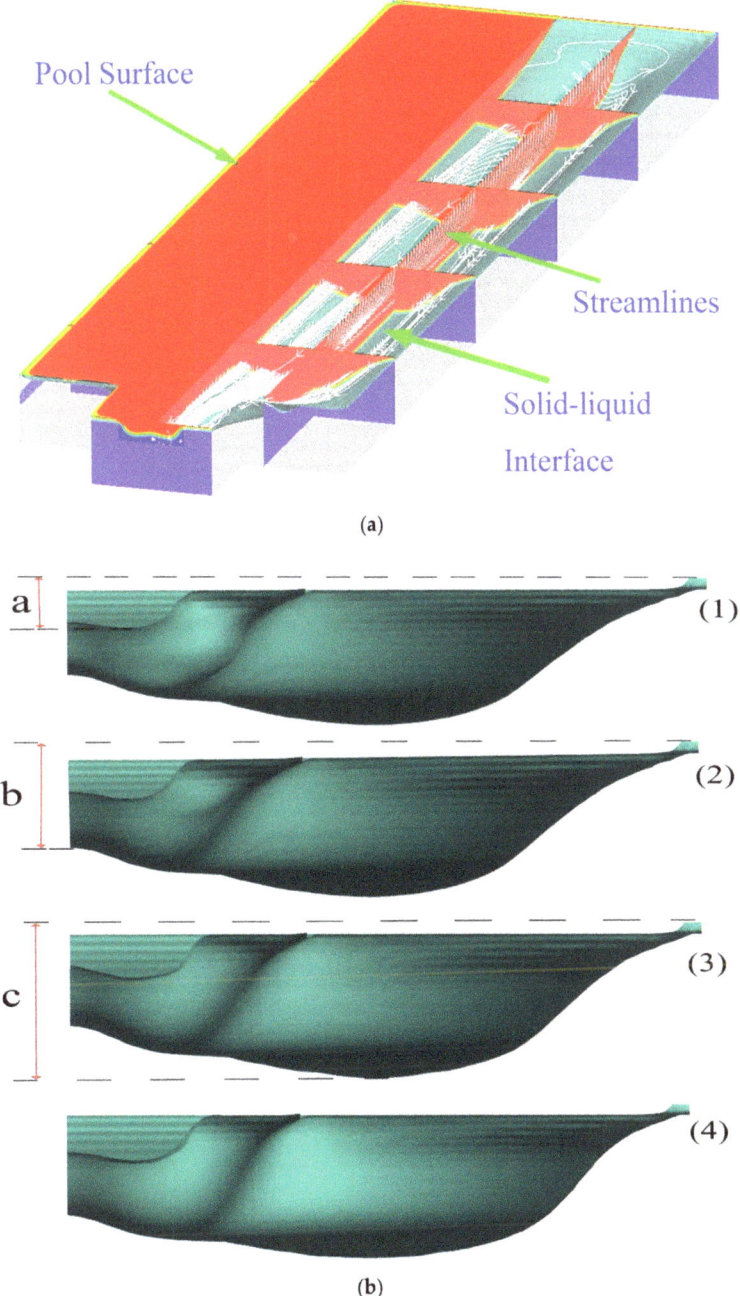

Figure 3. *Cont.*

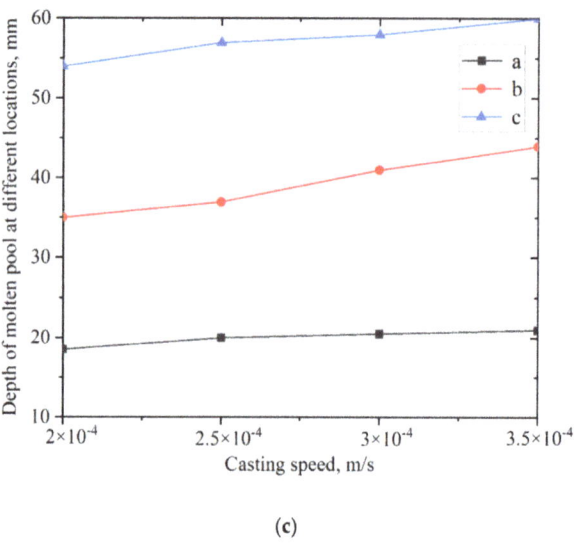

(c)

Figure 3. (a) Diagram of cold hearth for the pool surface, solid–liquid interface, and streamlines. (b) Contours of the solid–liquid interface at casting velocities of (1) 2×10^{-4} m/s, (2) 2.5×10^{-4} m/s, (3) 3×10^{-4} m/s, and (4) 3.5×10^{-4} m/s with a pouring temperature of 2273 K. Here, (a) is the deepest location below the inlet, (b) is the deepest location on the symmetry plane, and (c) is the deepest location on the outlet. (c) Depth of the molten pool at (a) the deepest location below the inlet, (b) the deepest location on the symmetry plane, and (c) the deepest location on the outlet.

3.1. Effect of Casting Velocities on the Solid–Liquid Interface

The influence of different casting velocities on the profile of the molten pool was studied. Figure 3b shows the molten pool morphology in four states using casting velocities of 2×10^{-4}, 2.5×10^{-4}, 3×10^{-4}, and 3.5×10^{-4} m/s with a pouring temperature of 2273 K. The depth values of the three characteristic positions were studied to quantitatively analyze and determine the influence of casting velocity on the molten pool. These three characteristic positions are the deepest locations below the inlet, on the symmetry plane, and on the outlet (Figure 3b). The specific values for these depths are shown in Figure 3c. The depth of the molten pool increased with an increase in the casting velocity, and the solid–liquid interface gradually moved down.

3.2. Effect of Casting Velocities on the Flow Field

By studying the influence of casting velocity on the profile of the molten pool, it was determined that the cold hearth was sensitive to velocity. To study the effect of the flow field on the cold hearth, the state of the cold hearth's velocity field was varied using casting velocities of 2×10^{-4}, 2.5×10^{-4}, 3×10^{-4}, and 3.5×10^{-4} m/s with a pouring temperature of 2273 K (Figure 4a). With an increase in the casting velocity, the surface velocity and the velocity inside the molten pool gradually increased. The velocity magnitude near the outlet was noticeably higher than those of the other positions in the cold hearth. A lower velocity magnitude with poorer fluidity and stirring effect of the melt was observed as the distance from the outlet increased; this was more likely to produce defects such as segregation.

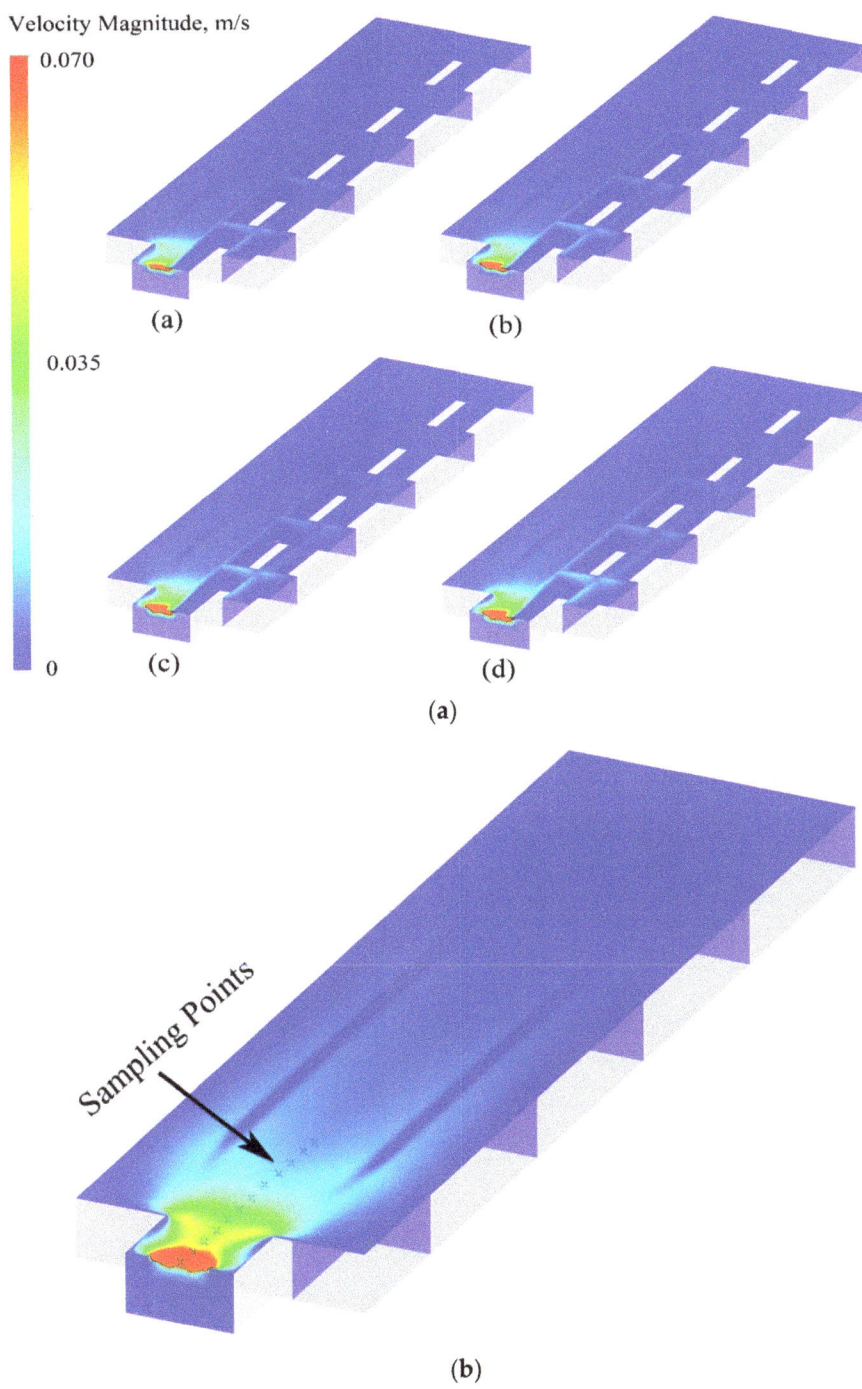

Figure 4. *Cont.*

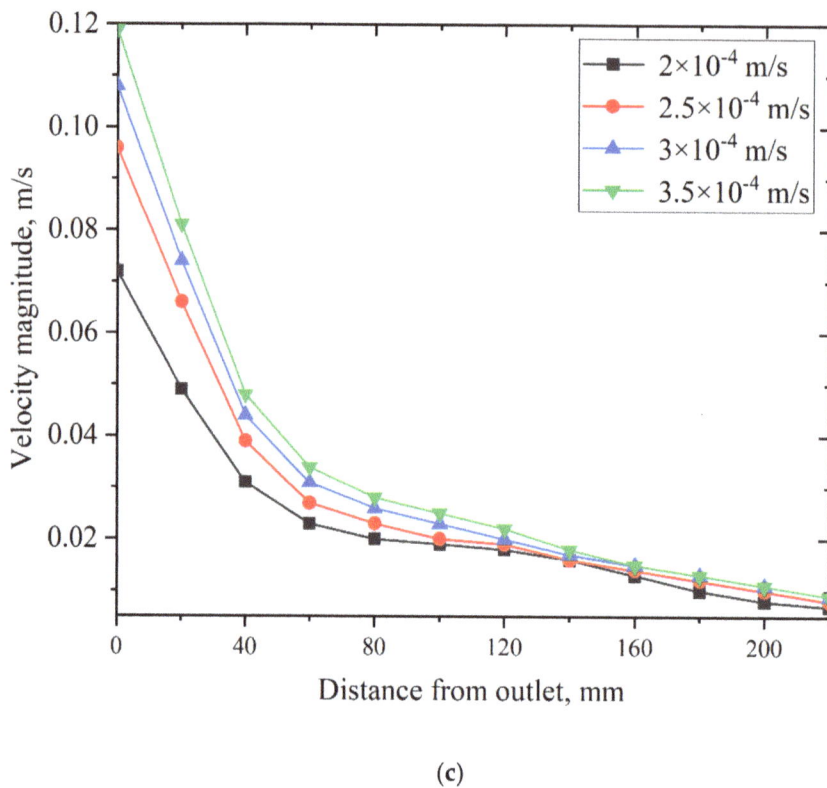

(c)

Figure 4. (a) Contours of the velocity magnitude at casting velocities of (a) 2×10^{-4}, (b) 2.5×10^{-4}, (c) 3×10^{-4}, and (d) 3.5×10^{-4} m/s with a pouring temperature of 2273 K. (b) The sampling points at the symcenter on the contour of the velocity magnitude. (c) Velocity magnitudes of sampling points at casting velocities of 2×10^{-4}, 2.5×10^{-4}, 3×10^{-4}, and 3.5×10^{-4} m/s with a pouring temperature of 2273 K.

To show the trend of the velocity magnitude more intuitively (Figure 4b), an area with a length of 220 mm was selected on the surface of the cold hearth, and 12 sampling points were evenly selected. A line chart was created showing the relationship between the velocity magnitude and the distance of these sampling points (Figure 4c). As the cast velocity increased, the velocity on the surface of the cold hearth increased. When the distance was between 0 and 60 mm away from the outlet, the velocity dropped rapidly. When the distance was between 60 to 120 mm away from the outlet, the velocity decreased more slowly.

3.3. Influence of Different Casting Velocities and Inclusion Specifications on the Trajectory of Inclusions

The trajectory of inclusions with densities of 3.5, 4.5, and 5.5 g/cm^3, and diameters of 10, 40, 70, and 100 μm at different casting velocities with a pouring temperature of 2273 K are shown in Figure 5. All these inclusions were enumerated and then classified according to different statistical methods. To study the movement behavior of the inclusions in the cold hearth, 20 inclusions were selected and flowed evenly into the cold hearth from the entrance; the reason why 20 inclusions were selected to be injected into the melt here is that this number of inclusions can fully reflect the movement law, and the trajectory is also clear, which is convenient for statistics. If the number of inclusions is higher, there will be a large number of trajectory coincidence, which is not conducive to statistics. By obtaining the flow trajectory of the inclusions, the trends of removing various inclusions in the cold hearth were shown, providing theoretical guidance for the actual production of the Ti-0.3Mo-0.8Ni alloy.

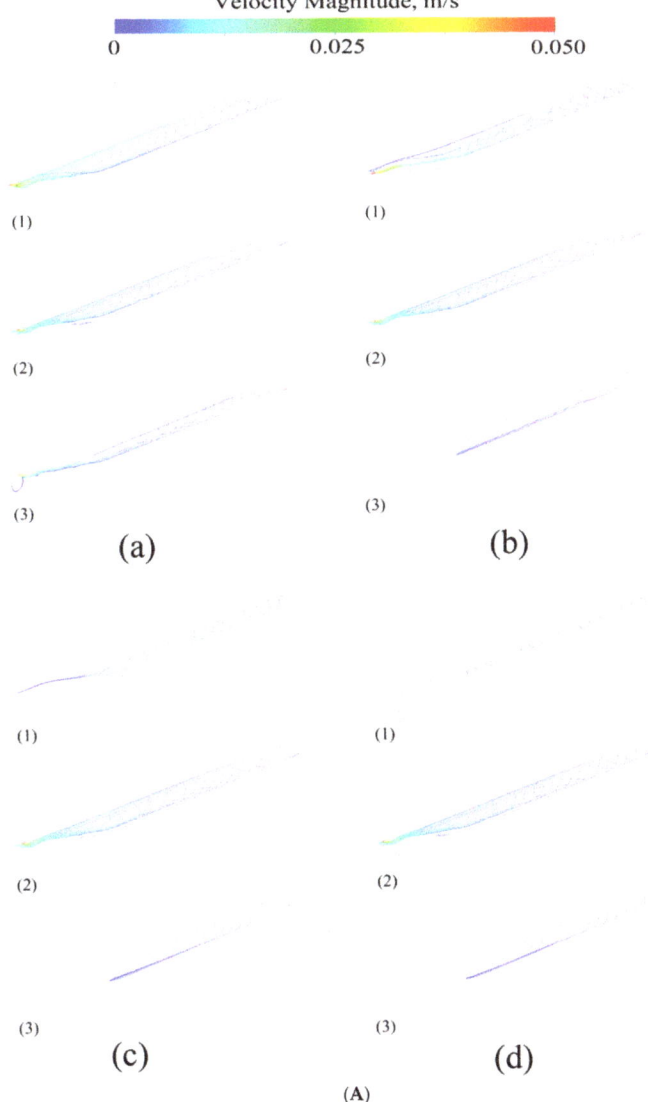

Figure 5. Cont.

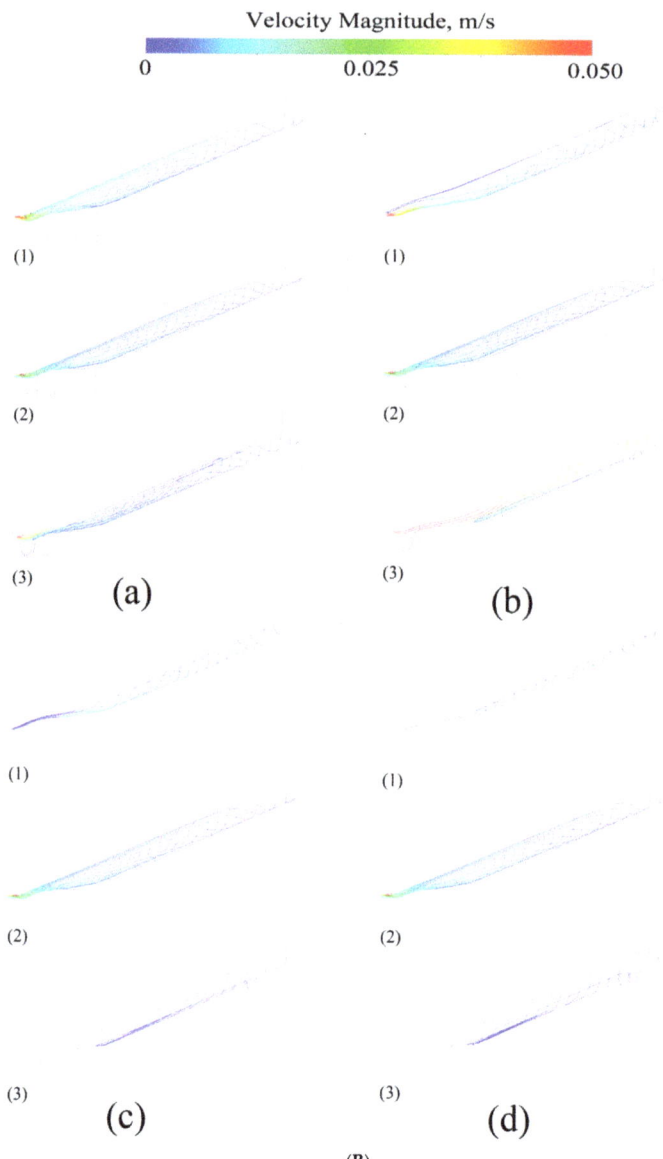

Figure 5. Cont.

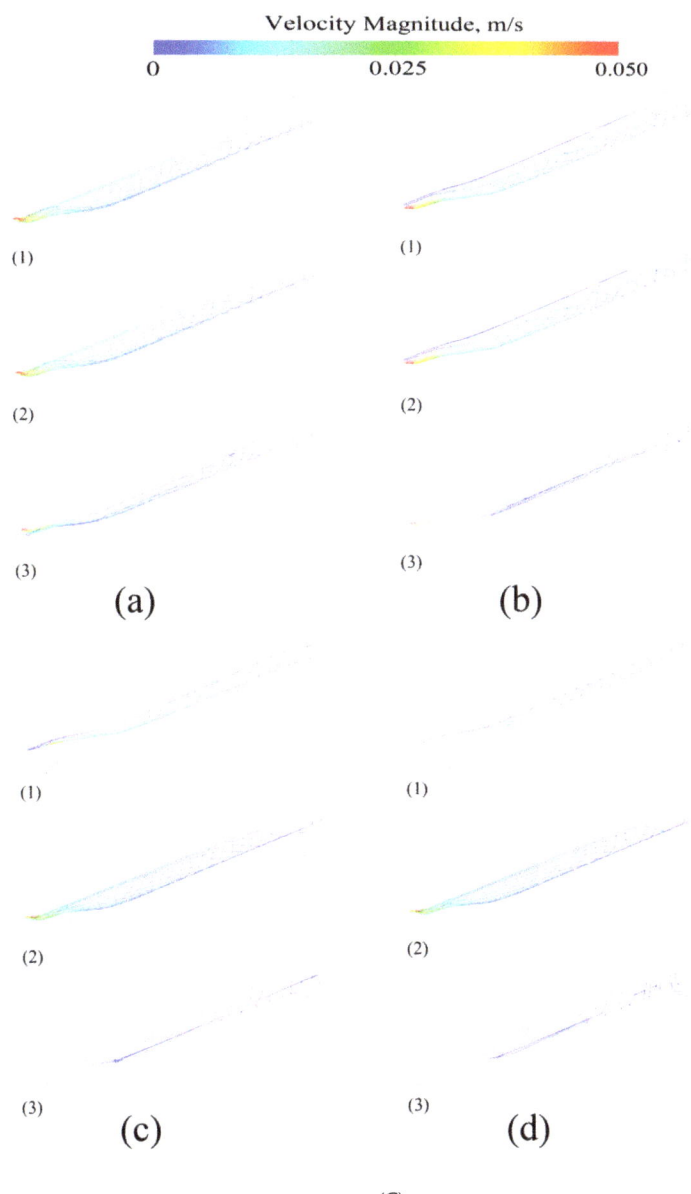

Figure 5. Cont.

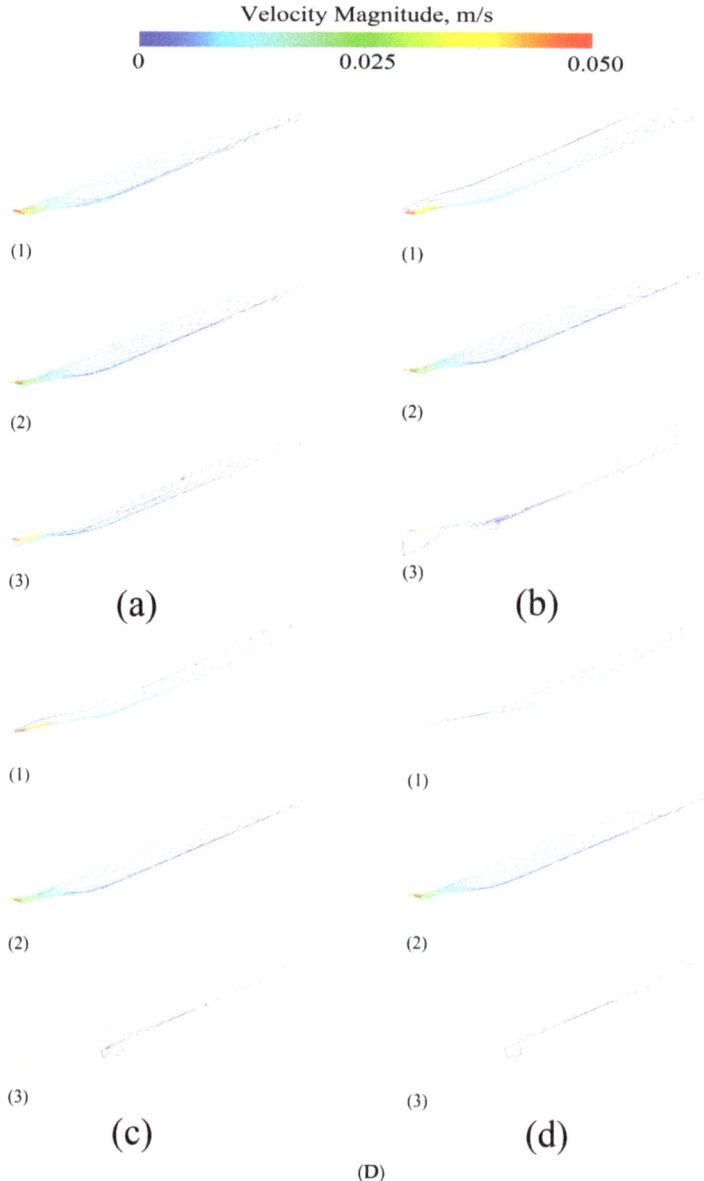

Figure 5. (**A**) The trajectory of inclusions with densities of (1) 3.5, (2) 4.5, and (3) 5.5 g/cm^3 and diameters of (a) 10, (b) 40, (c) 70, and (d) 100 μm at a casting velocity of 2×10^{-4} m/s with a pouring temperature of 2273 K. (**B**) The trajectory of inclusions with densities of (1) 3.5, (2) 4.5, and (3) 5.5 g/cm^3 and diameters of (a) 10, (b) 40, (c) 70, and (d) 100 μm at a casting velocity of 2.5×10^{-4} m/s with a pouring temperature of 2273 K. (**C**) The trajectory of inclusions with densities of (1) 3.5, (2) 4.5, and (3) 5.5 g/cm^3 and diameters of (a) 10, (b) 40, (c) 70, and (d) 100 μm at a casting velocity of 3×10^{-4} m/s with a pouring temperature of 2273 K. (**D**) The trajectory of inclusions with densities of (1) 3.5, (2) 4.5, and (3) 5.5 g/cm^3 and diameters of (a) 10, (b) 40, (c) 70, and (d) 100 μm at a casting velocity of 3.5×10^{-4} m/s with a pouring temperature of 2273 K.

The velocity information of the different types of inclusions is shown in Figure 5. Except those floating on the surface or deposited on the cold hearth, almost all other inclusions will increase their moving velocity with the melt flow. The trajectory of these inclusions was counted, the number of inclusions escaping from the outlet was quantitatively analyzed and compared, and the movement trend of inclusions was obtained.

3.4. The Escape Trend of Inclusions under Different Influencing Factors

To study the movement of different types of inclusions in the cold hearth, the escape quantity of inclusions was quantitatively analyzed. Figure 6A shows the number of inclusions escaping from the outlet with different diameters and densities at casting velocities of 2×10^{-4} m/s, 2.5×10^{-4}, 3×10^{-4}, and 3.5×10^{-4} m/s with a pouring temperature of 2273 K. The inclusions with a density of 4.5 g/cm^3 were very difficult to remove, because this density was close to that of the Ti-0.3Mo-0.8Ni alloy; therefore, it was easy to suspend in the melt and drift with the current. As the diameter of the inclusions with a density of 3.5 g/cm^3 increased, the number of escaped inclusions decreased and remained in the cold hearth; since the density was lower, the larger diameter inclusions tended to float on the surface of the cold hearth. The inclusions with a density of 5.5 g/cm^3 had a similar escaping trend to those with a density of 3.5 g/cm^3. Due to the increase in the diameter of the inclusions with a density of 5.5 g/cm^3, the gravity on these inclusions had a larger effect and caused them to sink more easily. As the casting velocity increased, the same trends were shown.

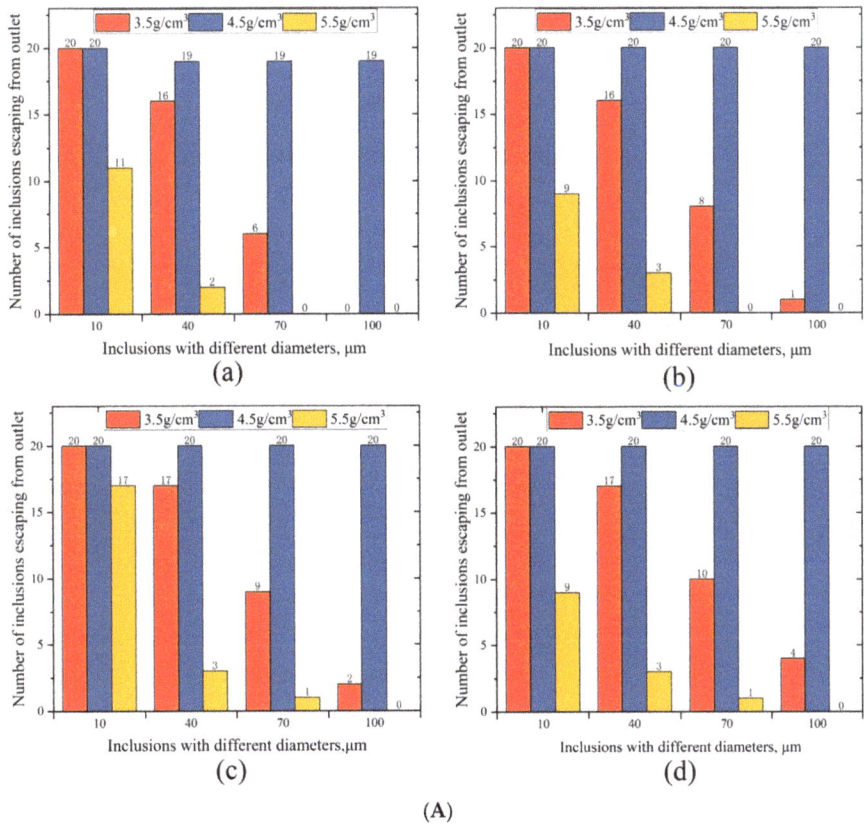

Figure 6. Cont.

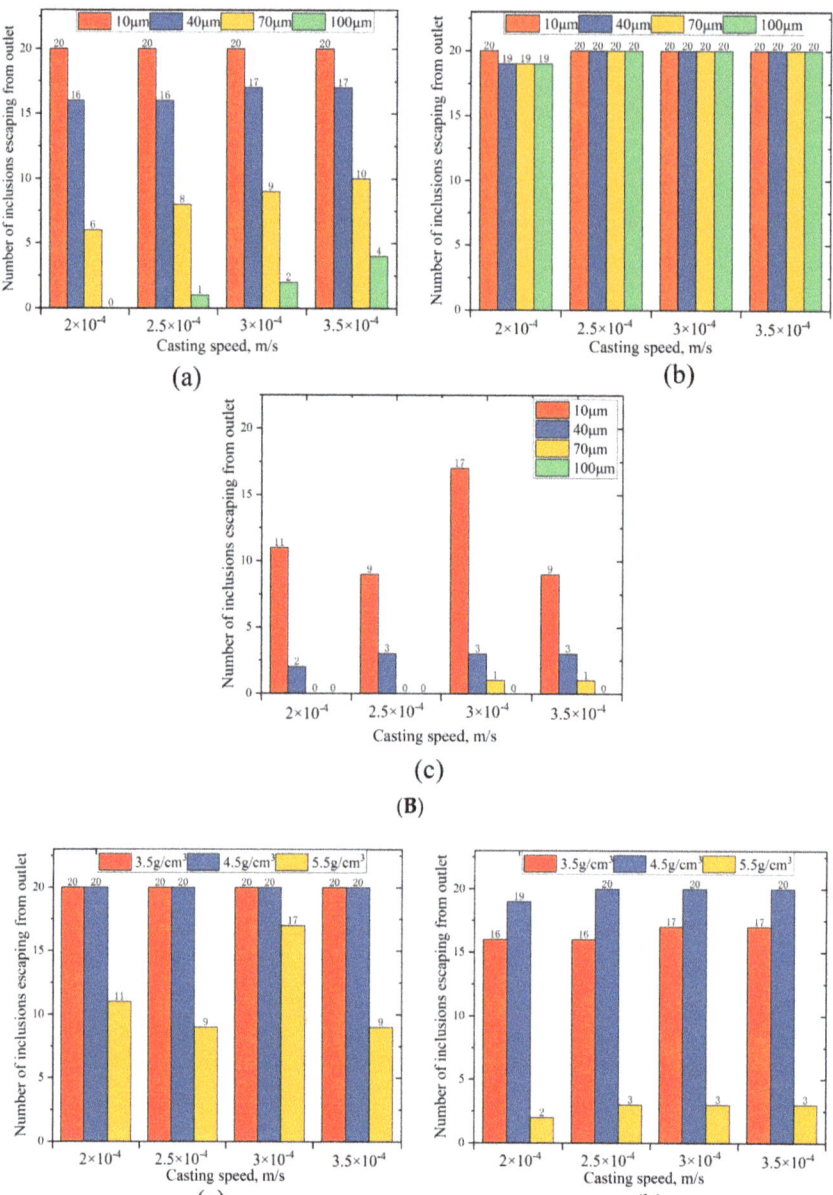

Figure 6. *Cont.*

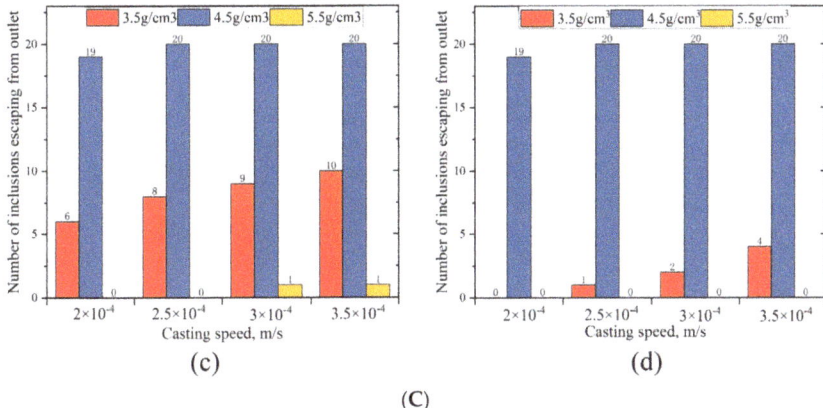

Figure 6. (**A**) Number of inclusions escaping from the outlet with different diameters and densities at casting velocities of (a) 2×10^{-4}, (b) 2.5×10^{-4}, (c) 3×10^{-4}, and (d) 3.5×10^{-4} m/s with a pouring temperature of 2273 K. (**B**) Number of inclusions escaping from the outlet with different diameters and casting velocities at densities of (a) 3.5, (b) 4.5, and (c) 5.5 g/cm^3 with a pouring temperature of 2273 K. (**C**) Number of inclusions escaping from the outlet with different densities and casting velocities at diameters of (a) 10, (b) 40, (c) 70, and (d) 100 μm with a pouring temperature of 2273 K.

The number of inclusions at a density of 3.5 g/cm^3 escaping the outlet with different casting velocities and diameters is shown in Figure 6B(a). As the diameter of the inclusions increased, the number of inclusions escaping gradually decreased; since the density was lower, the larger diameter inclusions tended to float on the surface of the cold hearth. However, as the casting velocity was increased, the number of escaped inclusions with diameters of 40, 70, and 100 μm showed a slight increase; as both the diameter of the inclusion and the casting velocity were increased, the flow force on the inclusion increased weakly causing this slight increase. The inclusions with a density of 4.5 g/cm^3 were difficult to remove at all diameters and casting velocities (Figure 6B(b)). When the density was 5.5 g/cm^3, the removal effect for inclusions with a diameter of 10 μm was typical; the removal of the inclusions with diameters of 40, 70, and 100 μm was optimal, because as the diameter increased, the gravity effect was greater, and they tended to sink.

Almost all of the inclusions with a diameter of 10 μm and a density of 3.5 g/cm^3 escaped. For inclusions with a density of 3.5 g/cm^3 and diameters of 40, 70, and 100 μm, as the casting velocity increased, the number of inclusions that escaped gradually increased, and the removal effect gradually became worse. With the increase in the diameter of the inclusion, the number of inclusions with densities of 3.5 and 5.5 g/cm^3 that escaped continuously decreased. The large diameter inclusions were removed more effectively, especially for the inclusions with a large diameter and large density.

4. Conclusions

The melt flow field and inclusions movement in the cold hearth for the Ti-0.3Mo-0.8Ni alloy were investigated by a three-dimensional model. The results of this study have highly significant outcomes for the production of a high-quality Ti-0.3Mo-0.8Ni alloy.

According to our calculations, the solid–liquid interface of the cold hearth appeared W-shaped. The depth of the molten pool increased with an increase in casting velocity, and the solid–liquid interface gradually moved down.

With an increase in the casting velocity, the melt velocity inside and the velocity on the surface of the cold hearth increased. From the outlet to about 60 mm away from the outlet, the velocity decreased rapidly, and then, after 60 mm, it decreased more slowly. The solidification shell collected the inclusions precipitated in the melt.

The moving trajectories of the various inclusions in the cold hearth were obtained and used to determine the removal effect of the inclusions. The study found that the inclusions with densities of 4.5 g/cm^3 were exceedingly difficult to remove at all diameters and casting velocities. As the diameter of the inclusions with a density of 3.5 g/cm^3 increased, the number of inclusions that escaped decreased; they floated on the pool surface and remained in the cold hearth. Almost all of the inclusions with a diameter of 10 μm and a density of 3.5 g/cm^3 escaped. As the casting velocity increased for the inclusions with a density of 3.5 g/cm^3 and diameters of 40, 70, and 100 μm, the removal effect gradually became worse.

The inclusions with a density of 5.5 g/cm^3 have a similar escaping trend as inclusions with a density of 3.5 g/cm^3; as the diameters increased, the gravity of inclusions became larger and caused them to sink more easily. For inclusions with a diameter of 10 μm, the removal effect was typical; the removal effect for inclusions with diameters of 40, 70, and 100 μm was optimal at the various casting velocities. In general, high- and low-density inclusions with large diameters were easier to remove and improved the removal efficiency of the inclusions. Suitable measures should be taken to promote the growth and polymerization of inclusions in metallurgical production.

Author Contributions: Methodology, X.L.; formal analysis, W.X.; writing—original draft preparation, Z.Z.; writing—review and editing, R.Z.; project administration, Z.L.; funding acquisition, R.Z. All authors have read and agreed to the published version of the manuscript.

Funding: This work was supported by the Major Science and Technology Projects of Yunnan Science and Technology Plan (No. 202002AB080001) and Science and Technology Major Project of Yunnan Province (No. 202202AG050007).

Data Availability Statement: Not applicable.

Conflicts of Interest: The authors declare no conflict of interest.

References

1. Lutjering, G.; Williams, J.C. *Titanium*; Springer: New York, NY, USA, 2003.
2. Shuster, R.E. *Modeling of Aluminum Evaporation during Electron Beam Cold Hearth Melting of Titanium Alloy Ingots*; University of British Columbia: Vancouver, BC, Cannada, 2017.
3. Harker, H.R. Experience with large scale electron beam cold hearth melting (EBCHM). *Vacuum* **1990**, *41*, 2154–2156. [CrossRef]
4. Santis, M.D.; Ferretti, A. Thermo-fluid-dynamics modeling of the solidification process and behaviour of nonmetallic inclusions in the continuous casting slabs. *ISIJ Intern.* **1996**, *36*, 673–680. [CrossRef]
5. Li, H.; Wen, J.; Zhang, J.M.; Wang, X.H.; Yasushi, S.; Mitsutaka, H. Simulation on cluster-agglomeration of inclusions in molten steel with DLA model. *J. Univ. Sci. Technol. Beijing* **2006**, *28*, 343–347.
6. Yu, S.; Long, M.J.; Zhang, M.Y.; Chen, D.F.; Xu, P.; Duan, H.M.; Yang, J. Effect of mold corner structures on the fluid flow, heat transfer and inclusion motion in slab continuous casting molds. *J. Manuf. Process.* **2021**, *68*, 1784–1802. [CrossRef]
7. Lei, S.W.; Zhang, J.M.; Zhao, X.K.; Dong, Q.P. Study of molten steel flow and inclusions motion behavior in the solidification processes for high speed continuous casting slab by numerical simulation. *Trans. Indian Inst. Met.* **2016**, *69*, 1193–1207. [CrossRef]
8. Liu, Z.Q.; Sun, Z.B.; Li, B.K. Modeling of quasi-four-phase flow in continuous casting mold using hybrid Eulerian and Lagrangian approach. *Metall. Mater. Trans. B Process Metall. Mater. Process. Sci.* **2017**, *48*, 1248–1267. [CrossRef]
9. Liu, Z.Q.; Li, B.K.; Jiang, M.F.; Tsukihashi, F. Euler-Euler-Lagrangian modeling for two-phase flow and particle transport in continuous casting mold. *ISIJ Int.* **2014**, *54*, 1314–1323. [CrossRef]
10. Yuan, X.L. *Editorial Board of China Aeronautical Materials Manual*; Standards Press of China: Beijing, China, 2002.
11. Li, Z.C. Research, Production and Application of Ti-0.3Mo-0.8Ni Alloy Abroad. *Rare Met. Mater. Eng.* **1984**, *5*, 75–81.
12. Zhu, Z.Z.; Li, X.M.; Zhou, R.F.; Huang, H.G.; Xiong, W.T.; Li, Z.L. Numerical simulation of molybdenum and nickel distribution in large-scale slab ingots of Ti-0.3 wt.% Mo-0.8 wt.% Ni alloys during electron beam cold hearth melting. *JOM* **2022**, *74*, 3811–3820. [CrossRef]
13. Truong, V.D.; Hyun, Y.T.; Won, J.W.; Lee, W.J.; Yoon, J.H. Numerical Simulation of the Effects of Scanning Strategies on the Aluminum Evaporation of Titanium Alloy in the Electron Beam Cold Hearth Melting Process. *Materials* **2022**, *15*, 820. [CrossRef]
14. Bellot, J.P.; Ablitzer, D.; Hess, E. Aluminum volatilization and inclusion removal in the electron beam cold hearth melting of Ti alloys. *Metall. Mater. Trans. B* **2000**, *31*, 845–854. [CrossRef]
15. Kroll-Rabotin, J.S.; Gisselbrecht, M.; Ott, B.; May, R.; Fröhlich, J.; Bellot, J.P. Multiscale simulation of non-metallic inclusion aggregation in a fully resolved bubble swarm in liquid steel. *Metals* **2020**, *10*, 517. [CrossRef]
16. Li, J.; Wu, M.H.; Ludwig, A. Simulation of macrosegregation in a 2.45-ton steel ingot using a three-phase mixed columnar-equiaxed model. *Int. J. Heat Mass Transf.* **2014**, *72*, 668–679. [CrossRef]

17. Chen, H.B.; Long, M.J.; Chen, D.F.; Liu, T.; Duan, H.M. Numerical study on the characteristics of solute distribution and the formation of centerline segregation in continuous casting slab. *Int. J. Heat Mass Transf.* **2018**, *126*, 843–853. [CrossRef]
18. Bennon, W.D.; Incropera, F.P. A continuum model for momentum, heat and species transport in binary solid-liquidphase change systems. *Int. J. Heat Mass Transf.* **1987**, *30*, 2161–2170. [CrossRef]
19. Brent, A.D.; Voller, V.R.; Reid, K.J. Enthalpy-porosity technique for modeling convection-diffusion phase change: Application to the melting of a pure metal. *Numer. Heat Transf.* **1988**, *13*, 297–318.
20. Hirt, C.W.; Nichols, B.D.; Hotchkiss, R.S. Volume of fluid (VOF) method for the dynamics of free boundaries. *J. Comput. Phys.* **1981**, *39*, 201–225. [CrossRef]
21. Launder, B.E.; Spalding, D.B. The numerical computation of turbulent flows. *Comput. Methods Appl. Mech. Eng.* **1974**, *3*, 269–289. [CrossRef]
22. Voller, V.R.; Prakash, C. A fixed-grid numerical modeling methodology for convection-diffusion mushy region phase change problems. *Int. J. Heat Mass Transf.* **1987**, *30*, 1709–1720. [CrossRef]
23. Thomas, B.G.; Yuan, Q.; Sivaramakrishnan, S.; Shi, T.B.; Vanka, S.P.; Assar, M.B. Comparison of four methods to evaluate fluid velocities in a continuous Slab casting mold. *ISIJ Int.* **2001**, *41*, 1262–1271. [CrossRef]
24. Waheed, M.A.; Nzebuka, G.C. Investigation of macrosegregation for different dendritic arm spacing, casting temperature, and thermal boundary conditions in a direct-chill casting. *Appl. Phys. A* **2020**, *126*, 725. [CrossRef]
25. Zhao, X.K.; Reilly, C.; Yao, L.; Maijer, D.M.; Cockcroft, S.L.; Zhu, J. A three-dimensional steady state thermal fluid model of jumbo ingot casting during electron beam re-melting of Ti-6Al-4V. *Appl. Math. Model.* **2014**, *38*, 3607–3623. [CrossRef]
26. Nzebuka, G.C.; Ufodike, C.O.; Egole, C.P. Influence of various aspects of low-Reynolds number turbulence models on predicting flow characteristics and transport variables in a horizontal direct-chill casting. *Int. J. Heat Mass Transf.* **2021**, *179*. [CrossRef]
27. Zhao, J.J. Numerical simulation of inclusion movement in the continuous casting slab mold. In Proceedings of the International Symposium On Clean Steel, Beijing, China, 17 September 2008.
28. Li, A.; Ahmadi, G. Dispersion and deposition of spherical particles from point sources in a turbulent channel flow. *Aerosol Sci. Technol.* **1992**, *16*, 209–226. [CrossRef]

MDPI
St. Alban-Anlage 66
4052 Basel
Switzerland
Tel. +41 61 683 77 34
Fax +41 61 302 89 18
www.mdpi.com

Crystals Editorial Office
E-mail: crystals@mdpi.com
www.mdpi.com/journal/crystals

www.ingramcontent.com/pod-product-compliance
Lightning Source LLC
LaVergne TN
LVHW070457100526
838202LV00014B/1740